COLUMBIA UNIVERSITY
LIBRARIES

TIME SERIES ANALYSIS:
Theory and Practice 6
Hydrological, Geophysical and Spatial Applications

TIME SERIES ANALYSIS:
Theory and Practice 6
Hydrological, Geophysical and Spatial Applications

Proceedings of the International Conference
held at Toronto, Canada, 10-14 August 1983

Edited by

O. D. ANDERSON
TSA&F, Nottingham, England

J. K. ORD
Pennsylvania State University, USA

and

E. A. ROBINSON
University of Tulsa, Oklahoma, USA

1985

NORTH-HOLLAND – AMSTERDAM · NEW YORK · OXFORD

© ELSEVIER SCIENCE PUBLISHERS B.V., 1985

All rights reserved. No part of this publication may be reproduced, stored in a retrieval system, or transmitted, in any form or by any means, electronic, mechanical, photocopying, recording or otherwise, without the prior permission of the copyright owner.

ISBN: 0 444 87683 9

Published by:
ELSEVIER SCIENCE PUBLISHERS B.V.
P.O. Box 1991
1000 BZ Amsterdam
The Netherlands

Sole distributors for the U.S.A. and Canada:
ELSEVIER PUBLISHING COMPANY, INC.
52 Vanderbilt Avenue
New York, N.Y. 10017
U.S.A.

Library of Congress Cataloging in Publication Data

Main entry under title:

Time series analysis.

   Proceedings from the 10th International Time Series Meeting held Aug. 10-14, 1983, Toronto, Canada.
  1. Hydrology--Statistical methods--Congresses.
2. Geophysics--Statistical methods--Congresses.
3. Space and time--Statistical methods--Congresses.
4. Time-series analysis--Congresses. I. Anderson, O. D. (Oliver Duncan), 1940-   . II. Ord, J. K.
III. Robinson, Enders A. IV. International Time Series Meeting (10th : 1983 : Toronto, Ont.)
GB656.2.S7T56 1985   551.48'028   84-25956
ISBN 0-444-87683-9 (Elsevier)

PRINTED IN THE NETHERLANDS

*To
Luisa
Janice
and
Helen*

CONTENTS

O.D. ANDERSON
Introduction — 1

Part 1: HYDROLOGICAL TIME SERIES

K.W. HIPEL, A.I. McLEOD and W.-K. LI
Causal and Dynamic Relationships between Natural Phenomena — 13

R.M. THOMPSTONE, K.W. HIPEL and A.I. McLEOD
Grouping of Periodic Autoregressive Models — 35

D.A. CLUIS
Information Content of the Mean and Rescaled Range for
Short Samples Derived from AR(2) Processes — 51

P.A. MORETTIN, A.R. DE MESQUITA and J.G.C. DA ROCHA
Rainfall at Fortaleza in Brazil Revisited — 67

H. KOPPELMAN and T. SCHILPEROORT
Identification and Parameter Estimation of Phosphate Balance
Models in Drinking Water Reservoirs — 87

Part 2: GEOPHYSICAL TIME SERIES

E.A. ROBINSON and S. TREITEL
The Right-Half Autocorrelation Theorem — 105

L.R. LINES and S. TREITEL
A Review of Nonlinear Regression and its Applications to
Geophysical Inverse Problems — 133

C.E. SCHMID
ARMA Maximum Entropy Spectral Analysis Using Iterative
Prewhitening — 179

I.R. MUFTI
Recent Developments in Seismic Migration 193

F. EL-HAWARY
An Approach to Seismic Information Extraction 223

Part 3: SPATIAL TIME SERIES

L.A. AROIAN
Time Series in M Dimensions: Past, Present, and Future 241

R.J. PERRY and L.A. AROIAN
Autoregressive Models in M Dimensions, $M=1$, Theory
and Examples 263

D.A. GRIFFITH, R.P. HAINING and R.J. BENNETT
Estimating Missing Values in Space-Time Data Series 273

D.S. STOFFER
Maximum Likelihood Fitting of STARMAX Models to
Incomplete Space-Time Series Data 283

A.D. CLIFF and J.K. ORD
Forecasting the Spread of an Epidemic 297

INTRODUCTION

About the Meeting

The 10th International Time Series Meeting (ITSM) took place in Toronto (Canada) 10-14 August, 1983, at the Sheraton Centre, immediately prior to the 1983 Joint Statistical Meetings (American Statistical Association, Biometric Society, Institute of Mathematical Statistics and the Statistical Society of Canada) which were held at the same location. The ITSM was a Special Topics Conference, with theme "Hydrological, Geophysical and Spatial Time Series".

This introduction lists the participants, provides the technical programme, and prints abstracts of those papers which were presented, but are not included in these Proceedings. It also acknowledges all individuals, who helped make the Meeting a success, and gives biographical sketches for those authors who filed appropriate details.

Conference Organisation

Oliver D. Anderson (UK), Convenor & General Chairman	TSA&F, Nottingham
Richard B. Anderson (Canada), Registration	York University, Downsview, Ontario
Leo A. Aroian (USA), Session Organiser, Spatial Stream	Union College, Schenectady, New York
J. Keith Ord (USA), Spatial Stream Leader	Pennsylvania State University
Enders A. Robinson (USA), Geophysical Stream Convenor	University of Tulsa, Oklahoma

Other Participants (making 58 people in all, from 11 countries)

Arminé G. Aroian (USA)	Guest, Schenectady, New York
Piero Barone (Italy)	Institute Mauro Picone, Rome
Christine Baufays (Belgium)	Guest, Namur
Pierre Baufays (Belgium)	Namur University
Denis Bolduc (Canada)	University of Montreal, Quebec
Jean-Pierre Chanut (Canada)	University of Quebec, Rimouski
Keewhan Choi (USA)	Centers for Disease Control, Atlanta, Georgia
Daniel A. Cluis (Canada)	University of Quebec, Sante-Foy
Eivind Damsleth (Norway)	Norwegian Computing Center, Oslo
Peter R. Defize (Netherlands)	Institute TNO, The Hague
Janice A. Derr (USA)	Guest, Pennsylvania State University
Ferial El-Hawary (Canada)	Technical University of Nova Scotia, Halifax
Mo El-Hawary (Canada)	Guest, Technical University of Nova Scotia, Halifax
Ömer L. Gebizlioglu (Turkey)	Middle East Technical University, Ankara
Pierre Goupillaud (USA)	Systems, Science and Software, La Jolla, California
Daniel A. Griffith (USA)	State University of New York, Buffalo
Zhi-Gang Han (China)	Heilongjiang University, Harbin
Jon Helgeland (Norway)	Norwegian Computing Center, Oslo
Keith W. Hipel (Canada)	University of Waterloo, Ontario
Urban Hjorth (Sweden)	Linköping Institute of Technology
Shiv G. Kapoor (USA)	University of Illinois, Urbana-Champaign
Myron J. Katzoff (USA)	Bureau of Census, Washington DC

Larry R. Lines (USA)	Amoco Production Co, Tulsa
Donald J. Malec (USA)	Bureau of Census, Washington DC
A. Ian McLeod (Canada)	University of Western Ontario, London
Priya Ranjan Mohanty (Canada)	University of Manitoba, Winnipeg
Marc Moore (Canada)	Ecole Polytechnique, Montreal, Quebec
Pedro A. Morettin (Brazil)	University of São Paulo
Ishard Mufti (USA)	Superior Oil, Houston
Trygve S. Nilsen (Norway)	University of Bergen
France Paradis (Canada)	Guest, Montreal, Quebec
Anne M. Perry (USA)	Guest, Schenectady, New York
Robert J. Perry (USA)	Schenectady, New York
Charles H. Proctor (USA)	North Carolina State University, Raleigh
David L. Quigg (USA)	Bradley University, Peoria, Illinois
Tom Schilperoort (Netherlands)	Delft Hydraulics Laboratory
Charles Schmid (USA)	Honeywell, Seattle
Laurel Sharp (USA)	Guest, Columbia, South Carolina
Will Sharp (USA)	Guest, Columbia, South Carolina
W. Ed Sharp (USA)	University of South Carolina, Columbia
Janice Singh-Stoffer (USA)	Guest, Pittsburgh, Pennsylvania
Lena Smolon (USA)	Guest, Schenectady, New York
William J. Smolon (USA)	Union College, Schenectady, New York
David S. Stoffer (USA)	University of Pittsburgh, Pennsylvania
Mathew Stoffer (USA)	Guest, Pittsburgh, Pennsylvania
Marcia Synnott (USA)	Guest, Columbia, South Carolina
Lucille Terry (USA)	Guest, Bowling Green State University, Ohio
W. Robert Terry (USA)	University of Toledo, Ohio
Robert M. Thompstone (Canada)	Alcan Smelters and Chemicals Ltd, Jonquière, Quebec
George Treviño (USA)	CHIRES Associates, Las Cruces, New Mexico
Dan Wartenburg (USA)	State University of New York, Stoney Brook
Terry L. Watt (USA)	University of Tulsa, Oklahoma
Zhi-Ming Wu (China)	Shanghai Jiao Tong University

Technical Programme (30 contributions in a single stream)

Thursday, 11 August

Session 1 Plenary 1   08.30-10.15

O.D. Anderson (UK) Welcome to 10th International Time Series Meeting (ITSM) and 4th American Conference
J.K. Ord (USA) Forecasting the Spread of an Epidemic (with *A.D. Cliff*, UK)
L.A. Aroian (USA) Time Series in M-Dimensions: Past, Present and Future

Session 2 Hydrological 1   10.45-12.30

D. Cluis (Canada) Information Content of the Mean and Rescaled Range of Short Samples derived from AR(2) Processes
W.R. Terry (USA) A Markovian State Vector Representation for the Influence of Wind Speed and Atmospheric Stability on Sulfur Dioxide Concentration (with *A. Kumar*)
R.M. Thompstone (Canada) Grouping of Periodic Autoregressive Models (with *K.W. Hipel* & *A.I. McLeod*)

Session 3 Geophysical 1   14.30-16.15   *Organised by Professor E.A. Robinson*

E.A. Robinson (USA) A Bridge between Geophysical and Other Time Series Analysts
T.L. Watt (USA) Robust Homomorphic Filtering (with *J.B. Bednar*)
L.R. Lines (USA) A Review of Nonlinear Regression and its Applications to Problems in Geophysical Inverse Theory (with *S. Treitel*)

Introduction

Session 4 <u>Spatial 1</u> 16.45-18.30 *Organised by Professors J.K. Ord & L.A. Aroian*

D.A. *Griffith* (USA) Estimating Missing Values in Space-Time Data Series (with R.P. *Haining* & R.J. *Bennett*, UK)
D. *Wartenburg* (USA) Blocked Quadratic Variance or Spectral Density: Speed versus Power in Transect Analysis
W.E. *Sharp* (USA) Statistical Space Series on a Square Net (with L.A. *Aroian*)

Friday, 12 August

Session 5 <u>Hydrological 2</u> 08.30-10.15

P.A. *Morettin* (Brazil) Rainfall at Fortaleza in Brazil Revisited (with A.R. *de Mesquita*)
T. *Schilperoort* (Netherlands) Identification and Parameter Estimation of Models for the Phosphate Balance in Drinking Water Reservoirs (with H. *Koppelman*)
S. *Kapoor* (USA) A Vector Time Series Analysis of the Primary Pollutants which Cause Acid Rain (with W.R. *Terry*)

Session 6 <u>Geophysical 2</u> 10.45-12.30 *Organised by Professor E.A. Robinson*

F. *El-Hawary* (Canada) An Approach to Seismic Information Extraction
I.R. *Mufti* (USA) Recent Developments in Seismic Migration
P. *Goupillaud* (USA) The Layer-Cake Model Revisited

Session 7 <u>Spatial 2</u> 16.45-18.30 *Organised by Professors J.K. Ord & L.A. Aroian*

R.J. *Perry* (USA) An Investigation into the Suitability of a Space-Time Autoregressive Model for Forecasting Highway Condition Data
Ü.L. *Gebizlioglu* (Turkey) On the Modelling, Estimation and Hypothesis Testing for Spatial ARMA Processes
D.L. *Quigg* (USA) ARMA Model Identification: Time Series in M-Dimensions

Saturday, 13 August

Session 8 <u>Mixed Session</u> 08.30-10.15

G. *Treviño* (USA) A Method for Approximating the Mean-Value of Nonstationary Random Data
C.E. *Schmid* (USA) ARMA Maximum Entropy Spectral Analysis Using Iterative Prewhitening
D.J. *Malec* (USA) Spatial Analysis of Census Data (with M.J. *Katzoff*)

Session 9 <u>Spatial 3</u> 10.45-12.30

U. *Hjorth* (Sweden) Space-Time Modeling of a Moving Field
P. *Barone* (Italy) A Class of STAR Models for Stationary Processes Smoothly Varying in Space
D.S. *Stoffer* (USA) Maximum Likelihood Fitting of STARMAX Models to Incomplete Space-Time Series Data

Session 10 <u>Plenary 2</u> 16.45-18.30

K.W. *Hipel* (Canada) Causal and Dynamic Relationships between Natural Phenomena (with W.-K. *Li*, Hong Kong, & A.I. *McLeod*)
E.A. *Robinson* (USA) The Right-Half Autocorrelation Theorem (with S. *Treitel*)
O.D. *Anderson* (UK) Closing Remarks.

Edited Abstracts (for presented papers not included in these Proceedings)

1. P. BARONE (Italy)
A Class of STAR models for Stationary Processes Smoothly Varying in Space

In biological environments, one often measures the same phenomenon from different spatial locations and within a given time interval. The detrended series, corresponding to each site, is generally correlated with those at all other sites, and may exhibit a time dependent correlation structure which varies smoothly over space. To model such series, a general multivariate AR model was simplified by partitioning the parameter matrix, for each lag, into subsets and then imposing deterministic constraints on each subset. This gives rise to a STAR model with specific weighting matrices.

2. Ö.L. GEBIZLIOGLU (Turkey)
On the Modelling, Estimation and Hyopthesis Testing for Spatial ARMA Processes

This paper attempts to provide a theoretical development for the representation of purely spatial series by ARMA family models. With an appropriate sampling scheme it is shown that, using the correlation structure of an observed spatial process, a suitable model can be fit with parameters estimated by minimum variance prediction error, least squares, or maximum likelihood methods. A means of hypothesis testing the representation is presented.

3. P.L. GOUPILLAUD (USA)
The Layer-Cake Model Revisited

After 25 years, the layer-cake (1-D) model of the stratified earth is today one of the major tools for seismic exploration. Being a lattice model, it can be viewed as a Markov chain or an ARMA system; therefore standard spectral estimation methods can be applied to both the forward and inverse problems. The application of statistical concepts to seismic exploration, introduced by E.A. Robinson 30 years ago, has not yet produced its full potential return. An attempt is made to explain why, and suggestions are offered to show where, in the author's opinion, research efforts should be focused.

4. U. HJORTH (Sweden)
Space-Time Modeling of a Moving Field

A stationary or periodically stationary field is moving relatively to a reference coordinate system. A covariance model is derived for the multivariate time series observed at a given set of points in the reference system. Applications to weather observations are indicated.

5. S.G. KAPOOR & W.R. TERRY (USA)
*A Vector Time Series Analysis of the Primary Pollutants which Cause Acid Rain**

This paper evaluates two automatic methods for modeling time series: (1) the Data Dependent Systems approach of Pandit and Wu, and (2) Akaike's Markovian State Vector representation. These are compared with respect to the form of the family of models each uses, and the means by which they select an appropriate model. Results obtained from modeling atmospheric concentrations of nitrogen and sulfur dioxides are discussed.

6. D.K. QUIGG (USA)
ARMA Model Identification: Time Series in M-Dimensions

Time series in M-dimensions form a class of models developed to study data

---

* Paper to appear in <u>Time Series Analysis: Theory and Practice 7</u>.

dependent upon both time and space via an extension of Box-Jenkins analysis. The identification and subsequent analysis of ARMA models is crucial. In this paper we extend to time series in M-dimensions the generalized partial autocorrelation array, the GPAC. We study the theoretical and sample properties of the GPAC, the autocorrelation function, and the partial autocorrelation function, for such series with M = 1; and comment upon correct model order identification using these means.

7. W.R. TERRY & A. KUMAR (USA)
A Markovian State Vector Representation for the Influence of Wind Speed and Atmospheric Stability on Sulfur Dioxide Concentration

Considering the physical phenomena suggested that wind speed and atmospheric stability are important determinants for the maximum sulfur dioxide (SO_2) concentration in the immediate vicinity of a major pollutant source. Since all these variables vary continuously and randomly in time, their dynamic behavior was conceptually represented by a stochastic differential equation in which maximum SO_2 concentration is the dependent variable and wind speed and atmospheric stability are independent variables. Since such a stochastic differential equation has an equivalent state vector representation, the work of Akaike was used to determine the most appropriate Markovian state vector representation for the system. Implications of the resulting model are discussed.

8. G. TREVIÑO (USA)
*A Method for Approximating the Mean-Value of Nonstationary Random Data**

By "nonstationary" is meant data whose mean-value, $\mu(t)$, is a definite function of time, in contrast to "stationary" data where $\mu(t)$ is constant. The method is formulated by first assuming that while the functional form of $\mu(t)$, $0 \leq t \leq T$, is some polynomial in t, $\mu(t) = \sum_{n=0}^{\infty} a_n t^n$, this can be approximated linearly in a "small enough" time interval. The defining slope and intercept are then determined from the available data. The slope is computed following Treviño (1982), while the intercept is computed using a "shifted-average" scheme, the application of which is new.

9. D. WARTENBURG (USA)
Blocked Quadrat Variance or Spectral Density: Speed versus Power in Transect Analysis

Field ecologists collecting survey data are often confronted with a dilemma in data analysis: should one analyze transect data by quick, easy to compute, blocksize ANOVA techniques or should one use spectral techniques which require computer processing? It has been shown elsewhere that blocksize approaches are equivalent to considering spectra of filtered series, and thus only use a portion of the information available through spectral techniques. Blocksize techniques are recommended for quick field analysis, and to focus attention on features of interest, but rigorous spectral approaches are valuable in providing a more complete picture of the variation. Examples using actual data are given.

These Proceedings

Regrettably, only about half of the papers offered for this Conference have eventually made the present book. Some were not accepted for final presentation, others could not be presented; and the remainder failed to be

---
* Paper to appear in <u>Time Series Analysis: Theory and Practice 7</u>.

completed, or were not revised satisfactorily. We share the disappointment of the unpublished authors, but our aim to hold standards high must be maintained.

Although some contributions span more than one of the Special Topic areas of Time Series Analysis, which were considered at the Conference, the published papers are ordered within the three (interconnected) streams: Hydrological, Geophysical, and Spatial.

The Referees

We are most grateful to the following 52 specialists (from 9 countries) who helped with the refereeing, and in the reviewing process, for this volume:

A. Abakuks (UK)	A.J. Girling (UK)	C.A. Oprian (UK)
M.M. Ali (USA)	P.J. Green (UK)	J.K. Ord (USA)
A.P. Andersen (Australia)	D.A. Griffith (USA)	P.E. Pfeifer (USA)
O.D. Anderson (UK)	E.J. Hannan (Australia)	A.E. Raftery (Ireland)
L.A. Aroian (USA)	J.R.M. Hosking (UK)	E. Redfern (UK)
F.G. Ball (UK)	I.T. Jolliffe (UK)	E. Renshaw (UK)
M.S. Bartlett (UK)	D.A. Jones (UK)	E.A. Robinson (USA)
B. Bass (USA)	R.H. Jones (USA)	N.D. Singpurwalla (USA)
R.J. Bennett (UK)	C.A. Kang (USA)	D.N. Smith (UK)
J.A. Cadzow (USA)	K.V. Mardia (UK)	H. Spliid (Denmark)
R.T. Clarke (UK)	F.H.C. Marriott (UK)	D. Sprevak (UK)
D.M. Cooper (UK)	R. Marshall (UK)	P. Stoica (Romania)
C. Craggs (UK)	R.J. Martin (UK)	W.R. Terry (USA)
A.H. El-Shaarawi (Canada)	R.J. Meinhold (USA)	S. Treitel (USA)
D.F. Findley (USA)	P.A. Morettin (Brazil)	G. Treviño (USA)
R.F. Galbraith (UK)	J. Newton (UK)	A.M. Yaglom (USSR) &
M.F. Gard (USA)	P.M. North (UK)	S. Zachary (UK).
A. Gerstenkorn (USA)		

The Authors

OLIVER D. ANDERSON is a graduate of the UK Universities of Cambridge, London, Birmingham and Nottingham; with first class honours degrees in Mathematics and Economics, and master's degrees in Mathematics and Statistics.

He has worked in Industry, for Government, and as a Consultant Statistician; and taught in Schools, Colleges and Universities. He has lectured in over 20 countries and published some 200 items. He is an active member of a dozen professional societies (in England and abroad) concerned with Education, Mathematics, Statistics, Operations Research, Management Science, Economics and Econometrics; and, in 1979, was honoured by election to the International Statistical Institute. More recently he has become involved with Management, Marketing and Information Science.

Dr LEO A. AROIAN has been Research Professor of Management and Administration at Union College, Schenectady, New York, USA, since 1974; and, before that, Professor from 1968, after considerable previous experience in both industry and academia. He has received much recognition during his outstanding career and many honours, including election to Fellowship of the American Statistical Association.

ANDREW D. CLIFF is University Lecturer in Geography, and a Fellow of Christ's College, at Cambridge University, England. His research interests include spatial diffusion processes and related areas of quantitative geography. His publications include <u>Spatial Processes: Models and Applications</u> with Keith Ord, and <u>Locational Analysis</u> with Peter Haggett and Alan Frey.

Introduction

DANIEL A. CLUIS is a graduate of the French universities of Paris and Grenoble, with a doctorate in Statistical Fluid Mechanics. In 1970, he joined the Institut National de la Recherche Scientifique (INRS-Eau), a research center of the University of Quebec devoted to multidisciplinary water research and graduate studies. His research concerns water quality modelling, network rationalization and the application of statistical techniques to the hydrosciences.

AFRANIO R. DE MESQUITA has an MPhil (1972) in Statistics from the University of Southampton, UK; and an MS (1969) and PhD (1978), in Oceanography, from the University of São Paulo, Brazil, where he is currently an Associate Professor in the Department of Physical Oceanography.

FERIAL EL-HAWARY received her BSc (1967) and MSc (1971) in Electrical Engineering from the Universities of, respectively, Alexandria (Egypt) and Alberta (Canada); and, in 1981, was awarded a PhD in Ocean Engineering by the Memorial University of Newfoundland. Currently, she is a Research Assistant Professor at the Technical University of Nova Scotia, and has published several papers in "modeling and seismic signal processing for identification of ocean subsurface features". Dr El-Hawary is a member of Sigma Xi, IEEE, and the Marine Technology Society.

DANIEL A. GRIFFITH graduated from Indiana University, Pennsylvania, and the University of Toronto; his degrees include a BS in mathematics, and an MA and PhD in Geography. He has taught in Schools, Polytechnical Institutes, and Universities, and has had a dozen research projects funded. He has directed two NATO Advanced Studies Institutes, has lectured in 7 countries, and has published some 32 items, including two edited books. He is an active member of the Association of American Geographers and the Regional Science Association, and serves on the editorial board of Geographical Analysis. In 1980 he was selected a Nystrom Competition finalist for his doctoral dissertation work.

KEITH W. HIPEL is an Associate Professor, and Associate Chairman for Undergraduate Studies, within the Department of Systems Design Engineering at the University of Waterloo. He obtained his Bachelors and PhD degrees in Civil Engineering, and his Master's degree in Systems Design Engineering, from the University of Waterloo.

As well as being a member of the American Water Resources Association, the American Geophysical Union and the Association of Professional Engineers of Ontario, Dr Hipel is an Associate Editor of Water Resources Bulletin and Chairman of the Surface Runoff Committee of the Hydrology Section in the American Geophysical Union. Besides doing engineering consulting in operational research, Dr Hipel has taught special courses and seminars in foreign countries such as Brazil, China, Japan, Singapore and the United States. His present research interests include the use and development of stochastic modelling techniques in the geophysical sciences and the incorporation of political and non-quantitative considerations into the systems design of large scale engineering projects.

Dr Hipel has co-authored two textbooks and co-edited two volumes on stochastic modelling, and has published his research in various international journals of engineering and operational research.

A. IAN McLEOD is an Assistant Professor in the Department of Statistical and Actuarial Sciences at the University of Western Ontario. He holds a PhD in Statistics from the University of Waterloo, for which he won the doctoral gold medal prize in 1978. His main research interests are in the theory and application of time series models, stochastic hydrology and statistical algorithms.

PEDRO A. MORETTIN, MA (1971) and PhD (1972) in Statistics, from the University of California at Berkeley, is Chairman of the Statistics Department, University of São Paulo, Brazil. He is also Editor (Theory and Methods) for <u>Estadistica</u>, journal of the Inter-American Statistical Institute.

IRSHAD R. MUFTI received his PhD in geophysics in 1966 from Clausthal Technische Universitaet, West Germany. He has served as senior research scientist, Amoco Production Co (1971-1980); research physicist, Geological Survey of West Germany (1963-1967); and geophysicist, Geological Survey of Pakistan (1958-1962). He has been actively engaged in various research and exploration projects dealing with seismics, geoelectrics, gravity and aeromagnetics. He holds two patents dealing with geoelectric and thermal remote-sensing along boreholes.

Dr Mufti joined Superior Oil in 1980 as research consultant. His current interests include the application of numerical and iterative techniques to the processing and interpretation of seismic data. He is a member of SEG, EAEG and the Geophysical Society of Houston.

J. KEITH ORD is Professor of Management Science and Statistics at the Pennsylvania State University, USA. Previously he was Reader in Statistics at Warwick University, England. His main research interests lie in forecasting and in spatial processes. His publications include <u>Spatial Processes: Models and Applications</u> with Andrew Cliff, and <u>The Advanced Theory of Statistics, Volume 3</u>, with the late Sir Maurice Kendall and Alan Stuart.

ROBERT J. PERRY is a graduate of Rensselaer Polytechnic Institute, the Graduate School of Public Affairs at the State University of New York, and Union College, Schenectady; with a bachelor's degree in Chemical Engineering, masters degrees in Public Administration and Applied Statistics, and a doctorate in Administrative and Engineering Systems. He has been employed by the New York State Department of Transportation since 1962; and, at present, is a Principal Civil Engineer in the Construction Division. He is a licensed Professional Engineer and a member of Sigma Xi, the American Statistical Association, and the American Society of Public Administration. Prior to 1977, Dr Perry was a member of the American Society for Testing and Materials' Committee D4, "Road and Paving Materials", and the Transportation Research Board Committee A2G04, "Adhesives".

ENDERS A. ROBINSON has made enormous contributions to signal processing, in geophysics; and to time series analysis, in general. He has published numerous research papers and books on these subjects, which combine outstanding scholarship with exceptional readability. A giant in both theory and practice, Dr Robinson is rightly regarded as the Father of Deconvolution, which he has nurtured and developed for over 30 years. Enders is currently a Distinguished Professor at the University of Tulsa, Oklahoma.

TOM SCHILPEROORT received an MSc (1977) in electrical engineering and mathematical engineering from the Twente University of Technology, Netherlands; and is currently project engineer at the Delft Hydraulics Laboratory. His main research interests are in time series analysis, system identification and parameter estimation, and adaptive filtering and control. Within the framework of consulting activities he has carried out applied work on many different hydrological and hydraulic problems; including optimization of monitoring networks, spectral analysis of waves and turbulence, optimization of depth of channels, identification of water quality models, and prediction of water levels and wave energy in estuaries.

CHARLES E. SCHMID received a BSEE (1963) from Cornell University, an MSEE (1968) from the University of Connecticut, and a PhD (1977) in electrical engineering from the University of Washington. He has been with Honeywell since 1966, where he works on various aspects of underwater acoustics and digital signal processing. In 1980 he spent 3 months with Honeywell ELAC in Keil, West Germany. Most

Introduction

recently, he has been working on sonar trainers in conjunction with Honeywell's training organization in West Covina, California. Dr Schmid is a member of Tau Beta Pi and belongs to the Acoustical Society of America (presently serving as chairman of its N.W. Chapter).

W. EDWIN SHARP, BA (1958), MA (1960) and PhD (1964), from the University of California at Los Angeles, is currently Associate Professor, Department of Geology, University of South Carolina, Columbia, USA. He has also worked as a Research Scientist (1964-66) in Johannesburg, South Africa, and held a post-doctoral Fellowship from the Royal Norwegian Council for Scientific and Industrial Research (1976-77). Dr Sharp has published 35 papers and a number of maps, monographs and reports.

ROBERT M. THOMPSTONE is Senior Coordinator of the Hydraulic Resources Group Power Operations in Quebec for Alcan Smelters and Chemicals Ltd, where he has been employed for the last nine years. Prior to that, he spent two years with a private consulting firm specializing in water resources management. Mr Thompstone holds a BASc in civil engineering and an MASc in management sciences from the University of Waterloo, and a BSpAdm in business administration from the University of Quebec at Chicoutimi. He is also currently a part-time PhD candidate in the Department of Systems Design Engineering, University of Waterloo.

Acknowledgements

I would like to thank all participants for coming, speakers for presenting their work, referees for assessing it, and authors for preparing final copy for publication; and, especially, the Organising Committee and the other Editors, for all their extensive and extended efforts. As usual, I think everyone will agree that Inez van der Heide, at North-Holland, has done an excellent job in preparing this volume for print.

Looking Forward

The Proceedings for the final 1983 ITSM are also in press:

Time Series Analysis: Theory and Practice 7
General Interest ITSM, Toronto (Canada) 18-21 August, 1983.

Although there will be no ITSMs held in 1984, we have begun to plan for 1985. Prospective authors should write to the first Editor, at the address below, for details on submitting abstracts and papers.

OLIVER D. ANDERSON
TSA&F, 9 Ingham Grove, Lenton Gardens, Nottingham NG7 2LQ, England
July 1984

Part 1
Hydrological Time Series

CAUSAL AND DYNAMIC RELATIONSHIPS BETWEEN NATURAL PHENOMENA

Keith W. Hipel
Department of Systems Design Engineering, University of Waterloo
Waterloo, Ontario, Canada, N2L 3G1

A. Ian McLeod
Department of Statistical and Actuarial Sciences, The University of Western Ontario
London, Ontario, Canada N6A 5B9

Wai Keung Li
Department of Statistics, University of Hong Kong, Hong Kong

Flexible statistical procedures are employed to investigate possible causal relationships among a large variety of geophysical time series and to construct transfer function-noise models to link two time series when meaningful relationships are found. The properties of the cross-correlation function of the residuals from stochastic models fitted to two time series are employed for revealing the type of causality that may be present. Contrary to a previous suggestion, it is found that annual sunspot numbers do not significantly affect the yearly flows of the Volga River in Russia. Other causality studies demonstrate that temperatures for certain months of the year can significantly affect the annual flows of rivers and also the price of wheat. Upon detecting significant causal connections between two phenomena, a transfer function-noise model can be developed by following the identification, estimation and diagnostic check stages of model construction.

1. INTRODUCTION

Is it possible to substantiate the claim of a Soviet hydrologist (Smirnov, 1969) that yearly sunspot numbers have a significant affect upon the annual flows of the Volga River? What is the influence of temperature upon the price of wheat? In other words, how and when can one say that one phenomenon definitely causes another?

The foregoing kinds of questions have been baffling scientists for decades and previously some research had been carried out to attempt to answer them. For example, Brillinger (1969) and Rodriguez-Iturbe and Yevjevich (1968) employed cross-spectral and other statistical methods to investigate relationships between natural time series. However, comprehensive statistical tools are now available to assist in solving causality problems and these useful techniques have yet to be applied to a large variety of geophysical data sets. Consequently, the purpose of this paper is to present flexible statistical procedures for formally answering causality questions and then to apply the methodologies to a wide range of natural time series. In particular, Granger's (1969) definition of causality is first defined and then it is explained how a cross-correlation analysis of the residuals from the stochastic models fitted to two series, can be employed to detect causal relationships (Pierce and Haugh, 1977). The information from the cross-correlation analysis can then be used to design a transfer function-noise model to explicitly describe the mathematical relationship between the two data sequences (Haugh and Box, 1977; Box and Jenkins, 1970, Ch. 11). In the section on applications, a large number of interesting cross-correlation studies are carried out to detect possible causal relationships between many different phenomena. The time series studied include sunspot numbers, annual and monthly temperatures, seven annual river flow series, Beveridge wheat price indices, and tree ring widths.

2. CAUSALITY

2.1 Definition

Wiener (1956) originally formulated a definition of causality between two time series, which is suitable for empirical detection and verification of meaningful relationships. More recently, Granger (1969) presented a formal definition of causality while Pierce and Haugh (1977) expanded upon the work of Granger (1969) and gave a comprehensive survey regarding research on causality in temporal systems. Other research which is related to Granger's (1969) definition of causality can be found by referring to the appropriate statistical literature (see for example Jenkins and Watts (1968), Haugh (1972, 1976), Haugh and Box (1977), and McLeod (1979)).

Granger (1969) defined causality between two time series in terms of predictability. A variable X causes another variable Y, with respect to a given universe or information set that includes X and Y, if present Y can be better predicted by using past values of X than by not doing so, all other relevant information (including the past of Y) being used in either case. This definition of causality does not require the system to be linear but when it is, linear predictions are compared. To be more specific, let X_t and Y_t be two time series and let A_t for $t = 0, \pm 1, \pm 2, \ldots$, be the given information set that includes at least X_t and Y_t. Allow $\bar{A}_t = \{A_s : s < t\}$, $\tilde{A}_t = \{A_s : s \leq t\}$ and in a similar fashion define $\bar{X}_t, \tilde{X}_t, \bar{Y}_t,$ and $\tilde{Y}_t$. Given the information set A_t, let $P_t(Y|A_t)$ be the minimum mean square error one step ahead predictor of Y_t and denote the resulting mean square error by $\sigma^2(Y|A)$. According to Granger (1969) X causes Y if

$$\sigma^2(Y|\bar{A}_t) < \sigma^2(Y|\bar{A}_t - \bar{X}_t) \tag{1}$$

while X causes Y instantaneously if

$$\sigma^2(Y|\bar{A}_t, \tilde{X}_t) < \sigma^2(Y|\bar{A}_t) \tag{2}$$

Causality from Y to X can be defined in the same way. Feedback occurs when X causes Y and Y also causes X.

2.2 Cross-Correlation

To ascertain the type of causality relationship that exists between X and Y, the properties of the cross-correlations are examined for the prewhitened series. When prewhitening discrete time series such as X_t or Y_t, the first step is to consider suitable transformations to form the transformed series, x_t and y_t. The reasons for transforming the series include stabilizing the variance, improving the normality assumption, eliminating trends, removing seasonality, and getting rid of nonstationarity. The selected transformations should allow x and y to be related causally in the same manner as X and Y when considering Granger's (1969) definition of causality. In practice, causality is preserved by many of the common types of transformations. For example, often the given series may be transformed by a Box-Cox transformation (Box and Cox, 1964) and following this the data may be differenced (Box and Jenkins, 1970, Chs. 4 and 9) to render the data stationary. When dealing with seasonal geophysical series the data may be transformed using a Box-Cox transformation and subsequent to this the seasonality may be removed by invoking an appropriate deseasonalization technique. For instance, when modelling an average monthly river flow series, often the series is first transformed by taking natural logarithms and then each data point is deseasonalized by subtracting out the monthly mean and dividing this by the monthly standard deviation. A Box-Cox transformation such as natural logarithms should not alter causality relationships for series consisting of all positive values, since the manner in which one series affects the predictability of another will not be changed by a strictly monotonic transformation that preserves the same relative position of every data point in the series. Deseasonalizing each time series is equivalent to removing a periodic component to eliminate seasonality where the periodic component is ultimately due to hydrologic factors such as precipitation and temperature. Because the deseasonalization parameters are estimated from the historical data and are assumed to be the same in the future, the deseasonalization should not alter the causality relationship existing in the original series when entertaining Granger causality. However, the periodic portion still constitutes one of the components needed to form the overall seasonal series.

The second step in the prewhitening procedure is to fit appropriate stochastic models to the x_t and y_t series in order to obtain white noise residuals. For instance, when the transformed series are nonseasonal it may be suitable to fit autoregressive-moving average (ARMA) models (Box and Jenkins, 1970) to x_t and y_t such that

$$\phi_x(B)(x_t - \mu_x) = \theta_x(B) u_t \qquad (3)$$

and

$$\phi_y(B)(y_t - \mu_y) = \theta_y(B) v_t \qquad (4)$$

where μ_x is the theoretical mean of the x_t series; B is the backward shift operator defined by $Bx_t = x_{t-1}$ and $B^k x_t = x_{t-k}$ where k is a positive integer; $\phi_x(B) = 1 - \phi_{x,1} B - \phi_{x,2} B^2 - \cdots - \phi_{x,p_x} B^{p_x}$ is the nonseasonal autoregressive (AR) operator of order p_x such that the roots of the characteristic equation $\phi_x(B) = 0$ lie outside the unit circle for nonseasonal stationarity and the $\phi_{x,i}$, $i = 1, 2, \ldots, p_x$, are the nonseasonal AR parameters; $\theta_x(B) = 1 - \theta_{x,1} B - \theta_{x,2} B^2 - \cdots - \theta_{x,q_x} B^{q_x}$ is the nonseasonal moving average (MA) operator of order q_x such that the roots of $\theta_x(B) = 0$ lie outside the unit circle for invertibility and θ_i, $i = 1, 2, \ldots, q_x$, are the nonseasonal MA parameters; u_t is white noise (also called innovation or disturbance) that has a mean of zero and variance of σ_u^2; and similar definitions to μ_x, $\phi_x(B)$, $\theta_x(B)$, and u_t hold for μ_y, $\phi_y(B)$, $\theta_y(B)$, and v_t, respectively. To indicate the orders of the AR and MA operators of the models in (3) or (4), the notation ARMA(p,q) is employed. Because of the linear nature of the operators in (3) and (4), this insures that u and v are causally related in the same way as x and y.

Subsequent to prewhitening of the time series, the cross correlation function (CCF), at lag k between the u_t and v_t series in (3) and (4), respectively, can be considered using

$$\rho_{uv}(k) = E(u_t \cdot v_{t+k}) / [E(u_t^2) E(v_t^2)]^{1/2} \qquad (5)$$

Due to the form of (5), the values of the CCF can range from negative one to positive one. Unlike the autocorrelation function (ACF), the CCF is not usually symmetric about lag zero and therefore the properties of $\rho_{uv}(k)$ must be examined for $k = 0, \pm 1, \pm 2 \ldots$. In addition to reflecting the type of linear dependence between u and v and consequently between X and Y, $\rho_{uv}(k)$ gives the kind of causality relationship between these variables for linear systems.

As explained by Pierce and Haugh (1977), there are many possible types of causal interactions between X and Y which can be characterized by the properties of $\rho_{uv}(k)$. Using the results of Pierce and Haugh (1977, p. 276, Table 3) some of the important causal relationships are categorized according to the restrictions on $\rho_{uv}(k)$ in Table 1. Due to the findings of Price (1979) and also Pierce and Haugh (1979), any of the relationships in Table 1 which involve instantaneous causality are only valid when there is no feedback. The entries in Table 1 are self-explanatory. For example, when there is unidirectional causality from X to Y, $\rho_{uv}(k) \neq 0$ for some $k > 0$, $\rho_{uv}(k) = 0$ for all $k < 0$, and $\rho_{uv}(0)$ may either be zero or else have some real non-zero value. For the case where Y does not cause X at all, there is no instantaneous causality between X and Y since $\rho_{uv}(0) = 0$.

When checking for the type of causality between two given time series the estimated CCF of the model residuals must be examined to ascertain which values are significantly different from zero. Suppose that two sequences x_t and y_t are given for $t = 1, 2, \ldots, n$. By utilizing (3) and (4) or other appropriate linear models, the two series can be prewhitened to obtain the estimated innovation series or residuals, $\hat{u}_t$ and $\hat{v}_t$. The residual CCF at lag k between $\hat{u}_t$ and $\hat{v}_t$ is defined by

$$r_{uv}(k) = c_{\hat{u}\hat{v}}(k) / [c_{\hat{u}}(0) \, c_{\hat{v}}(0)]^{1/2} \tag{6}$$

where

$$c_{\hat{u}\hat{v}}(k) = \begin{cases} n^{-1} \sum_{t=1}^{n-k} \hat{u}_t \hat{v}_{t+k} & k \geq 0 \\ n^{-1} \sum_{t=1-k}^{n} \hat{u}_t \hat{v}_{t+k} & k < 0 \end{cases}$$

is the estimated cross covariance function at lag k between the residual series;

$c_{\hat{u}}(0) = n^{-1} \sum_{t=1}^{n} \hat{u}_t^2$ is the estimated variance of the $\hat{u}_t$ sequence; and

$c_{\hat{v}}(0) = n^{-1} \sum_{t=1}^{n} \hat{v}_t^2$ is the estimated variance of the $\hat{v}_t$ series.

TABLE 1. Causal Relationships Between Two Variables

RELATIONSHIP	RESTRICTIONS ON $\rho_{uv}(k)$
X causes Y	$\rho_{uv}(k) \neq 0$ for some $k > 0$
Y causes X	$\rho_{uv}(k) \neq 0$ for some $k < 0$
Instantaneous Causality	$\rho_{uv}(0) \neq 0$
Feedback	$\rho_{uv}(k) \neq 0$ for some $k > 0$ and for some $k < 0$
X causes Y but not instantaneously	$\rho_{uv}(k) \neq 0$ for some $k > 0$ and $\rho_{uv}(0) = 0$
Y does not cause X	$\rho_{uv}(k) = 0$ for all $k < 0$
Y does not cause X at all	$\rho_{uv}(k) = 0$ for all $k \leq 0$
Unidirectional causality from X to Y	$\rho_{uv}(k) \neq 0$ for some $k > 0$ and $\rho_{uv}(k) = 0$ for either (a) all $k < 0$ or (b) all $k \leq 0$
X and Y are only related instantaneously	$\rho_{uv}(0) \neq 0$ and $\rho_{uv}(k) = 0$ for all $k \neq 0$
X and Y are independent	$\rho_{uv}(k) = 0$ for all k

The residual CCF can be plotted against lag k for $k \simeq -n/4$ to $k \simeq n/4$. In order to plot confidence limits the distribution of the residual CCF must be known. Assuming that the x_t and y_t series are independent (so $\rho_{uv}(k) = 0$ for all k), Haugh (1972, 1976) showed that for large samples $r_{\hat{u}\hat{v}}(k)$ is normally independently distributed with a mean of zero and variance of $1/n$. Consequently, to obtain the approximate 95% confidence limits a line equal to $1.96 \, n^{-1/2}$ can be plotted above and below the zero level for the residual CCF. McLeod (1979) obtained the asymptotic distribution of the residual CCF for the general case where the x_t and y_t series do not have to be independent of each other, and consequently more accurate confidence limits can be obtained by utilizing his results.

One reason why the residual CCF is examined rather than the CCF for the x_t and y_t series, is that it is much easier to interpret the results from a plot of $r_{\hat{u}\hat{v}}(k)$. This is because when both the x_t and y_t series are autocorrelated, the estimates of the CCF for x_t and y_t can have high variance and the estimates at different lags can be highly correlated with one another (Bartlett, 1935). In other words, the distribution of the estimated CCF for x_t and y_t is more complex than the distribution of $r_{\hat{u}\hat{v}}(k)$. Monte Carlo studies executed by Stedinger (1981), demonstrate the advantages of prewhitening two series before calculating their CCF. Additionally, from an intuitive point of view it makes sense to examine the residual CCF. Certainly, if the "driving mechanisms" or residuals between two series are significantly correlated, then meaningful relationships would exist between the original series.

From an examination of the residual CCF, the type of relationship existing between X and Y can be ascertained by referring to the results in Table 1. Suppose, for example, the X variable is precipitation and the Y variable is river flow. From a physical understanding of hydrology it is obvious that precipitation causes river flow. This knowledge would be mirrored in a plot of the residual CCF for these two series. For $k \geq 0$ there would be at least one value of $r_{\hat{u}\hat{v}}(k)$ which is significantly different from zero. However, all values of the residual CCF for $k < 0$ would not be significantly different from zero. In situations where the type of causality between two series is not known (for instance, do sunspots cause river flows), an examination of the residual CCF can provide valuable insight into the problem (see Application Section).

Formal tests of significance may also be derived when examining causal relationships (see, for example, McLeod (1979) and Pierce (1977)). Suppose that it is known a piori that Y does not cause X so that $\rho_{uv}(k) = 0$ for $k < 0$ (for instance river flow does not cause precipitation). Consequently, one may wish to test the null hypothesis that X does not cause Y and hence $\rho_{uv}(k) = 0$ for $k = 0, 1, 2, \ldots, M$, where M is a suitably chosen lag such that after M time periods it would be expected there would not be a relationship between the x_t and y_t series. The statistic

$$Q_M = n^2 \sum_{k=0}^{M} \frac{1}{n-k} r_{uv}^2(k) \qquad (7)$$

is then approximately distributed as $\chi^2(M+1)$. A significantly large value for Q_M would mean that the hypothesis should be rejected and therefore X causes Y.

A limitation of the methods explained in this section is that they are only useful when describing the relationships between two time series. If three or more time series are mutually related, then analyzing them only two at a time may lead to spurious relationships. Consequently, further research on causality between linear systems is still required. Nevertheless, in many situations bivariate causality studies are of direct interest to the practitioner (see, for example, the Applications Section of this paper).

When sufficient data are available, an alternative approach for detecting causal linear relationships is to work in the frequency domain rather than the time domain by employing the coherence function. An advantage of this procedure is that it can be extended for handling multiple-input and multiple-output systems (Bendat and Piersol, 1980).

3. DYNAMIC MODELS

3.1 Description

As explained in the previous section, the properties of the residual CCF can be employed in conjunction with a physical understanding of the problem to ascertain the types of causal relationships that exist between two time series. When meaningful connections are detected a stochastic model can then be developed that mathematically describes the formal links between the series. This stochastic model can then be utilized for applications such as forecasting and simulation.

When there is feedback between two series (i.e., there are some values of $r_{\hat{u}\hat{v}}(k)$ which are significantly different from zero at both positive and negative lags), multivariate models must be considered. For the applications considered later in this paper multivariate models are not required. Therefore, multivariate model building will not be discussed but the reader may wish to refer to the appropriate statistical literature for a description of these models and how to use them in practice (see, for instance, Tiao and Box (1981), Hillmer and Tiao (1979), and Li and McLeod (1981)).

As shown in Table 1, for the situation where one variable causes another, the residual CCF must have one or more values which are significantly different from zero, where all the large magnitudes appear entirely at either non-negative or else non-positive lags. For instance, when X causes Y but not instantaneously, at least one value of $r_{\hat{u}\hat{v}}(k)$ must be significantly different from zero for positive lags and small magnitudes must exist at all non-positive lags. Transfer function-noise models can be designed to mathematically model the connections between two series when it is known that one data set definitely causes another and therefore one sequence acts as an input to cause the output series. These models can also be developed when there are multiple input or covariate series that cause the single output series.

Consider the situation where there is no feedback and X causes Y. A transfer function-noise model that relates the x_t and y_t series can be written mathematically as (Box and Jenkins, 1970, Ch. 10)

$$y_t - \mu_y = \nu(B)(x_t - \mu_x) + N_t \qquad (8)$$

where $\nu(B) = \frac{\omega(B)}{\delta(B)} = \nu_0 + \nu_1 B + \nu_2 B^2 + \ldots$ is the transfer function with weights $\nu_0, \nu_1, \nu_2, \ldots$, which are called the impulse response function of the system; $\omega(B) = \omega_0 - \omega_1 B - \omega_2 B^2 - \ldots - \omega_s B^s$ is the operator in the numerator of the transfer function such that the roots of $\omega(B) = 0$ lie outside the unit circle and ω_i, $i = 1, 2, \ldots, s$, are the parameters of $\omega(B)$; $\delta(B) = 1 - \delta_1 B - \delta_2 B^2 - \ldots - \delta_r B^r$ is the operator in the denominator of the transfer function and for stability the roots of $\delta(B) = 0$ lie outside the unit circle and δ_i, $i = 1, 2, \ldots, r$, are the parameters of $\delta(B)$; and $N_t = \frac{\theta(B)}{\phi(B)} a_t$ is the ARMA noise term with a white noise sequence denoted by a_t which has a mean of zero and variance σ_a^2, and where $\theta(B)$ and $\phi(B)$ are the MA and AR operators, respectively, that are defined in the same fashion as the operators in (3) or (4).

When the parameters for the $\omega(B)$ and $\delta(B)$ operators are known, as is the case when they are estimated from the given data, the ν_j coefficients can be determined by utilizing the equation

$$\delta(B) \nu_j = -\omega_j \quad \text{for } j = 1, 2, \ldots \qquad (9)$$

where B operates on the subscript j and therefore $B^k \nu_j = \nu_{j-k}$; $\nu_0 = \omega_0$; and $\nu_j = 0$ for $j < 0$.

3.2 Model Construction

No matter what type of model is being fitted to a given data set it is recommended to follow the identification, estimation and diagnostic check stages of model development (Box and Jenkins, 1970). When designing a transfer function-noise model, the number of parameters required in the $\nu(B)$ operator and N_t term in (8) must be identified. Three procedures for model identification are the empirical approach that has been used when modelling hydrological time series

(Hipel et al., 1977b,c, 1975, 1982; Baracos et al., 1981; McLeod et al., 1983) the technique of Haugh and Box (1977), and the method of Box and Jenkins (1970) which is based upon suggestions by Bartlett (1935). The latter two methodologies rely heavily upon the results of cross-correlation studies and often the first procedure can be used in conjunction with either the second or third approaches. For a detailed description of how to apply the foregoing identification procedures and also the relative advantages and limitations of each method, the reader can refer to the text of Hipel and McLeod (1984).

Following the identification of one or more plausible transfer function-noise models, maximum likelihood estimates (MLE) are obtained for all the model parameters. Within the Application Section of this paper, the method of McLeod (1977) is employed for getting MLE's of the parameters although other recommended maximum likelihood procedures include those of Ansley (1979) and Ljung and Box (1979). When parameters are estimated for a number of possible models, a convenient method for choosing the most appropriate model is to select the model that has the minimum value of the Akaike information criterion (AIC) (Akaike, 1974) defined by

$$AIC = -2\ln ML + 2k \qquad (10)$$

where ML denotes maximum likelihood and k is the number of independently adjusted parameters in the model. Within the hydrological literature, the method of employing the AIC in conjunction with the three stages of model development has been clearly explained (Hipel and McLeod, 1984; Hipel, 1981) and the efficacy of the AIC has been confirmed by a wide range of stochastic modelling applications (see for example McLeod et al. (1977, 1983), McLeod and Hipel (1978), Hipel and McLeod (1984), Baracos et al. (1981), and Hipel (1981)). If a suitable range of models are considered, it has been found in practice that the model possessing the minimum AIC value also satisfies diagnostic tests of the model residuals.

The innovation sequence, a_t, is assumed to be independently distributed and a recommended procedure for checking the whiteness assumption is to examine a plot of the residual ACF along with confidence limits. The residual ACF, $r_{\hat{a}\hat{a}}(k)$, can be calculated by replacing both $\hat{u}_t$ and $\hat{v}_t$ by $\hat{a}_t$ in (6). Since $r_{\hat{a}\hat{a}}(k)$ is symmetric about lag zero, the residual ACF is only plotted against lags for $k = 1$ to $k \simeq n/4$ and the method of McLeod (1978) can be employed to calculate confidence limits. If the residuals are correlated, this suggests some type of model inadequacy. To ascertain the source of the error in the model, the CCF for the $\hat{u}_t$ and $\hat{a}_t$ sequences can be studied (leave $\hat{u}_t$ as $\hat{u}_t$ and replace $\hat{v}_t$ by $\hat{a}_t$ in (6) to estimate $r_{\hat{u}\hat{a}}(k)$). Because the $\hat{u}_t$ and $\hat{a}_t$ series are assumed to independent of one another, the estimated values of $r_{\hat{u}\hat{a}}(k)$ should not be significantly different from zero where one standard error is approximately $n^{-1/2}$

when the CCF is normally distributed. When a plot of $r_{\hat{u}\hat{a}}(k)$ from $k \simeq -n/4$ to $k \simeq n/4$ along with chosen confidence limits indicate whiteness while significant correlations are present in $r_{\hat{a}\hat{a}}(k)$, the model inadequacy is probably in the noise term, $\hat{N}_t$. The form of the residual ACF for the $\hat{a}_t$ series should suggest appropriate modifications to the noise structure. However, if both $r_{\hat{a}\hat{a}}(k)$ and $r_{\hat{u}\hat{a}}(k)$ possess one or more significant values, where $r_{\hat{u}\hat{a}}(k)$ only has large values at non-negative lags, this could mean that the transfer function is incorrect and the noise term may or may not be suitable. When feedback is indicated by significant values of $r_{\hat{u}\hat{a}}(k)$ at negative lags, a multivariate model should be considered rather than a transfer function-noise model. Whenever problems arise in the model building process, suitable model modifications can be made from information at the diagnostic check and identification stages. Subsequent to estimating the model parameters for the new model, the modelling assumptions should be checked to see if further changes are necessary.

Besides being independently distributed, the a_t sequence is assumed to possess constant variance (homoscedasticity) and follow a normal distribution. Tests are available for checking the homoscedastic and normality suppositions (see for example Hipel et al. (1977a) and McLeod et al. (1977)), and in practice it has been found that suitable Box-Cox transformations of the Y_t and/or X_t series can often correct heteroscedasticity and non-normality in the residuals. The Box-Cox transformation for the Y_t series is given as

$$y_t = \begin{cases} \lambda^{-1}[(Y_t + c)^\lambda - 1] & \lambda \neq 0 \\ \ln(Y_t + c) & \lambda = 0 \end{cases} \qquad (11)$$

where the constant c is usually assigned a magnitude which is just large enough to make all the entries in the Y_t series to be positive.

4. APPLICATIONS

4.1 Data

For a long time hydrologists have been attempting to ascertain the impact of exogenous forces upon specific hydrological and meteorological phenomena. In many instances, the great complexity of the physical problem at hand has precluded the development of suitable physical or statistical models to realistically describe the situation. Consequently, a wide range of phenomena are now studied in order to detect and model meaningful dynamic relationships.

The time series investigated are listed in Table 2. Except for monthly temperatures from the English Midlands, all of the data sets consist of annual values. The sunspot numbers, annual and monthly temperatures, seven river flow series in m^3/s where each average yearly flow is calculated for the water year from October 1st of one year to September 30th of the next year, and Beveridge wheat price indices, are obtained from articles by Waldemeier (1961), Manley (1953, pp. 255-260), Yevjevich (1963), and Beveridge (1921), respectively. The tree ring widths given in units of 0.01 mm are for Bristlecone Pine and were received directly from V.C. LaMarche at the Laboratory of Tree Ring Research, University of Arizona, Tuscon, Arizona. The length and accuracy of the tree ring series make it a valuable asset in cross-correlation studies for determining the effects of external variables such as temperature and amount of sunlight. The reason for considering the Beveridge wheat price index data is that the series could be closely related to climatic conditions and therefore may be of interest to hydrologists and climatologists. For example, during years when the weather is not suitable for abundant grain production the price of wheat may be quite high.

TABLE 2. Time Series Used in the Causality Studies

DATA SET	LOCATION	PERIOD	LENGTH
Sunspots	Sun	1700-1960	261
Annual Temperatures	English Midlands	1723-1970	248
12 Monthly Temperature Sequences	English Midlands	1723-1970	248 per month
St. Lawrence River	Ogdensburg, New York, USA	1860-1957	97
Volga River	Gorkii, USSR	1877-1935	58
Neumunas River	Smalininkai, USSR	1811-1943	132
Rhine River	Basle, Switzerland	1807-1957	150
Gota River	Sjotorp-Vanersburg, Sweden	1807-1957	150
Danube River	Orshava, Romania	1837-1957	120
Mississippi River	St. Louis, Missouri, USA	1861-1957	96
Beveridge Wheat Price Index	England	1500-1869	370
Tree Ring Widths	Campito Mountain, California, USA	1500-1969	470

4.2 Prewhitening

When checking for causality, the time series under investigation must first be prewhitened. Table 3 describes the types of models which were used to prewhiten the series from Table 2. In all cases, the models were determined by following the three stages of model construction in conjunction with the AIC and in some instances the most appropriate models are constrained models where some of the model parameters are omitted. For example, the best ARMA model for the sunspot series is a constrained ARMA (9,0) model where ϕ_3 to ϕ_8 are left out of the model and the original data is transformed by a square root transformation where $\lambda = 0.5$ in (11) and $c = 1$ due to some zero values in the series. Using the format in (4), the estimated sunspot model is written in difference equation form as

$$(1 - 1.245B + 0.524B^2 - 0.192B^9)(y_t - 10.652) = a_t \qquad (12)$$

where

$$y_t = (1/0.5)[(Y_t + 1.0)^{0.5} - 1.0]$$

Notice for the Beveridge wheat price indices that the data are transformed using a natural logarithmic transformation where $\lambda = 0$ and $c = 0$ in (11). The transformed data are then differenced once to remove nonstationarity by using

TABLE 3. Models Used to Get Residuals for the CCF Studies

DATA SET	ARMA (p,q) MODEL	$\hat{u}_t$	$\hat{v}_t$
Sunspots	(9,0) without ϕ_3 to ϕ_8, $\lambda = 0.5$ and $c = 1$	✓	
Annual Temperature	(2,0) without ϕ_1	✓	✓
12 Monthly Temperature Sequences	(0,0) for all months	✓	✓
St. Lawrence River	(3,0) without ϕ_2		✓
Volga River	(0,0)		✓
Neumunas River	(0,1), $\lambda = 0$		✓
Rhine River	(0,0)		✓
Gota River	(2,0)		✓
Danube River	(0,0)		✓
Mississippi River	(0,1)		✓
Beveridge Wheat Price Index	(8,1) without ϕ_3 to ϕ_7, $\lambda = 0$, and series is differenced once		✓
Tree Ring Widths	(4,0) without ϕ_2		✓

$$x_t = \ln X_{t+1} - \ln X_t$$

for $t = 1,2,3,\ldots,n-1$. Following this, identification procedures reveal that an ARMA (8,1) without ϕ_3 to ϕ_7 should be fitted to x_t where the estimated model is given as

$$(1 - 0.729B + 0.346B^2 + 0.119B^8)x_t = (1 - 0.783B)u_t \qquad (13)$$

The reader should bear in mind that only the family of ARMA models were entertained when selecting the best model to describe each data set in Table 3. In certain instances it may be appropriate to also consider other types of models. For example, Akaike (1978) noted that because of the nature of sunspot activity a model based on some physical consideration of the generating mechanism may produce a better fit to the sunspot series than an ARMA model. For modelling the sunspot series, Granger (1978) utilized a bilinear model. Whatever the case, for each time series in Table 3 extensive diagnostic checking was executed to insure that the best ARMA model was ultimately chosen.

4.3 Causality Studies

Following prewhitening, (6) is employed to calculate the residual CCF for two specified residual series. In the third and fourth columns of Table 3 check marks indicate when the residuals of a given series are used as $\hat{u}_t$ and/or $\hat{v}_t$, respectively, in (6). Whenever two series are cross-correlated, the residual values are used for the time period during which the $\hat{u}_t$ and $\hat{v}_t$ data sets overlap. The sunspot residuals could possibly affect all the other series in Table 3 and therefore the sunspot residuals are separately cross-correlated with each of the remaining series in Table 3. For the monthly temperature data, each monthly sequence is considered as a separate sample when the residual CCF is calculated between the sunspot series as $\hat{u}_t$ and a given monthly temperature data set as $\hat{v}_t$. However, it is also possible that temperature can affect the phenomena listed below the temperature series in Table 3. For example, April temperatures may influence tree ring growth in the Northern Hemisphere since the month of April is when the growing season begins after the winter months. Consequently, the residual series for the annual temperature data set and the 12 monthly temperature sequences are each cross-correlated with the residual series of the data given below the temperatures.

In many situations it may not be known whether or not one phenomenon definitely causes another. Although the direction of suspected causality is often known a priori due to a physical understanding of the problem, proper statistical methods must be employed to ascertain if the available evidence confirms or denies the presence of a significant causal relationship. Consider, for example, determining whether or not sunspots and river flows are causally related. Obviously, it is only physically possible for sunspots to cause river flows and not vice versa.

Based upon ad hoc graphical procedures comparing annual flows of the Volga River in the USSR with yearly sunspot numbers, Smirnov (1969) postulated that sunspots unequivocally affect river flows. However, when the residual CCF is used to scientifically detect causality, the results do not support Smirnov's strong claim. In Figure 1, the residual CCF along with the 95% confidence limits are presented for the residuals from the ARMA model fitted to the annual flows of the Volga River at Gorkii, USSR, and the residuals from the ARMA model fitted to the annual sunspot numbers (refer to Table 2 for a description of these data sets and to Table 3 for the types of models fitted to the two time series). As can be seen, there are no significant values of the CCF at lag zero and the smaller positive lags. If sunspot activity did affect the Volga flows it would be expected that this would happen well within the time span of a few years. Therefore, the absence of significant values of the CCF from lags 0 to 2 or 3 indicates that the current information does not support the hypothesis that sunspots cause the Volga River flows. The slightly large magnitudes at lags 5 and 11 are probably due to chance. Nevertheless, it is possible, but highly unlikely, that the value at lag 11 could be due to the fact that the best ARMA model could not completely remove the

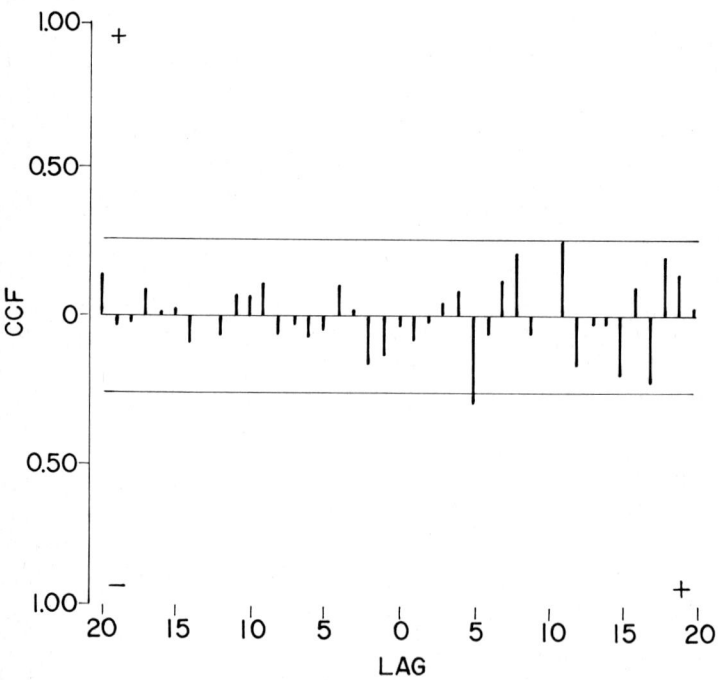

Fig. 1. Residual CCF for the sunspot numbers and the Volga River flows.

periodicity present in the sunspot series. Previously, Granger (1957) found that the periodicity of sunspot data follows a uniform distribution with a mean of about 11 years. However, the sunspot model in Table 3 was originally designed by McLeod et al. (1977) to account for this and these authors subjected the model to rigorous diagnostic checks to demonstrate that the periodicity was not present in the model residuals and none of the values of the residual ACF were significantly different from zero, even at lag 11.

Besides the annual flows of the Volga River, no meaningful causality relationships are detected when the sunspot residuals are separately cross-correlated with the other river flow residuals and also the remaining residuals series which are considered as $\hat{v}_t$ in Table 3. As emphasized earlier, if correct statistical procedures are not followed it would not be possible to reach the aforesaid conclusions regarding the causality relationships between the sunspots and the other phenomena. For example, in Figure 2 it can be seen that the values of the CCF calculated for the given annual sunspot and Gota River flows series are large in magnitude at negative and positive lags (recall that the 95% confidence limits in Figure 2 are derived for independent series). Furthermore, the cyclic nature

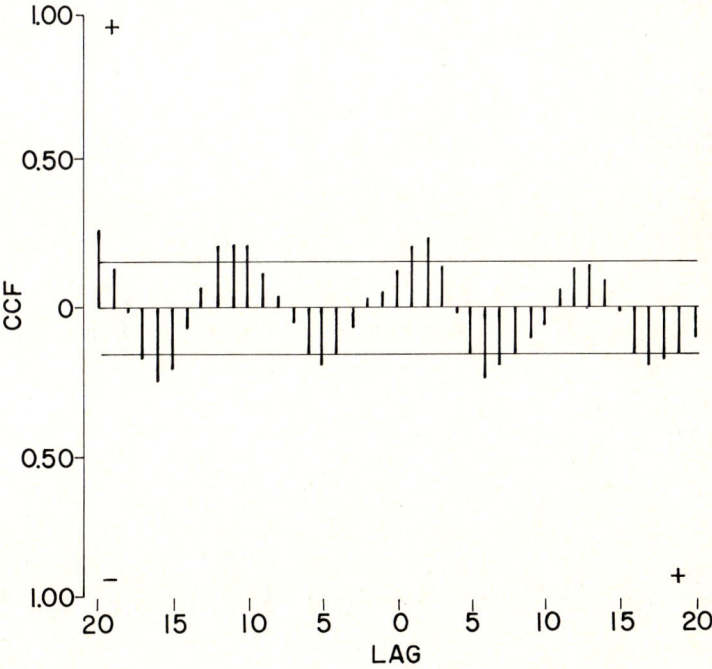

Fig. 2. CCF for the given sunspot numbers and the Gota River flows.

of the sunspot data is portrayed by the sinusoidal characteristics in the graph. To uncover the underlying causal relationship between the series it is necessary to examine the residual CCF. As just noted, the residual CCF for the sunspot and Gota River series does not reveal that sunspot numbers affect the flows of the Gota River.

For the case where the $\hat{u}_t$ sequence, as represented by the residuals of the annual temperature data, is cross-correlated with each of the last nine $\hat{v}_t$ series in Table 3, no meaningful relationships are found. However, some significant values of the residual CCF are discovered when each monthly temperature series is cross-correlated separately with each residual sequence for the river flows and also the Beveridge wheat price indices. Table 4 shows the lags at which the residual CCF possesses large values when $\hat{u}_t$ is a designated monthly temperature series and $\hat{v}_t$ is either the annual river flow or Beveridge wheat price index residuals. Since it would be expected from a physical viewpoint that a given monthly temperature data set would have the most effect upon the other time series in the

TABLE 4. Residual CCF Results of Monthly Temperatures and Other Series

$\hat{v}_t$	MONTHLY TEMPERATURES $\hat{u}_t$	LAGS FOR LARGER VALUES OF RESIDUAL CCF
St. Lawrence River	February	1
Volga River	February April May July	2 2 0 and 1 2
Neumunas River	May July December	0 2 2
Rhine River	October	0
Gota River	June July August September	0 0 0 0
Danube River	September October	0 0
Mississippi River	December	1
Beveridge Wheat Price Index	February November December	0 0 0

current year or perhaps one or two years into the future, large values of the residual CCF are only indicated in Table 4 when they occur at lags 0 to 2. As an illustrative example, consider the graph of the residual CCF for the August temperatures and the Gota River residuals which is shown in Figure 3. As can be seen, the large negative correlation at lag zero extends well beyond the 95% confidence limits. When the Q_M statistic in (7) is calculated for lags 0 to 2, the estimated value for the residual CCF in Figure 3 is 26.6. Because this value is much larger than the tabulated $\chi^2(3)$ value of 7.8 for the 5% significance level, one must reject the null hypothesis that the August temperatures do not affect the annual flows of the Gota River.

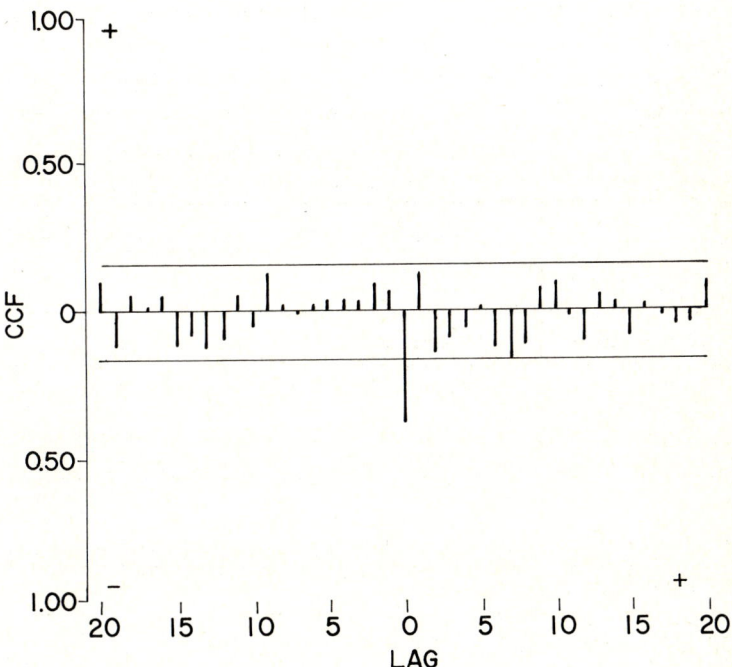

Fig. 3. Residual CCF for the August temperatures and the Gota River flows.

4.4 Dynamic Modelling

The residual CCF studies can be considered as part of the "exploratory" data analysis stage where simple graphical and statistical tools are employed for detecting important statistical characteristics of the data (Tukey, 1977). At the "confirmatory" data analysis step, the dynamic model in equation (8) can be utilized to formally model and confirm the mathematical relationships which are discovered at the exploratory data analysis stage. Examples of dynamic model building are available in the hydrological literature (McLeod et al., 1983; Hipel et al., 1982; Hipel and McLeod, 1984) and in this paper a dynamic model is now presented for formally describing the relationships between the annual Gota River flows and the monthly temperatures.

From Table 4, it can be seen that there are significant values of the residual CCF at lag zero between the flows of the Gota River and each of four monthly temperature series. This relationship is displayed graphically in Figure 3 for the case of the residual CCF for the Gota River flows and the August temperatures. By following the model construction phases outlined in Section 3.2, an appropriate dynamic model can be developed. When all four temperature series are used as covariate series in a transfer function-noise model, the transfer function parameter estimates for June, July and September are not significantly different from zero and can therefore be left out of the model. Whereas the residual CCF can only be used for pairwise comparisons, the transfer function-noise model can be employed to ascertain the most meaningful relationship when there are multiple covariate series and a single response or output series. For the case where only the August temperatures are used as a covariate series, the parameter estimates and standard errors are listed in Table 5 while the difference equation for the model which follows the format of equation (8) is written as

$$Y_t - 535.464 = -23.869(X_t - 15.451) + (1 - 0.672B + 0.329B^2)^{-1} a_t \qquad (14)$$

where Y_t represents the annual flows of the Gota River and X_t stands for the monthly temperatures. Besides portraying the dynamic relationship between the Gota River flows and August temperatures, the model in equation (14) can be employed for applications such as forecasting and simulation.

TABLE 5. Parameter Estimates for the August Temperature - Gota River Flow Transfer Function-Noise Model

PARAMETER	MLE	STANDARD ERROR
ω_0	-23.869	4.285
ϕ_1	0.672	0.077
ϕ_2	-0.329	0.077
σ_a	5.714×10^3	

It would be expected that temperature could significantly affect tree ring growth. As noted by La Marche (1974), because Bristlecone Pines are located at the upper treeline on mountains, temperature is a key factor in controlling growth. However, this growth would only be sensitive to local temperature conditions and the temperatures recorded in the English Midlands are probably not representative of the temperatures at Campito Mountain in California. If local temperatures were available, the results of the residual CCF between the temperature and the tree ring widths could be utilized to design a dynamic model to connect the two series. Because the tree ring series would be much longer than any existing temperature series, the estimated transfer function-noise model connecting the two series could be used to back-forecast the missing values of the temperature series.

5. CONCLUSIONS

Comprehensive procedures are now available for detecting causal relationships between two time series. The results of Table 4 demonstrate that monthly temperatures can significantly affect annual river flows and also the price of wheat. However, no meaningful links are found between the annual sunspot numbers and the other phenomena designated by $\hat{v}_t$ in Table 3. In particular, the statistical evidence from Figure 1 cannot support the claim (Smirnov, 1969) that sunspots significantly affect the annual flows of the Volga River. While some of the findings of this paper may be somewhat interesting, it is also informative to note the types of results that Pierce (1977) discovered in the field of econometrics. Using residual CCF studies, Pierce (1977) found that numerous economic variables which were generally regarded by economists as being strongly interrelated were in fact independent or else only weakly correlated. These conclusions are of course based upon the information included in the time series which Pierce analyzed. If it were possible to improve the design of the data collection scheme for a causality study, this would of course enhance the conclusions reached at the analysis stage. Certainly, it is necessary that a sufficiently wide range of values of the relevant variables appear in the sample, in order to increase the probability of detecting relationships which do actually exist in the real world. However, as is the case in economics and also in geophysics, the experimenter has little control over the phenomena which produce the observations and must therefore be content with the data that can be realistically collected. Perhaps "God" may have a switch that can greatly vary the number of sunspots that appear on the sun so that "mortal man" can assess beyond a shadow of a doubt whether or not sunspots can significantly affect river flows.

Given the available information, it is essential that this data be properly analyzed. For example, if a sample CCF were calculated for the x_t and y_t

series, "spurious" correlations may seem to indicate that the variables are causally related (see Figure 2, for instance). However, an examination of the residual CCF for the two series may clearly reveal that based upon the given data no meaningful relationships do in fact exist between the two phenomena. It is of course possible that no significant correlations may appear in the residual CCF even though two variables are functionally related. This is because correlation is only a measure of linear association and nonlinear relationships that contain no linear component, may be missed. To minimize the occurrence of this type of error, the fitted ARMA models that are used to prewhiten the series are subjected to stringent diagnostic checks. In this way, any problems that arise due to the use of these linear models will be detected prior to examining the residual CCF.

Subsequent to the revelation of causality using the residual CCF, a dynamic model from (8) can be built to mathematically describe the formal connections between the x_t and y_t series. An inherent advantage of the transfer function-noise models discussed in this paper is that well developed methodologies are available for use at the three stages of model construction. For instance, at the identification step the results of the residual CCF study that detected the causal relationship in the first place, can be utilized to design the dynamic model (Haugh and Box, 1977). When the y_t series has been altered by one or more external interventions, intervention components can be introduced into the transfer function-noise model to account for possible changes in the mean level (Box and Tiao, 1975; Hipel et al., 1975, 1977b,c; Baracos et al., 1981; McLeod et al., 1983).

ACKNOWLEDGEMENTS

The authors greatly appreciate the useful comments of the editor and referees which enhanced the quality of the paper.

REFERENCES

AKAIKE, H. (1974). A new look at the statistical model identification. IEEE Transactions on Automatic Control 19, 716-723.

AKAIKE, H. (1978). On the likelihood of a time series model. In Proceedings of the International Conference on Time Series Analysis and Forecasting (convened by O.D. Anderson for the Institute of Statisticians), Cambridge University, England, July, 1978, The Statistician 27, 217-235.

ANSLEY, C.F. (1979). An algorithm for the exact likelihood of a mixed autoregressive-moving average process. Biometrika 66, 59-65.

BARACOS, P.C., HIPEL, K.W. and MCLEOD, A.I. (1981). Modelling hydrologic time series from the Arctic. Water Resources Bulletin 17, 414-422.

BARTLETT, M.S. (1935). Stochastic Process. Cambridge University Press, London.

BENDAT, J.S. and PIERSOL, A.G. (1980). Engineering Applications of Correlation and Spectral Analysis, Wiley, New York.

BEVERIDGE, W.H. (1921). Weather and harvest cycles. Economics Journal 31, 429-552.

BOX, G.E.P. and COX, D.R. (1964). An analysis of transformations. Journal of Royal Statistical Society, Series B 26, 211-252.

BOX, G.E.P. AND JENKINS, G.M. (1970). Time Series Analysis: Forecasting and Control. Holden-Day, San Francisco.

BOX, G.E.P. and TIAO, G.C. (1975). Intervention analysis with applications to economic and environmental problems. Journal of American Statistical Association 70, 70-79.

BRILLINGER, D.R. (1969). A search for a relationship between monthly sunspot numbers and certain climatic series. Bulletin of International Statistical Institute 43, 293-307.

GRANGER, C.W.J. (1957). A statistical model for sunspot activity. Astrophysics Journal 126, 152-158.

GRANGER, C.W.J. (1969). Investigating causal relations by econometric models and cross-spectral methods. Econometrica 37, 424-438.

GRANGER, C.W.J. (1978). New classes of time series models. In Proceedings of the International Conference on Time Series Analysis and Forecasting (convened by O.D. Anderson for the Institute of Statisticians), Cambridge University, England, July, 1978. The Statistician 27, 237-253.

HAUGH, L.D. (1972). The Identification of Time Series Interrelationships with Special Reference to Dynamic Regression. Ph.D. Thesis, Department of Statistics, University of Wisconsin, Madison, Wisconsin, USA.

HAUGH, L.D. (1976). Checking the independence of two covariance-stationary time series: a univariate residual cross-correlation approach. Journal of American Statistical Association 71, 378-385.

HAUGH, L.D. and BOX, G.E.P. (1977). Identification of dynamic regression (distributed lag) models connecting two time series. Journal of American Statistical Association 72, 121-130.

HILLMER, S.C. and TIAO, G.C. (1979). Likelihood function of stationary multiple autoregressive moving average models. Journal of American Statistical Association 74, 652-660.

HIPEL, K.W. (1981). Geophysical model discrimination using the Akaike information criterion. IEEE Transactions on Automatic Control 26, 358-378.

HIPEL, K.W., LENNOX, W.C., UNNY, T.E. and MCLEOD, A.I. (1975). Intervention analysis in water resources. Water Resources Research 11, 855-861.

HIPEL, K.W. and MCLEOD, A.I. (1984). Time Series Modelling for Water Resouces and Environmental Engineers. Elsevier, Amsterdam, in press.

HIPEL, K.W., MCLEOD, A.I. and LENNOX, W.C. (1977a). Advances in Box-Jenkins modelling, 1, model construction. Water Resources Research 13, 567-575.

HIPEL, K.W., MCLEOD, A.I. and MCBEAN, E.A. (1977b). Stochastic modelling of the effects of reservoir operation. Journal of Hydrology 32, 97-113.

HIPEL, K.W., MCLEOD, A.I. and NOAKES, D.J. (1982). Fitting dynamic models to hydrological time series. In Time Series Methods in Hydroscience. Eds: A.H. El-Shaarawi and S.R. Esterby, Elsevier, Amsterdam, 110-129

HIPEL, K.W., MCLEOD, A.I., UNNY, T.E. and LENNOX, W.C. (1977c). Intervention analysis to test for changes in the mean level of a stochastic process. In Stochastic Processes in Water Resources Engineering, Eds: L. Gottschalk, G. Lindh and L. Maré. Water Resources Publications, Fort Collins, Colorado, 1: 93-113.

JENKINS, G.M. and WATTS, D.G. (1968). Spectral Analysis and its Applications. Holden-Day, San Francisco.

LA MARCHE, V.C. JR. (1974). Paleoclimatic inferences from long tree-ring records. Science 183, 1042-1048.

LI, W.K. and MCLEOD, A.I. (1981). Distribution of the residual autocorrelations in multivariate ARMA time series models. Journal of Royal Statistical Society Series B 43, 231-239.

LJUNG, G.M. and BOX, G.E.P. (1979). The likelihood function of stationary autoregressive-moving average models. Biometrika 66, 265-270.

MANLEY, G. (1953). The mean temperatures of central England (1698-1952). Quarterly Journal of Royal Meteorological Society 79, 242-261.

MCLEOD, A.I. (1977). Improved Box-Jenkins estimators. Biometrika 64, 531-534.

MCLEOD, A.I. (1978). On the distribution of residual autocorrelations in Box-Jenkins models. Journal of Royal Statistical Society Series B 4, 296-302.

MCLEOD, A.I. (1979). Distribution of the residual cross-correlation in univariate ARMA time series models. Jorunal of American Statistical Association 74, 849-855.

MCLEOD, A.I. and HIPEL, K.W. (1978). Preservation of the rescaled adjusted range, 1, a reassessment of the hurst phenomenon. Water Resources Resarch 14, 491-508.

MCLEOD, A.I., HIPEL, K.W. and CAMACHO, F. (1983). Trend assessment of water quality time series, Water Resources Bulletin 19, 537-547.

MCLEOD, A.I., HIPEL, K.W. and LENNOX, W.C. (1977). Advances in Box-Jenkins modelling, 2, applications. Water Resources Research 13, 577-586.

PIERCE, D.A. (1977). Relationships - and the lack thereof - between economic time series, with special reference to money and interest rates. Journal of American Statistical Association 72, 11-21.

PIERCE, D.A. and HAUGH, L.D. (1977). Causality in temporal systems. Journal of Econometrics 5, 265-293.

PIERCE, D.A. and HAUGH, L.D. (1979). The characterization of instantaneous causality, a comment. Journal of Econometrics 10, 257-259.

PRICE, J.M. (1979). The characterization of instantaneous causality, a correction. Journal of Econometrics 10, 253-256.

RODRIGUEZ-ITURBE, I. and YEVJEVICH, V. (1968). The Investigation of Relationships Between Hydrologic Time Series and Sunspot Numbers. Hydrology Paper No. 26, Colorado State University, Fort Collins, Colorado.

SMIRNOV, N.P. (1969). Causes of long-period streamflow fluctuations. Bulletin of All-Union Geographic Society (Izvestiya VGO) 101, 443-440.

STEDINGER, J.R. (1981). Estimating correlations in multivariate streamflow models. Water Resources Research 17, 200-208.

TIAO, G.C. and BOX, G.E.P. (1981). Modelling multiple time series with applications. Journal of American Statistical Association 76, 802-816.

TUKEY, J.W. (1977). Exploratory Data Analysis. Addison-Wesley, Reading, Massachusetts.

WALDEMEIER, M. (1961). The Sunspot Activity in the Years 1610-1960. Schulthas and Company, Zurich, Switzerland.

WIENER, N. (1956). The theory of prediction. In Modern Mathematics for Engineers, Series 1. Ed: E.F. Beckenback. Chapter 8. McGraw-Hill, New York.

YEVJEVICH, V.M. (1963). Fluctuation of Wet and Dry Years, 1, Research Data Assembly and Mathematical Models. Hydrology Paper No. 1, Colorado State University, Fort Collins, Colorado.

GROUPING OF PERIODIC AUTOREGRESSIVE MODELS

Robert Marshal Thompstone
Alcan Smelters and Chemicals Ltd., P.O. Box 1500,
Jonquière, Québec, G7S 4L2, Canada

Keith William Hipel
Dept. of Systems Design Engineering, University of Waterloo,
Waterloo, Ontario, N2L 3G1, Canada

Angus Ian McLeod
Dept. of Statistical and Actuarial Sciences, University of Western Ontario,
London, Ontario, N6A 5B9, Canada

Periodic autoregressive (PAR) models are described and recent developments in their identification, parameter estimation and diagnostic checking stages reviewed. Combining individual autoregressive models for various seasons, to obtain a single model for all seasons in a given group, is then examined. After grouping, parameters of the more parsimonious PAR models (designated as "PPAR" models) are estimated and diagnostically checked, and the PAR and PPAR models compared. The techniques are illustrated using monthly and quarter-monthly hydrological time series (ie, the year is divided into 12 or 48 periods).

1. INTRODUCTION

Seasonal hydrological time series exhibit an autocorrelation structure which depends on not only the time lag between observations but also the season of the year (Moss and Bryson, 1974). Periodic autoregressive (PAR) models attempt to preserve this seasonally-varying autocorrelation structure by fitting a separate and different autoregressive (AR) model to each season of the year. However, one could reasonably question the necessity of going to the extreme of having a different model for each and every season. The purpose of this paper is to present a methodology for identifying groups of seasons with similar autocorrelation structures and to examine the consequences of fitting different AR models to each group rather than to each individual season. As will be seen later, the grouping of seasons results in a reduction of the overall number of AR parameters, and for this reason the resulting models are labelled herein as parsimonious periodic autoregressive (PPAR) models. Note that other researchers have argued that PAR models for monthly data become far too prodigal in parameters (see, for example, Anderson, 1978, p. 75).

The following section of this paper describes in greater detail the overall class of PAR models, and reviews recent developments in the identification, parameter estimation and diagnostic checking of such models. The subsequent section formally introduces the PPAR class of models, and presents a methodology for identifying groups of seasons with similar AR characteristics. Next, PAR and PPAR models are fit to three monthly and three quarter-monthly hydrological time series. The models are compared using the log-likelihood ratio of the residuals, the Akaike information criterion (AIC) (Akaike, 1974) and Schwarz's approximation of the Bayes information criterion (BIC) (Schwarz, 1978). The models are then employed in a forecasting experiment in which one step ahead forecast errors are compared using three years of data not considered in their identification and estimation. It is concluded that PPAR models offer an attractive alternative to PAR models.

2. PERIODIC AUTOREGRESSIVE (PAR) MODELS

Let z_t, $t = 1, 2, \ldots$, be a seasonal time series with period s. The time index, t, may be regarded as a function of the year, T (T = 1, 2, ..., N), and the season, m (m = 1, 2, ..., s). Thus the time index may be written as $t = (T-1)s + m$. The PAR $(p_1, p_2, \ldots, p_s)$ model may be written as

$$\phi^{(m)}(B)(z_t^{(\lambda)} - \mu_t) = a_t \qquad (1)$$

where $\phi^{(m)}(B) = 1 - \phi_1^{(m)} B - \ldots - \phi_{p_m}^{(m)} B$ is the AR operator of order p_m for season m, B is the backward shift operator on t, $\mu_t = \mu^{(m)}$ is the mean for season m, and $a_t \sim NID(0, \sigma^{2(m)})$. The superscript m obeys modulo arithmetic, ie, $\mu^{(1)} \equiv \mu^{(s+1)} \equiv \mu^{(-s+1)}$.

The superscript λ is the exponent of an appropriate Box-Cox transformation. The Box-Cox transformation (Box and Cox, 1964) is given by

$$z_t^{(\lambda)} = \begin{cases} \lambda^{-1}[(z_t + \text{constant})^\lambda - 1] & \lambda \neq 0 \\ \ln(z_t + \text{constant}) & \lambda = 0 \end{cases} \qquad (2)$$

Models similar to this family of seasonal models have previously been employed by other researchers (see Yevjevich, 1972; Clarke, 1973; Rao and Kashyap, 1974; Tao and Delleur, 1976; Croley and Rao, 1977; Anderson, 1978; McLeod and Hipel, 1978b; Pagano, 1978; Sen, 1978; Parzen and Pagano, 1979; Troutman, 1979; Salas, Boes and Smith, 1982). However, additional results to be used at the identification, estimation and diagnostic check stages of model development have recently been presented (McLeod and Hipel, 1984; Noakes, McLeod and Hipel, 1983, 1984) and are fully incorporated in the results presented herein. Thus the identification, estimation and verification of PAR models will be briefly reviewed.

Two procedures for obtaining parameter estimates for PAR models may be employed. In the first approach, efficient conditional maximum likelihood estimates of the AR parameters are obtained directly from the multiple linear regression of

$z_t^{(\lambda)}$ on $z_{t-1}^{(\lambda)}, z_{t-2}^{(\lambda)}, z_{t-3}^{(\lambda)}, \ldots, z_{t-p_m}^{(\lambda)}$.

The maximum order of autoregression for each season, p_m, need not be equal for each season and subset autoregression (McClave, 1975) can be used where specific autoregressive parameters are constrained to zero.

In the second scheme, the Yule-Walker equations are formulated and solved (McLeod and Hipel, 1983 a,b) to obtain estimates of the model parameters. For a given parameter in the same model, usually there is very little difference between the estimates when the multiple linear regression and Yule-Walker estimates are used. In the Yule-Walker approach, p_m may vary from season to season but all of the AR parameters are estimated. The order of the AR model fit to each season (p_m) may either be determined using some automatic selection criterion such as the AIC or by examining plots of the partial autocorrelation function (PACF) for each season (Sakai, 1982; McLeod and Hipel, 1983a). Tjostheim and Paulsen (1983) have recently discussed the bias of some commonly-used estimates of the parameters of time series models.

Noakes et al. (1983,1984) used thirty monthly river flow time series to compare these procedures for identifying and estimating PAR models. They concluded that probably the best procedure with respect to forecasting is to use the PACF to identify the order of AR model for each season, and to use an efficient estimation procedure, such as one of those just mentioned, to estimate the parameters. Consequently, this approach has been adopted herein. The fact that the forecasting experiments of Noakes et al (1983,1984) also showed that this type of PAR model forecasts significantly better than several other types of stochastic models commonly advocated for seasonal hydrological series underlines the importance of such models.

In the estimation procedure adopted herein, the seasonal mean parameter, $\mu^{(m)}$, is estimated by

$$\hat{\mu}^{(m)} = \frac{1}{N} \sum_{i=1}^{N} z_{(i-1)s+m}^{(\lambda)} \qquad m = 1,2,\ldots,s \qquad (3)$$

The residuals, $\hat{a}_t$, are calculated from (1) by setting initial values of the residuals to zero and the residual variance, $\sigma^{2(m)}$, is then estimated by

$$\hat{\sigma}^{2(m)} = \frac{1}{N} \sum_{i=1}^{N} a_{(i-1)s+m}^2 \qquad m = 1,2,\ldots,s \qquad (4)$$

It should be noted that the parameters for the mth season (ie, $\mu^{(m)}$, $\sigma^{2(m)}$, $\phi_1^{(m)}, \phi_2^{(m)}, \ldots, \phi_{p_m}^{(m)}$), can be estimated entirely independently of the parameters of any other season. This is because the Fisher large sample information matrix is block diagonal (Pagano, 1978; McLeod and Hipel, 1978b) and hence the estimates of the parameters in different seasons are statistically independent. With respect to diagnostic checking, the adequacy of a fitted PAR model can be checked by examining the seasonal residual autocorrelations which are given by

$$\hat{r}^{(m)}(j) = \frac{\frac{1}{N}\sum_{i=1}^{N}\hat{a}_{(i-1)s+m} \cdot \hat{a}_{(i-1)s+m-j}}{\hat{\sigma}^{(m)} \cdot \hat{\sigma}^{(m-j)}} \tag{5}$$

McLeod and Hipel (1983a) have shown that if the model is adequate, $\hat{r}^{(m)}(j)$ is approximately normal with mean zero and variance less than $1/N$. Furthermore, the Portmanteau statistics (Li and McLeod, 1981)

$$Q_{m,\ell} = N \sum_{i=1}^{\ell} \hat{r}^{2(m)}(j) + \frac{\ell(\ell+1)}{2N} \tag{6}$$

are asymptotically independent for $m = 1, 2, \ldots, s$ and follow χ^2 distributions with $\ell - n_m$ degrees of freedom where n_m is the total number of parameters for season m. A large value of $Q_{m,\ell}$ indicates inadequacy of the model for the mth season. An overall test for model adequacy can be calculated using

$$Q_\ell = \sum_{m=1}^{s} Q_{m,\ell} \tag{7}$$

where Q_ℓ is χ^2 on $\sum_{m=1}^{s}(\ell - n_m)$ degrees of freedom if the overall PAR model is adequate.

3. PARSIMONIOUS PERIODIC AUTOREGRESSIVE (PPAR) MODELS

The PAR models described in the previous section attempt to preserve the seasonally-varying autocorrelation structure of a time series by fitting a separate AR model to each season of the year. However, one could reasonably question the necessity of going to the extreme of having a different model for each and every season. Assuming the s seasons are grouped into G groups of one or more seasons with similar AR characteristics, the parsimonious periodic autoregressive model PPAR $(p_1, p_2, \ldots, p_G)$ may be defined in a manner analogous to the PAR model as

$$\phi^{(g)}(B)(z_t^{(\lambda)} - \mu_t) = a_t \tag{8}$$

where $\phi^{(g)}(B) = 1 - \phi_1^{(g)} B - \ldots - \phi_{p_g}^{(g)} B^{p_g}$ is the AR operator of order p_g for group g, $\mu_t = \mu^{(m)}$ is the mean for season m, and $a_t \sim NID(0, \sigma^{2(g)})$. Notice from equation (8) that within a given group each seasonal mean is preserved by the parameter $\mu^{(m)}$.

In order to identify an appropriate grouping of seasons, the approach examined herein involves first fitting PAR models to the time series in question as outlined in the previous section. One then attempts to find seasons for which the AR models are "compatible". The equation of season m_2 is said to be compatible with that of season m_1, if the residuals obtained when the equation fit to season m_2 is applied to season m_1 are not significantly different from the residuals obtained from the equation fit to season m_1. In order to formally test for compatibility, define $a_T(m_1, m_2)$ to be the residuals obtained when the model fit to season m_2 is applied to season m_1 using equation (1) with initial values set to zero. These residuals can be used to estimate $\sigma^2(m_1, m_2)$, the residual variance when the model for season m_2 is applied to season m_1.

Consider the null hypothesis

$$H_o : \sigma^2(m_1, m_2) = \sigma^2(m_1, m_1)$$

Assuming that $(a_T(m_1, m_2), a_T(m_1, m_1))$ are jointly normally distributed with mean zero and are independent for successive values of T, a test developed by Pitman (1939) can be used to test this null hypothesis. Let

$$S_T = a_T(m_1, m_2) + a_T(m_1, m_1) \qquad (9)$$

and

$$D_T = a_T(m_1, m_2) - a_T(m_1, m_1) \qquad (10)$$

Pitman's test is then equivalent to testing if the correlation, r, between S_T and D_T is significantly different from zero. Thus, provided N > 25, H_o would be accepted at the 5% level of significance if $|r| < 1.96/\sqrt{N}$.

In practice, the residuals may not satisfy exactly the assumptions of a joint normal distribution with mean zero and independence for successive T. However, these assumptions are probably a sensible first approximation. The assumption of independence seems reasonable because, with annual periodicity, the residuals are chronoligically one year apart. Furthermore, the mean of zero is assured for the case of $a_T(m_1, m_1)$ due to the method of fitting the model. Pitman's test has often been used for testing the equality of variances of paired samples (Snedecor and Cochran, 1980, p. 190). It was pointed out in Lehmann (1959, p. 208, problem 33) that in this situation the test is unbiased and uniformly most powerful.

The above definition of compatibility of equations can be extended to mutual compatibility, ie, equations for seasons m_1 and m_2 are mutually compatible if, at a given level of significance, one would accept the following two hypotheses:

$$\sigma^2(m_2, m_1) = \sigma^2(m_1, m_1)$$

$$\sigma^2(m_1, m_2) = \sigma^2(m_2, m_2)$$

Thus, the criteria adopted herein for identifying seasons in the same group is that each pair of seasons in the group must be mutually compatible at a given level of significance and have the same order of AR model. In addition, seasons are not grouped together unless they are chronologically adjacent. Once the groups have been identified, the parameters are estimated using maximum likelihood estimation. Diagnostic checking invloves first calculating the residuals from (8) by setting initial values to zero, and then examining the seasonal residual autocorrelations and related Portmanteau test statistics as per equations (5) to (7). Note that occasionally the grouping given by these rules is not unique; in these cases, a subjective intervention was made to select one of the groupings which did respect the rules. Thompstone (1983, Ch. 3) shows that the Fisher large sample information matrix for parameters of the PPAR model is block diagonal and thus estimates of parameters in different groups are statistically independent. He also establishes confidence limits for the seasonal residual autocorrelation function of PPAR models by expanding on earlier work of McLeod (1978).

4. COMPARISON OF PAR AND PPAR MODELS

The PAR and PPAR models fit to a given series can be compared using the Akaike information criterion (AIC) (Akaike, 1974), Schwarz's approximation of the Bayes information criterion (BIC) (Schwarz, 1978; Neftci, 1982), and the log-likelihood ratio of the residuals. In general, the AIC is defined by

$$\text{AIC} = -2 \log (\text{maximum likelihood}) + 2 (\text{number of free parameters}) \quad (11)$$

and Schwarz's approximation of the BIC is defined by

$$\text{BIC} = -2 \log (\text{maximum likelihood}) + (\text{number of free parameters}) \log (\text{number of entries in the series}) \quad (12)$$

where log represents the natural logarithm.

Consider first the maximized log likelihood. For the case of the PAR model, it is given by Noakes et al (1983) as

$$L_{PAR} = -1/2 \sum_{m=1}^{s} N \log \sigma^2(m) + (\lambda-1) \sum_{t=1}^{s.N} \log z_t \quad (13)$$

For the case of the PPAR model, it is

$$L_{PPAR} = -1/2 \sum_{g=1}^{G} N_g \log \sigma^2(g) + (\lambda-1) \sum_{t=1}^{s.N} \log z_t \quad (14)$$

where N_g is the product of the number of seasons in group g and N, the number of years of data. The number of free parameters is the sum of the number of seasonal means (s for both PAR and PPAR models), the number of residual variances (s for the PAR model and G for the PPAR model), the total number of AR parameters, and, if λ is estimated, one for the Box-Cox transformation.

The AIC and BIC both incorporate the principles of model parsimony (Box and Jenkins, 1970; Ledolter and Abraham, 1981) and good statistical fit. The model which has the lower AIC (or BIC) is the preferred model. The AIC has been used previously in stochastic hydrology for determining the orders of various kinds of seasonal and nonseasonal models (see, for example, Hipel et al, 1977; McLeod et al, 1977; Hipel, 1981; Hipel and McLeod, 1983).

The log-likelihood ratio (Rao, 1973, p. 448) can be used to test the null hypothesis that there is no significant difference in the residuals of the two models. It may be expressed as

$$R = -2 [L_{PPAR} - L_{PAR}] \qquad (15)$$

and, assuming the null hypothesis is true, follows asymptotically a chi-squared distribution with the number of degrees of freedom equal to the difference in the number of free parameters in the PAR and PPAR models respectively (ie, the difference in the number of AR parameters and residual variances).

5. APPLICATIONS

PAR and PPAR models were fit to three monthly hydrological series and three quarter-monthly hydrological series, and the resulting models were compared using the three criteria described in the previous section. The series were based on (1) inflows to reservoirs of the hydroelectric system operated by Alcan Smelters and Chemicals Ltd in the Province of Quebec, Canada (Thompstone et al, 1980), (2) flows of the Saugeen River measured at Walkerton, Ontario, Canada (supplied by the Water Survey of Canada), and (3) flows of the Rio Grande measured at Furnas, Minas Gerais, Brazil (supplied by Mr. Paulo Holanda Sales at Electrobras, Brasil). The monthly series consisted of Alcan system inflows, 1943-79, Saugeen river flows, 1915-76, and Rio Grande flows, 1931-75; the quarter monthly flows consisted of Alcan system inflows, 1943-79, Saugeen river flows, 1915-76, and Rio Grande flows, 1931-72. Note that quarter-monthly data consists of flows averaged from the 1st to the 7th, from the 8th to the 15th, from the 16th to the 22nd, and from the 23rd to the end of the month, ie, periods of approximately one week each.

Grouping of seasons within the PPAR models was performed using three levels of significance in the Pitman test. Tables 1, 2 and 3 compare the PAR and PPAR models with levels of significance for grouping at 50%, 20% and 5% respectively. In general, as the level of significance increases, fewer seasons are considered to have "incompatible" models and thus there is more grouping, ie, a smaller

Table 1 : Comparison of PAR and PPAR models with 50% level of significance for grouping criterion

Season length		1/4 month			month		
Series		Alcan System	Rio Grande	Saugeen	Alcan System	Rio Grande	Saugeen
AIC	PAR	19 706.07	20 074.47	12 034.83	5 028.78	5 482.81	3 274.61
	PPAR	19 684.76	20 051.88	12 021.46		5 477.01	3 272.73
BIC	PAR	20 577.73	20 921.41	12 934.58	5 229.47	5 658.77	3 445.25
	PPAR	20 463.22	20 786.64	12 825.23		5 635.80	3 406.48
Log likelihood	PAR	-9 694.04	-9 886.23	-5 867.42	-2 465.39	-2 700.41	-1 600.30
	PPAR	-9 700.38	-9 894.94	-5 876.73		-2 701.51	-1 607.37
Number of AR parameters	PAR	62	54	53	24	16	12
	PPAR	53	44	45		14	8
Number of residual variances	PAR	48	48	48	12	12	12
	PPAR	40	38	40		10	8
Log-likelihood ratio		12.68	17.42	18.62		2.2	14.14
Degrees of freedom		17	20	16		4	8
Significant at 5% level		no	no	no		no	no
Standard error of forecasts	PAR	0.2575	0.2213	0.4070	0.3554	0.2301	0.4267
	PPAR	0.2567	0.2235	0.4075		0.2289	0.4267
Significant at 5% level		no	no	no		no	no

Table 2: Comparison of PAR and PPAR models with 20% level of significance for grouping criterion

Season length		1/4 month			month		
Series		Alcan System	Rio Grande	Saugeen	Alcan System	Rio Grande	Saugeen
AIC	PAR	19 706.07	20 074.47	12 034.83	5 028.78	5 482.81	3 274.61
	PPAR	19 673.59	20 053.91	12 020.65		5 473.23	3 265.29
BIC	PAR	20 577.73	20 921.41	12 934.58	5 229.47	5 658.77	3 445.25
	PPAR	20 331.44	20 665.28	12 620.49		5 614.85	3 389.82
Log likelihood	PAR	-9 694.04	-9 886.23	-5 867.42	-2 465.39	-2 700.41	-1 600.30
	PPAR	-9 716.80	-9 917.96	-5 910.33		-2 703.61	-1 605.65
Number of AR parameters	PAR	62	54	53	24	16	12
	PPAR	42	33	28		12	7
Number of residual variances	PAR	48	48	48	12	12	12
	PPAR	29	27	23		8	7
Log-likelihood ratio		45.52	63.46	85.82		6.40	10.70
Degrees of freedom		39	42	50		8	10
Significant at 5% level		no	yes	yes		no	no
Standard error of forecasts	PAR	0.2575	0.2213	0.4070	0.3554	0.2301	0.4267
	PPAR	0.2583	0.2221	0.4096		0.2321	0.4344
Significant at 5% level		no	no	no		no	no

Table 3 : Comparison of PAR and PPAR models with 5% level of significance for grouping criterion

Season length		Alcan System	1/4 month Rio Grande	Saugeen	Alcan System	month Rio Grande	Saugeen
Series							
AIC	PAR	19 706.07	20 074.47	12 034.83	5 028.78	5 482.81	3 274.71
	PPAR	19 683.43	20 146.94	12 040.33		5 477.45	3 268.97
BIC	PAR	20 577.73	20 921.41	12 934.58	5 229.47	5 658.77	3 445.25
	PPAR	20 280.98	20 634.91	12 580.18		5 597.61	3 375.05
Log likelihood	PAR	-9 694.04	-9 886.23	-5 867.42	-2 465.39	-2 700.41	-1 600.30
	PPAR	-9 732.72	-9 986.47	-5 930.16		-2 710.72	-1 611.49
Number of AR parameters	PAR	62	54	53	24	16	12
	PPAR	36	22	23		9	5
Number of residual variances	PAR	48	48	48	12	12	12
	PPAR	24	16	18		6	5
Log-likelihood ratio		77.36	200.48	125.48		20.62	22.38
Degrees of freedom		50	64	60		13	14
Significant at 5% level		yes	yes	yes		no	no
Standard error of forecasts	PAR	0.2575	0.2213	0.4070	0.3554	0.2301	0.4267
	PPAR	0.2552	0.2227	0.4098		0.2321	0.4297
Significant at 5% level		no	no	no		no	no

number of groups. A Box-Cox transformation with $\lambda = 0.0$ was used in all cases (ie, the data were transformed by taking their natural logarithms) since previous experience with seasonal hydrological time series has shown that this is usually the preferred transformation (Hipel and McLeod, 1984).

Table 1 shows that, when a 50% level of significance is used as the grouping criterion, the PPAR model is better than the PAR model according to the AIC and BIC for all 5 examples for which grouping was suggested. Note that with the 50% criterion (and thus with the 20% and 5% criteria), no grouping was suggested for the Alcan system monthly data. Table 1 also shows that, according to the log-likelihood ratio, the difference between models is not significant at the 5% level. Thus these PPAR models are definitely preferable to the PAR models. As shown in Table 2, when a 20% level of significance is used for grouping seasons, the PPAR models again have lower AIC's and BIC's than the PAR models for the 5 examples with grouping. However, according to the log-likelihood ratio, the models for two of the quarter-monthly series (ie, the Rio Grande and the Saugeen River) are different at the 5% level of significance. In the case of the 5% significance level for the grouping criterion, Table 3 indicates that the AIC for the PPAR model is actually higher than for the PAR model for these two quarter-monthly series. The AIC is lower for the three other cases, and the BIC is lower in all 5 cases. The log-likelihood ratio indicated significant differences for the three quarter-monthly series, but not for the monthly series.

6. FORECASTING STUDY

Three years of additional data were available for each of the six time series described in the previous section. In order to further compare the PAR and PPAR models, one step ahead forecasts were calculated for each model for the additional three-year periods. The standard errors of the forecasts were calculated for the three-year period and are shown in Tables 1-3. The Pitman test described earlier was used to test to see if there were significant differences in the mean squared errors of the forecasts from PPAR models as opposed to forecasts from PAR models. In no case (even with the 5% level of significance for the grouping criterion) was the mean squared error of the PPAR model forecasts significantly different (at the 5% test level) from that of the PAR model. As shown by Noakes et al (1983, 1984), a nonparametric test could also have been used to compare MSE's of forecasts, but it gave essentially the same results as the Pitman test.

7. FOURIER SERIES ESTIMATION OF AR PARAMETERS

Some authors (for example, Jones and Brelsford, 1967) have suggested the use of Fourier series to estimate the AR parameters. This would, of course, be

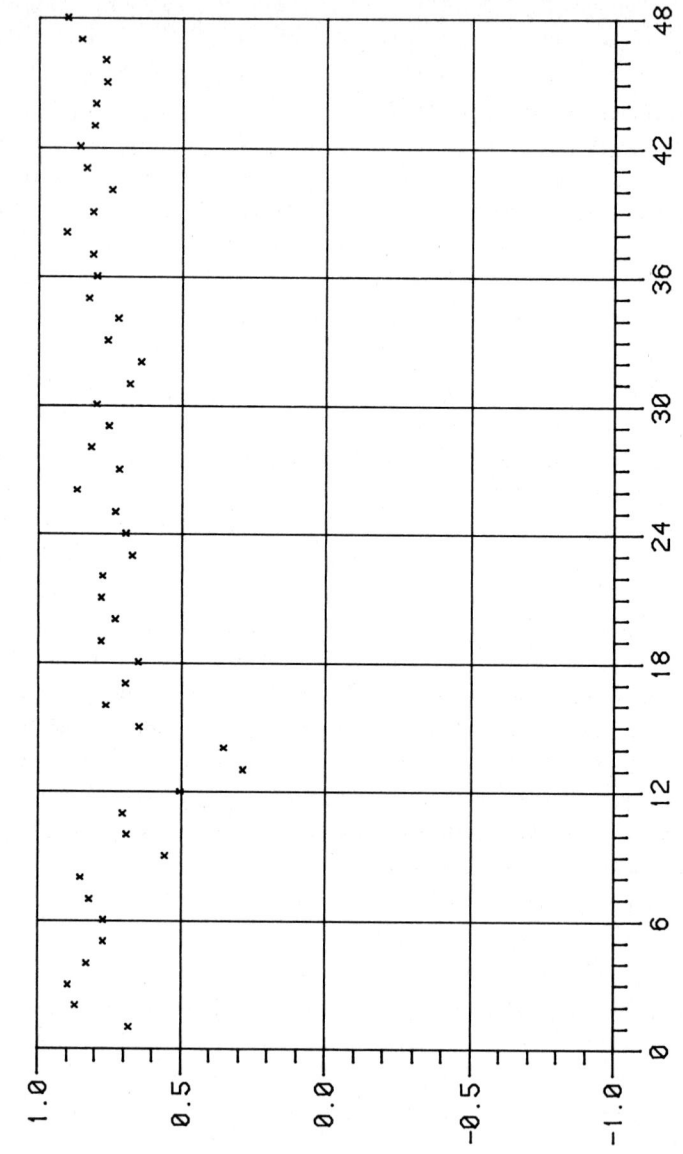

Figure 1 : First-Order Seasonal Autocorrelation of Log-Transformed Quarter-Monthly Flows, Saugeen River at Walkerton, 1915-76

appropriate for providing still more parsimonious models for certain time series. Consider, however, Figure 1, which shows a graph of first-order seasonal autocorrelations of the log-transformed quarter-monthly flows of the Saugeen River at Walkerton (1919-1976). No cyclic pattern which could be modeled by a low-ordered Fourier series is readily apparent. Because the correlation structure may change drastically during freshet, this behaviour is expected for river flow series. Furthermore, the approach to fitting PAR and PPAR models presented in this paper is sufficiently general to be applicable to series with or without a cyclic pattern in the AR parameters, and thus is most appropriate for the time series under consideration herein.

8. CONCLUSION

This paper has introduced a new type of stochastic model applicable to seasonal time series, namely the parsimonious periodic autoregressive or PPAR model. A technique was developed for identifying seasons with similar AR characteristics by first fitting PAR models to the series and then identifying seasons which are chronologically adjacent, have the same order of AR models, and for which the residuals are not significantly different when their PAR models are interchanged. Maximum likelihood estimates of AR parameters are then provided for each group of one or more seasons, and diagnostic checking is performed. Three monthly and three quarter-monthly hydrological time series were used to show that PPAR models are preferable to PAR models with respect to the AIC and BIC, and furthermore that there is no significant difference in the residuals or in one step ahead forecasts. Further research will be conducted to determine if PPAR models, when used to generate synthetic hydrological sequences, adequately preserve statistics which are of importance in the design and operation of water resource systems (McLeod and Hipel, 1978a,b).

ACKNOWLEDGEMENTS

The authors are greatful to Réjeanne Bergeron of Alcan Smelters and Chemicals Ltd for her assistance with certain computer aspects of this study, and to Donald J. Noakes of the University of Waterloo for initial versions of a number of computer programs used. Helpful comments of the referees and editor are very much appreciated. The work described here was financed in part by the Department of Industry, Trade and Commerce, Ottawa, Canada through the Program for Advancement of Industrial Technology.

REFERENCES

AKAIKE, H. (1974). A new look at the statistical model identification. <u>IEEE Transactions on Automatic Control</u>, AC-19, 716-723.

ANDERSON, O.D. (1978). A note on cyclic variation in the parameters of an AR(1) model, Metron, XXXVI (3-4), 73-77.

BOX, G.E.P. and COX, D.R. (1964). An analysis of transformations. Journal of the Royal Statistical Society, Series A, 127, 211-252.

BOX, G.E.P. and JENKINS, G.M. (1970). Time Series Analysis: Forecasting and Control. Holden-Day, San Francisco.

CLARKE, R.T. (1973). Mathematical models in hydrology, Irrigation and Drainage Paper, Food and Agricultural Organization of the United Nations, Rome.

CROLEY, T.E. and RAO, K.N.R. (1977). A manual for hydrologic time series deseasonalization and serial dependence reduction, Report No. 199, Iowa Institute of Hydraulic Research, University of Iowa, Iowa City, Iowa.

HIPEL, K.W. (1981). Geophysical model discrimination using the Akaike information criterion, IEEE Transactions on Automatic Control, AC-26, 358-378.

HIPEL, K.W. and McLEOD, A.I. (1984). Time Series Modelling for Water Resources and Environmental Engineers. Elsevier, Amsterdam (in press).

HIPEL, K.W., McLEOD, A.I. and LENNOX, W.C. (1977). Advances in Box-Jenkins modelling, 1, model construction, Water Resources Research, 13 (3), 567-575.

JONES, R.H. and BRELSFORD, W.M. (1967). Time series with periodic structure, Biometrika, 54(3), 403-408.

LEDOLTER, J. and ABRAHAM, B. (1981). Parsimony and its importance in time series forecasting, Technometrics, 23(4), 411-414.

LEHMANN, E.L. (1959). Testing Statistical Hypothesis. John Wiley and Sons, New-York.

LI, W.K. and McLEOD, A.I. (1981). Distribution of the residual autocorrelations in multivariate ARMA time series models, Journal of the Royal Statistical Society, Series B, 42(3), 231-239.

McCLAVE, J.T. (1975). Subset autoregression, Technometrics, 17, 213-220.

McLEOD, A.I. (1978). On the distribution of residual autocorrelations in Box-Jenkins models, Journal of the Royal Statistical Society, Series B, 40(3), 296-302.

McLEOD, A.I. and HIPEL, K.W. (1978a). Simulation procedures for Box-Jenkins models, Water Resources Research, 14(5), 969-975.

McLEOD, A.I. and HIPEL, K.W. (1978b). Developments in monthly autoregressive modelling, Technical Report 45-XM-011178, Dept. of Systems Design Engineering, University of Waterloo, Waterloo, Ontario, Canada.

McLEOD, A.I. and HIPEL, K.W. (1984). Periodic autoregression modelling, Unpublished manuscript, Department of Statistical and Actuarial Sciences, The University of Western Ontario, London, Ontario, Canada.

McLEOD, A.I., HIPEL, K.W. and LENNOX, W.C. (1977). Advances in Box-Jenkins modelling, 2, applications, Water Resources Research, 13(3), 577-586.

MOSS, M.E. and BRYSON, M.C. (1974). Autocorrelation structure of monthly streamflows, Water Resources Research, 10, 737-744.

NEFTCI, S.N. (1982). Specification of economic time series models using Akaike's criterion, Journal of the American Statistical Association, 77(379), 537-540.

NOAKES, D.J., McLEOD, A.I. and HIPEL, K.W. (1983). Forecasting experiments with seasonal hydrological time series models, Technical Report No. 117-XM-220283, Department of Systems Design Engineering, University of Waterloo, Waterloo, Ontario, Canada.

NOAKES, D.J., McLEOD, A.I. and HIPEL, K.W. (1984). Forecasting seasonal hydrological time series, Journal of Forecasting, to appear.

PAGANO, M. (1978). On periodic and multiple autoregression, Annals of Statistics, 6(6), 1310-1317.

PARZEN, E. and PAGANO, M. (1979). An approach to modelling seasonally stationary time series, Journal of Econometrics, 9, 137-153.

PITMAN, E.J.G. (1939). A note on normal correlation, Biometrika, 31, 9-12.

RAO, C.R. (1973). Linear Statistical Inference and its Applications. John Wiley and Sons Inc., New-York.

RAO, A.R. and KASHYAP, R.L. (1974). Stochastic modelling of river flows, IEEE Transactions on Automatic Control, AC-19, 874-881.

SAKAI, H. (1982). Circular lattice filtering using Pagano's method, IEEE Transactions on Acoustics, Speech and Signal Processing, 30(2), 279-287.

SALAS, J.D., BOES, D.C. and SMITH, R.A. (1982). Estimation of ARMA models with seasonal parameters, Water Resources Research, 18(4), 1006-1010.

SCHWARZ, G. (1978). Estimating the dimension of a model, Annals of Statistics, 6(2), 461-464.

SEN, Z. (1978). A mathematical model of monthly flow sequences, Hydrological Sciences Bulletin, 23(2), 223-229.

SNEDCOR, G.W. and COCHRAN, W.G. (1980). Statistical Methods, seventh edition. Iowa State University Press, Ames, Iowa.

TAO, P.C. and DELLEUR, J.W. (1976). Seasonal and nonseasonal ARMA models, ASCE Journal of the Hydraulics Division, 102, 1541-1559.

THOMPSTONE, R.M. (1983). Topics in Hydrological Time Series Modelling, PhD thesis, Dept. of Systems Design Engineering, University of Waterloo, Waterloo, Ontario, Canada.

THOMPSTONE, R.M., POIRE, A. and VALLEE, A. (1980). A hydrometeorological information system for water resources management, INFOR - Canadian Journal of Operational Research and Information Processing, 18, 258-274.

TJOSTHEIM, D. and PAULSEN, J. (1983). Bias of some commonly used time series estimates, Biometrika, 70(2), 389-399.

TROUTMAN, B.M. (1979). Some results in periodic autoregression, Biometrika, 66(2), 219-228.

YEVJEVICH, V.M. (1972). Structural analysis of hydrologic time series, Hydrology Paper No. 56, Colorado State University, Fort Collins, Colorado, U.S.A.

INFORMATION CONTENT OF THE MEAN AND RESCALED RANGE FOR SHORT SAMPLES DERIVED FROM AR(2) PROCESSES

Daniel A. Cluis
INRS-Eau, Université du Québec, C.P. 7500, Sainte-Foy, Québec, Canada G1V 4C7

Using a closed-form analytical expression for the variance of the mean in short samples derived from AR(2) processes, two related statistics, the Information Content of the mean and the Rescaled Range, are evaluated as functions of the first two autocorrelation coefficients, or the two model parameters, for samples of various lengths. The two statistics have been widely used for network design purposes (meteorology, hydrology and water quality studies) and in relation to the Hurst phenomenon. The results are discussed and compared with those of normal independent samples, and a sensitivity analysis is performed to explain possible discrepancies between data and the usual first order linear Markovian models.

1. INTRODUCTION

Consider the stochastic component of a stationary time-series once the trend and the cyclical fluctuations have been removed for both the mean and the variance. The AR(1) model has been widely used in hydrology and meteorology to represent the persistence of such residual series. Sometimes an AR(2) scheme shows a better fit to the observed data as for example in the case of daily discharges (BEARD, 1967; QUIMPO, 1967) and of daily water temperature (CLUIS, 1972). Two parameters of hydrological practical interest are studied here: the Information Content associated with the mean, a parameter widely used for temporal network design and trend assessment; and the Rescaled Range, a parameter associated with reservoirs and spillway design. Both statistics are closely related to the variance of the mean.

2. SOME RESULTS FOR THE AR(2) PROCESS

An AR(2) process X_t of mean zero may be expressed by the recurrence equation:

$$X_t = a\, X_{t-1} + b\, X_{t-2} + \varepsilon_t \tag{1}$$

where ε_t is an independently distributed random component with zero mean. Most of the results for this process can be expressed as a function of either a and b or of the first two population autocorrelation coefficients, ρ_1 and ρ_2. The two sets of coefficients are equivalent and related as:

$$\rho_1 = \frac{a}{1-b} \quad ; \quad \rho_2 = \frac{a^2}{1-b} + b \quad \text{or} \quad a = \frac{\rho_1(1-\rho_2)}{1-\rho_1^2} \quad ; \quad b = \frac{\rho_2 - \rho_1^2}{1-\rho_1^2}.$$

The relation between the variance of the AR(2) process and that of the random component is expressed as:

$$\frac{\operatorname{var} X}{\operatorname{var} \varepsilon} = \frac{1-b}{1+b} \cdot \frac{1}{[(1-b)^2 - a^2]} = \frac{(1-\rho_1)(1+\rho_1)}{(1-\rho_2)(1+\rho_2 - 2\rho_1^2)}.$$

The conditions for stationarity can then be deduced from these equations. The domain of stationarity lies in the area within the triangle $-1<b<1$; $a+b<1$; $b-a<1$, or alternatively within the parabola $\rho_1^2 < (\rho_2+1)/2$; $|\rho_1|<1$, $|\rho_2|<1$. The shape of the correlogram depends on the roots p and q of the characteristic equation $X^2 - aX - b = 0$. Two classes of process appear according to whether the roots are real or complex; the boundary between the two classes (Figure 1) is given by $a^2 + 4b = 0$ or is determined by: $\rho_1^2 (1-\rho_2)^2 + 4(\rho_2 - \rho_1^2)(1-\rho_1^2) = 0$. For a sample of length N from any stationary process, the variance of the mean, var $\overline{X}$, may be expressed (BAYLEY and HAMMERSLEY, 1946) by:

$$\text{var } \overline{X} = \frac{\text{var } X}{N} [1 + \frac{2}{N} \sum_{k=1}^{N-1} (N-k) \rho_k] = \frac{\text{var } X}{N} [1 + \frac{2}{N} \phi]. \tag{2}$$

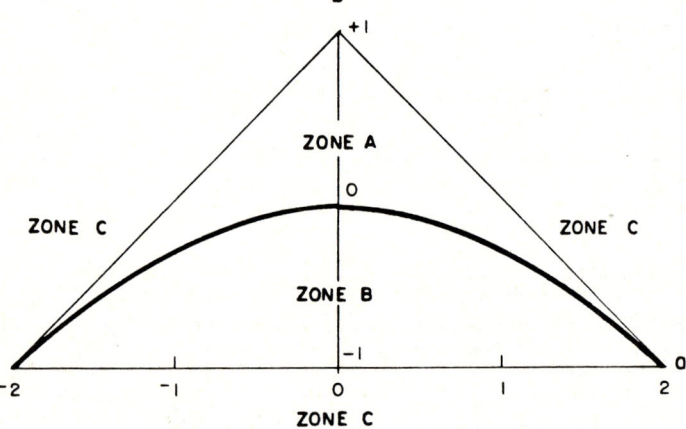

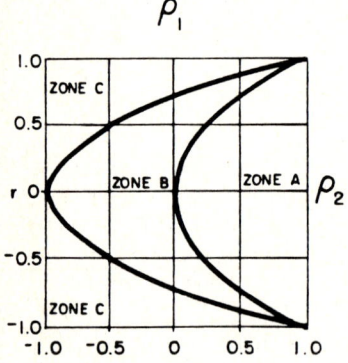

ZONE A: p and q real
ZONE B: p and q complex conjugates
ZONE C: non-stationary domain

Figure 1: Classification of the AR2 processes.

In the case of the AR(2) processes, the summation of equation (2) has lead to explicit expressions of ϕ for different shapes of correlograms. (See the Appendix).

3. INFORMATION CONTENT OF THE MEAN

To optimize the temporal sampling frequency of a data acquisition network, the concept of information content of the series is often used. FISHER (1921, 1925) has established that the Information Content of a statistical measure is inversely proportional to its variance. Applying this concept to the mean of a series, a most important practical statistic, one is lead to compare the variance of the mean for a sample of length N in the autocorrelated series to the variance of the mean for an independent series of the same length and variance:

$$I = \frac{(\text{var } X)/N}{\text{var } \overline{X}} = [1 + \frac{2}{N} \sum_{k=1}^{N-1} (N-k) \rho_k]^{-1} = [1 + \frac{2 \Phi}{N}]^{-1}. \tag{3}$$

This concept translates into the notion of an equivalent number N' of observations with the same information content as defined by FISHER, i.e. the same variance of the mean as N dependent observations, N' = N I. Some authors (REIHER and HUZZEN, 1967; QUIMPO, 1969) have noted cases in which AR(1) and AR(2) processes may induce Information Contents with values larger than one, which might seem surprising. Using a similar notion of "linkage coefficient", YULE (1945) noticed the same phenomenon for AR(2) series. Commenting on Yule's paper, KENDALL (1945) stated that such a situation "implies that for certain types of series the precision of the mean of N consecutive terms (as measured by its sampling variance) is less than that of an equal number of terms chosen from the series at random. Thus there may arise situations where it is better to choose a "clutch" of values than a random set even when the correlation between neighbouring values expressed by ρ_1 is positive, so far as concerns the estimation of the mean".

The calculation of the Information Content for AR(2) processes of lengths N = 10, 20, 50 and 100 is displayed in Table 1 for the plane (ρ_1, ρ_2) and in Table 2 for the plane (a, b); more complete results are available in CLUIS and BELLEMARE (1983). As the sample lengths increase, so does the Information Content for fixed parameters. Near the limits for stationarity, Information Content values greater than one are possible even if both ρ_1 and ρ_2 are positive, a situation representative of numerous geophysical time-series. This phenomenon is related to the large increase in the variance of the process around the boundaries for stationarity. For a very large sample, a common asymptotic expression called the Intrinsic Information Content of the AR(2) process as N → ∞ can be derived (Figures 2 and 3):

$$I = \frac{1-b}{1+b} \cdot \frac{1-a-b}{1+a-b} = \frac{(1 - \rho_1)(1 - \rho_2)}{(1 + \rho_1)(1 - 2\rho_1^2 + \rho_2)}. \tag{4}$$

Values larger than one appear in 3 of the 4 classes of process and in the case of major interest where both ρ_1 and ρ_2 are positive, this situation appears in a region where: $\rho_2 > 2\rho_1^2 - 1$ and $\rho_2 < \rho_1 (\rho_1^2 + \rho_1 - 1)$. In this last case, an alternative formulation of the Information Content such as the Information Content of the variance might be useful (BAYLEY and HAMMERSLEY, 1946).

The Markovian model is often used to represent the persistence of hydrometeorological phenomena. To assess the sensitivity to deviations from the Markovian model as demonstrated in the case of the Göta river in Sweden (YEVJEVICH, 1972, p 54), an AR(2) model of type $\rho_2 = \rho_1^2 + \Delta\rho_2$ is introduced into the expressions of the Information Content, which are then subjected to a Taylor expansion. Using $\Delta\rho_2 = 0.005$ as increment for ρ_2 in the expansion, numerical values of the relative sensitivity to small deviations were obtained as:

Table 1: Information Contents of samples of various lengths N = 10, 20, 50 and 100 derived from AR(2) processes with parameters (ρ_1, ρ_2)

N = 10

ρ_1 \ ρ_2	-0.8	-0.4	0	0.4	0.8
-0.8	*	*	*	14.18	3.55
-0.4	*	6.62	2.87	1.35	0.56
0	4.13	1.96	1	0.53	0.28
0.4	*	1.62	0.64	0.32	0.18
0.8	*	*	*	0.48	0.14

N = 20

ρ_1 \ ρ_2	-0.8	-0.4	0	0.4	0.8
-0.8	*	*	*	20.23	3.44
-0.4	*	8.26	3.12	1.32	0.42
0	6.44	2.13	1	0.47	0.18
0.4	*	1.84	0.63	0.27	0.11
0.8	*	*	*	0.51	0.09

N = 50

ρ_1 \ ρ_2	-0.8	-0.4	0	0.4	0.8
-0.8	*	*	*	30.04	3.43
-0.4	*	10.02	3.3	1.31	0.19
0	7.64	2.25	1	0.45	0.14
0.4	*	2.01	0.63	0.25	0.08
0.8	*	*	*	0.54	0.06

N = 100

ρ_1 \ ρ_2	-0.8	-0.4	0	0.4	0.8
-0.8	*	*	*	36.03	3.45
-0.4	*	10.78	3.37	1.3	0.33
0	8.27	2.29	1	0.44	0.12
0.4	*	2.08	0.63	0.24	0.07
0.8	*	*	*	0.55	0.05

* Outside the stationary domain.

Table 2: Information Contents of samples of various lengths N = 10, 20, 50 and 100 derived from AR(2) processes with parameters (a, b)

N = 10

b \ a	-0.75	-0.50	-0.25	0	0.25	0.5	0.75
-1.5	16.49	*	*	*	*	*	*
-1.	8.53	8	8.36	*	*	*	*
-.50	7.21	4.25	3.15	2.76	2.82	*	*
0	3.79	2.35	1.51	1	0.67	0.45	0.3
0.5	2.78	1.28	0.7	0.38	0.2	*	*
1	1.26	0.56	0.23	*	*	*	*
1.5	0.41	*	*	*	*	*	*

N = 20

b \ a	-0.75	-0.50	-0.25	0	0.25	0.5	0.75
-1.5	25.71	*	*	*	*	*	*
-1.	14.52	10.41	10.71	*	*	*	*
-.50	8.63	4.95	3.48	2.81	2.89	*	*
0	5.29	2.65	1.58	1	0.63	0.38	0.21
0.5	3.12	1.38	0.71	0.36	0.16	*	*
1	1.6	0.58	0.21	*	*	*	*
1.5	0.47	*	*	*	*	*	*

N = 50

b \ a	-0.75	-0.50	-0.25	0	0.25	0.5	0.75
-1.5	46.01	*	*	*	*	*	*
-1.	19.37	12.76	12.93	*	*	*	*
-.50	10.56	5.53	3.71	2.92	2.95	*	*
0	6.16	2.85	1.63	1	0.61	0.35	0.17
0.5	3.52	1.45	0.71	0.34	0.13	*	*
1	1.76	0.59	0.19	*	*	*	*
1.5	0.51	*	*	*	*	*	*

N = 100

b \ a	-0.75	-0.50	-0.25	0	0.25	0.5	0.75
-1.5	61.11	*	*	*	*	*	*
-1.	22.08	13.79	13.89	*	*	*	*
-.50	11.49	5.76	3.8	2.96	2.98	*	*
0	6.55	2.92	1.65	1	0.61	0.34	0.15
0.5	3.7	1.47	0.71	0.34	0.13	*	*
1	1.83	0.6	0.19	*	*	*	*
1.5	0.52	*	*	*	*	*	*

* Outside the stationary domain.

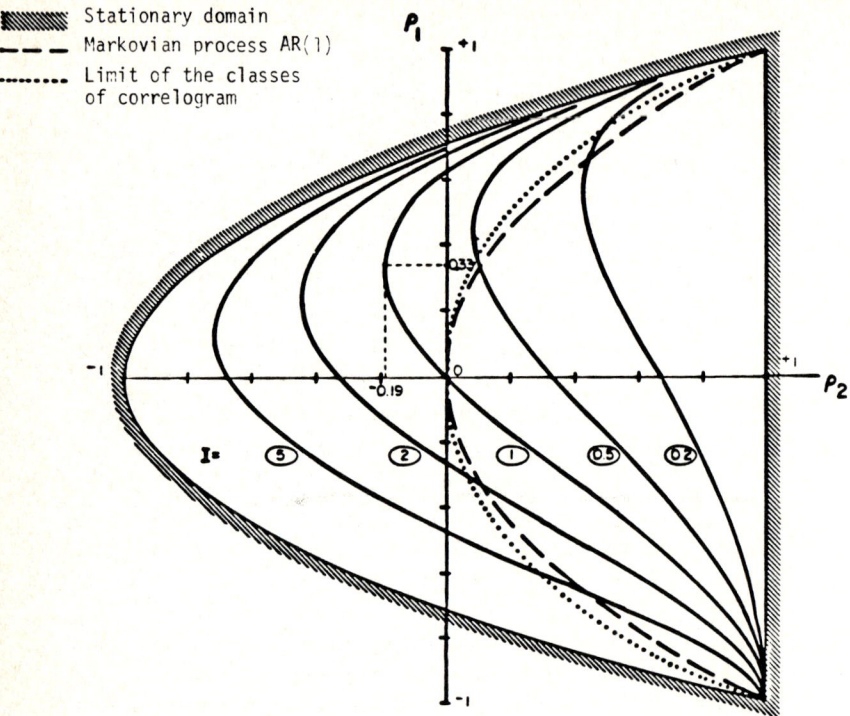

Figure 2: Intrinsic information content of an AR(2) process in the plane ρ_1, ρ_2.

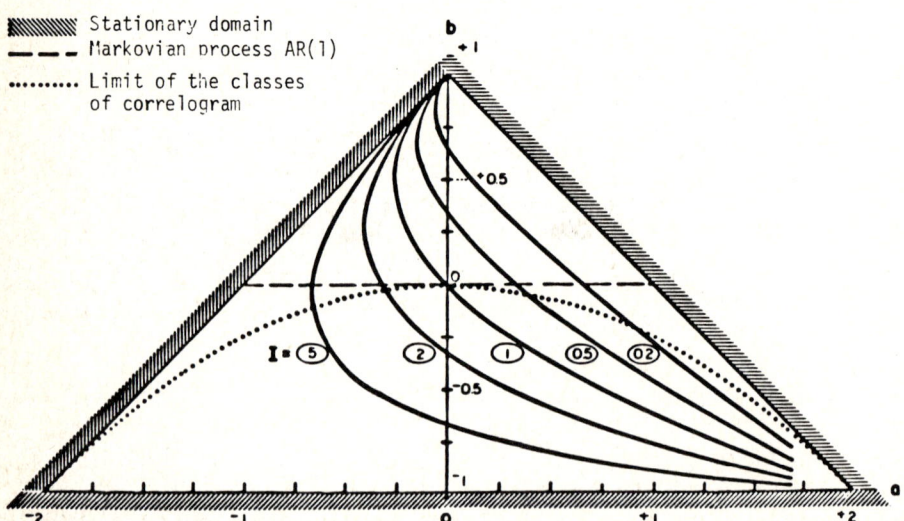

Figure 3: Intrinsic information content of an AR(2) process in the plane a,b.

$$\frac{\Delta I}{I} = \frac{\Delta \rho_2}{I} \frac{\partial I}{\partial \rho_2}. \tag{5}$$

The Information Contents of AR(1) processes for various values of ρ_1 and N are given in Table 3 and the sensitivity factors $\Delta I/(I \Delta \rho_2)$ are displayed in Table 4.

Table 3:   Information Content of AR(1) processes

N \ ρ_1	-0.8	-0.6	-0.4	-0.2	0	0.2	0.4	0.6	0.8
10	6.44	3.37	2.13	1.44	1	0.69	0.47	0.30	0.18
20	7.38	3.66	2.23	1.47	1	0.68	0.45	0.27	0.14
50	8.27	3.86	2.29	1.49	1	0.67	0.43	0.26	0.12
100	8.62	3.93	2.31	1.49	1	0.66	0.43	0.25	0.11

Table 4:   Sensitivity factors for the Information Content surrounding the Markovian case

N \ ρ_1	-0.8	-0.6	-0.4	-0.2	0	0.2	0.4	0.6	0.8
10	2.4	0.09	-0.73	-1.23	-1.6	-1.93	-2.28	-2.66	-3.04
20	2.14	-0.02	-0.88	-1.4	-1.8	-2.18	-2.63	-3.31	-4.56
50	2.28	-0.08	-0.98	-1.51	-1.92	-2.32	-2.83	-3.66	-5.76
100	2.36	-0.10	-1.01	-1.55	-1.90	-2.37	-2.89	-3.77	-6.12

The relative error in I resulting from the acceptance of an AR(1) instead of an AR(2) model may be very large. Taking $\rho_1 = 0.7$, $\rho_2 = 0.49$ and N = 100, we get I = 0.181; but, if $\rho_2 = 0.59$, then $\Delta \rho_2 = 0.10$ and $\Delta I/I = -0.46$, giving an overestimation of about 50%. This great sensitivity has major consequences in the field of network optimisation as well as in that of trend detection. It certainly questions the theoretical determination of the optimal sampling frequency of water resource data as established in hydrology (MATALAS and LANGBEIN, 1962), in meteorology (BROOKS and CARRUTHERS, 1953; HIRTZEL, QUON and COROTIS, 1982) as well as in water quality (LETTENMAIER, 1978; LOFTIS and WARD, 1980). This also raises questions concerning the accuracy of the trend detection techniques developed by LETTENMAIER (1967) and LETTENMAIER and BURGES (1977) which utilise the equivalent number of independent observations N' = NI, for use in the MANN-WHITNEY test.

4. EXPECTED RANGE OF HYDROLOGIC SEQUENCES

A large part of the theory of water storage and reservoir design has been developed using Range analysis. Let us now consider a sequence of random variables X_i of length N coming from a population of mean zero and finite variance var $X = \sigma^2$. If one considers the cumulative sum of the process

$S_i = \sum_{i=1}^{N} X_i$, one may define the time varying surpluses $S_N^+ = \max(0, S_1, S_2 \ldots S_N)$, deficits $S_N^- = \min(0, S_1, S_2 \ldots S_N)$ and ranges $R_n = S_N^+ - S_N^-$ of the process (Figure 4). For the study of small samples, the adjusted range R^*_N is defined to account for the sample mean by using $S^*_i = S_i - (i/N) S_N$.

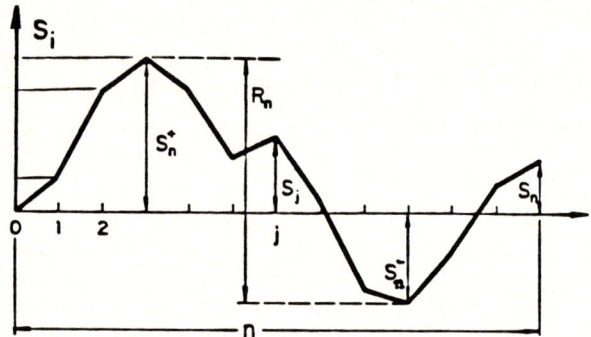

Figure 4: Definition of the surplus S_N^+, the deficit S_N^- and the range R_N.

HURST (1951) studied the storage required by the Great Lakes of the Nile River basin. He estimated the mean adjusted range R^* of cumulative departure of streamflow records using combinatorial analysis and the binomial expansion for approximating the normal probability density function. He then derived the asymptotic expected adjusted range as: $E(R^*_N) = \sigma \sqrt{N\pi/2}$.

Analyzing other large records of annual values of natural phenomena, he found that the observed adjusted ranges were not increasing with a power 0.5 of N, but as a higher power of order 0.72. Hurst's results have led many hydrologists to propose stochastic models in order to reproduce the departure from the square root law, generally called the Hurst phenomenon. FELLER (1951), assuming the hypothesis of independent normal random variables, obtained the asymptotic mean range as: $E(R_N) = 2\sqrt{2N/\pi} \simeq 1.596 N^{\frac{1}{2}}$.

ANIS and LLOYD (1953) studied the exact expected values of the surpluses S^+ and of the deficits S^- of independent normal variables with zero mean and unit variance:

$$E(R_N) = \sqrt{\frac{2}{\pi}} \sum_{i=1}^{N-1} i^{-\frac{1}{2}} \tag{6}$$

where $\text{var } S_i = \sum_{j=1}^{i} \text{var } X_j = i$, for $\text{var } X_j = 1$. If the variance of the process $\text{var } X_i = \sigma^2$ is not equal in value to one, Equation 6 should be multiplied by the standard deviation σ:

$$E(R_N) = \sqrt{\frac{2}{\pi}} \sum_{i=1}^{N} (\text{var } S_i)^{\frac{1}{2}}/i. \tag{7}$$

Using the data generation method, SATABUTRA (1967) studied the first-order Markov linear model while YEVJEVICH (1967) conducted the same kind of simulation with other linearly dependent normal variables including AR(1), AR(2) and MA(1) processes. They found that equation (7) very closely matched their results. Without definitive demonstration, YEVJEVICH (1967) suggested that this equation was valid for any linear dependence model such as the Markov and moving average schemes. As var $S_i = i^2$ var X_i, this equation may be expressed as:

$$E(R_N) = \sqrt{\frac{2}{\pi}} \sum_{i=1}^{N} (\text{var } \bar{X}_i)^{\frac{1}{2}}. \tag{8}$$

This equation confirms the close relationship between the expected range of a linear dependence process and the variance of its mean.

The rescaled range $E(R_N)/\sigma$ is obtained from the expression of the variance of the mean as:

$$E(R_N)/\sigma = \sqrt{\frac{2}{\pi}} \sum_{i=1}^{N-1} i^{-\frac{1}{2}} (1 + \frac{2\phi}{i})^{\frac{1}{2}}. \tag{9}$$

The results for AR(2) processes of lengths N = 10, 20, 50 and 100 are displayed in Tables 5 and 6. For given sample lengths, the rescaled ranges increase monotonically as the parameters of the processes ρ_1, ρ_2 or a, b increase. Taking both the range and rescaled range of a normal independent process as references, one can note the large variabilities of these statistics to changes in the two basic parameters of the AR(2) processes. Figures 5 and 6 in particular show large increases in these ranges when both ρ_1 and ρ_2 are positive.

For a given AR(2) process with parameters (ρ_1, ρ_2) or (a, b), a plot of the expected rescaled range $E(R_N)/\sigma_N$ as a function of the sample length N would show a decrease in power of N from 1 to about 0.5 for sample lengths varying from 10 to 100 (Figure 7). Of the 4 hypotheses put forward by O'CONNELL (1974) as possible explanations of the Hurst phenomenon (namely skewness, transience, non-stationarity and autocorrelation effects), our calculations for AR(2) processes show that transitory effects, with powers of N larger than 0.5 for small samples (N < 100), have effectively been obtained. This result does not exclude, of course, other effects also being present.

In a fashion similar to the Information Content of the mean, a numerical sensitivity analysis was performed for the expected range of deviations from the Markovian model:

$$\frac{\Delta[E(R_N)]}{E(R_N)} = \frac{\Delta\rho_2}{E(R_N)} \cdot \frac{\partial E(R_N)}{\partial \rho_2}. \tag{10}$$

The expected ranges of AR(1) processes for various values of ρ_1 and N are given in Table 7 and the sensitivity factors $\Delta[E(R_n)]/E(R_N) \Delta\rho_2$ are displayed in Table

Table 7: Expected range of AR(1) processes

N \ ρ_1	-0.8	-0.6	-0.4	-0.2	0	0.2	0.4	0.6	0.8
10	3.77	3.38	3.40	3.62	4.00	4.61	5.58	7.31	11.2
20	5.08	4.74	4.91	5.35	6.06	7.14	8.89	12.1	19.9
50	7.51	7.38	7.9	8.8	10.1	12.2	15.6	22.0	39.0
100	10.1	10.3	11.2	12.7	14.8	18.0	23.3	33.5	61.5

Table 5: Expected rescaled ranges of samples of various lengths N = 10, 20, 50 and 100 derived from AR(2) processes with parameters (ρ_1, ρ_2).

N = 10

ρ_2 \ ρ_1	-0.8	-0.4	0	0.4	0.8
-0.8	*	*	*	1.99	2.56
-0.4	*	2.42	2.91	3.49	4.29
0	2.77	3.38	4.01	4.07	5.49
0.4	*	3.82	4.77	5.63	6.44
0.8	*	*	*	5.66	7.22

N = 20

ρ_2 \ ρ_1	-0.8	-0.4	0	0.4	0.8
-0.8	*	*	*	2.49	3.69
-0.4	*	3.18	4.09	5.27	7.31
0	3.63	4.81	6.06	7.62	9.92
0.4	*	5.37	7.35	9.45	12.
0.8	*	*	*	8.57	13.6

N = 50

ρ_2 \ ρ_1	-0.8	-0.4	0	0.4	0.8
-0.8	*	*	*	3.31	5.91
-0.4	*	4.52	6.37	8.86	14.
0	5.18	7.59	10.2	13.7	20.5
0.4	*	8.32	12.5	17.5	25.8
0.8	*	*	*	14.2	29.5

N = 100

ρ_2 \ ρ_1	-0.8	-0.4	0	0.4	0.8
-0.8	*	*	*	4.11	8.43
-0.4	*	5.96	8.92	12.9	28.1
0	6.83	10.7	14.8	20.7	33.6
0.4	*	11.6	18.4	26.9	43.4
0.8	*	*	*	20.6	49.9

* Outside the stationary domain.

Table 6: Expected rescaled ranges of samples of various lengths N = 10, 20, 50 and 100 derived from AR(2) processes with parameters (a, b).

N = 10

b \ a	-0.75	-0.50	-0.25	0	0.25	0.5	0.75
-1.5	1.9	*	*	*	*	*	*
-1.	2.17	2.23	2.15	*	*	*	*
-.50	2.47	2.69	2.84	2.91	2.78	*	*
0	2.85	3.23	3.61	4.01	4.42	4.88	5.38
0.5	3.36	3.98	4.66	5.46	6.5	*	*
1	4.17	5.2	6.44	*	*	*	*
1.5	5.98	*	*	*	*	*	*

N = 20

b \ a	-0.75	-0.50	-0.25	0	0.25	0.5	0.75
-1.5	2.35	*	*	*	*	*	*
-1.	2.75	2.9	2.81	*	*	*	*
-.50	3.21	3.63	3.97	4.15	4	*	*
0	3.78	4.52	5.26	6.06	6.98	8.11	9.56
0.5	4.57	5.76	7.1	8.86	11.4	*	*
1	5.85	7.93	10.8	*	*	*	*
1.5	9.01	*	*	*	*	*	*

N = 50

b \ a	-0.75	-0.50	-0.25	0	0.25	0.5	0.75
-1.5	3.03	*	*	*	*	*	*
-1.	3.75	4.1	4	*	*	*	*
-.50	4.54	5.41	6.12	6.57	6.4	*	*
0	5.5	7	8.5	10.1	12.2	14.4	19.2
0.5	6.82	9.2	11.9	15.8	22.4	*	*
1	9.02	13.3	20.1	*	*	*	*
1.5	14.8	*	*	*	*	*	*

N = 100

b \ a	-0.75	-0.50	-0.25	0	0.25	0.5	0.75
-1.5	3.67	*	*	*	*	*	*
-1.	4.77	5.37	5.26	*	*	*	*
-.50	5.93	7.38	8.53	9.24	9.11	*	*
0	7.34	9.73	12.1	14.8	18.1	22.8	30.9
0.5	9.27	13.0	17.5	23.8	35.4	*	*
1	12.4	19.3	30.7	*	*	*	*
1.5	21.3	*	*	*	*	*	*

* Outside the stationary domain.

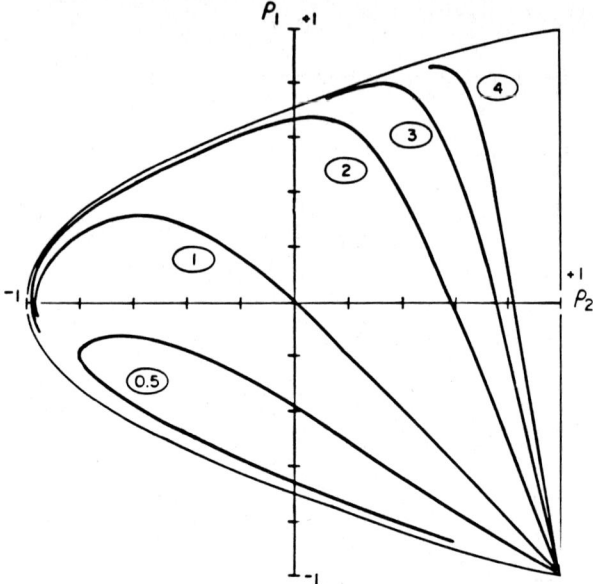

Figure 5: Expected relative ranges $E[R(\rho_1,\rho_2)]/E[R(0,0)]$ for samples of length N = 100.

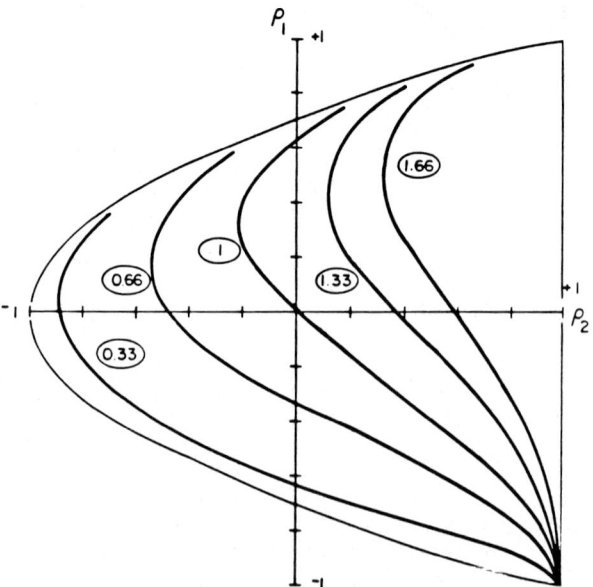

Figure 6: Expected relative rescaled ranges $\{E[R(\rho_1,\rho_2)]/\sigma(\rho_1,\rho_2)\}/\{E[R(0,0)]/\sigma(0,0)\}$ for samples of length N = 100.

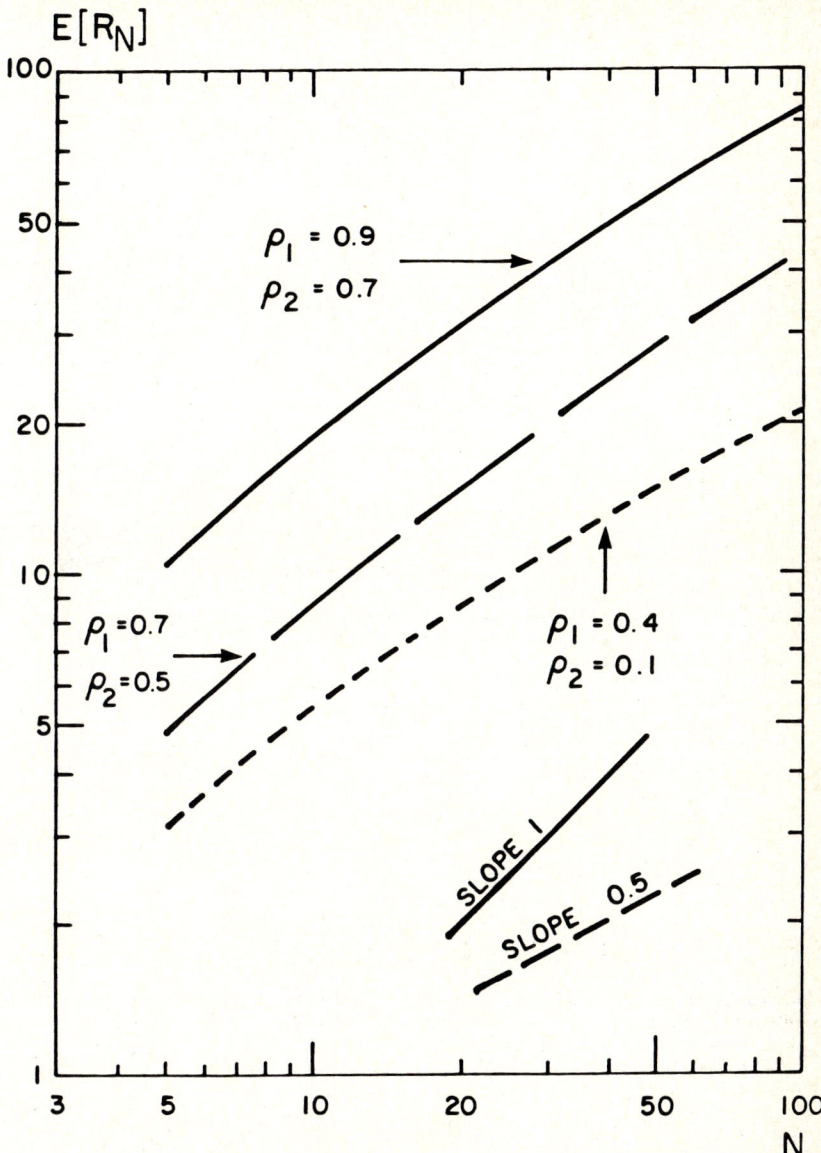

Figure 7: Relationship between the expected range $E[R_N]$ and the sample length N for various combinations of ρ_1 and ρ_2.

8. For the considered statistic, the relative error resulting from the acceptance of an AR(1) instead of an AR(2) model may be quite large. Taking $\rho_1 = 0.7$, $\rho_2 = 0.49$ and $N = 100$, then $E(R_N) = 43.3$; but if $\rho_2 = 0.59$, then $\Delta\rho_2 = 0.10$ and $\Delta E(R_N)/E(R_N) = 0.28$, giving an underestimation of about 30%.

Table 8: Sensitivity factors for the Expected Range surrounding the Markovian case

N \ ρ_1	-0.8	-0.6	-0.4	-0.2	0	0.2	0.4	0.6	0.8
10	1.17	0.43	0.30	0.32	0.40	0.55	0.80	1.26	2.57
20	1.07	0.46	0.39	0.44	0.56	0.74	1.04	1.59	3.08
50	0.95	0.51	0.49	0.57	0.71	0.92	1.26	1.92	3.75
100	0.86	0.54	0.55	0.65	0.79	1.01	1.38	2.10	4.12

5. CONCLUSIONS

The analytical expression of the variance of the mean of any AR(2) process has been obtained and appears to be very sensitive to small variations of the two basic coefficients as deviations contribute to the whole correlogram. Thus the two practical statistics studied here, the Information Content of the mean and the Rescaled Range, exhibit the same large sensitivity, particularly to deviations from the Markovian case. If these statistics are to be used at all for practical design purposes, care must be taken in the determination of the underlying theoretical model from small historical samples, as even insignificant (given the sample length) large-lag autocorrelation coefficients may influence them. In particular, blind automatic model fitting performed by computer packages should be avoided.

APPENDIX

The variance of the mean of any stationary process:

$$\text{var } \bar{X} = \frac{\text{var } X}{N} [1 + \frac{2}{N} \sum_{k=1}^{N-1} (N-k) \rho_k] = \frac{\text{var } X}{N} [1 + \frac{2 \Phi}{N}]$$

can be explicitly obtained for an AR(2) process, using the characteristic equation. Putting $a = p + q$ and $b = pq$, two cases may arise:

case 1: If p and q are real and both have absolute values less than one ($|a| < 2$; $|b| < 1$), then the autocorrelation coefficients ρ_k may be written (QUIMPO, 1969):

$$\rho_k = \frac{p(1-q^2)}{(p-q)(1+pq)} p^{|k|} + \frac{q(1-p^2)}{(q-p)(1+pq)} q^{|k|} = C_1 p^{|k|} + C_2 q^{|k|}.$$

The summation of this expression gives:

$$\Phi = [\frac{1+b}{1-a-b} - \frac{1}{1-b}] N + \frac{1}{(1-b)(1-a-b)^2} [-2b(1-b) - a(1+b^2)] + \frac{\psi_1 - \psi_2}{p-q}].$$

where $\psi_1 = p^{n+2} (1-q^2)(1-q)^2$ and $\psi_2 = q^{n+2} (1-p^2)(1-p)^2$.

case 2: If p and q are complex conjugates and the value of pq is less than one (b<1), KENDALL (1951) has shown that the ρ_k may be written as:

$$\rho_k = \frac{C^k \sin(k\theta + \psi)}{\sin \psi} \text{ where } C = \sqrt{-b}; \cos\theta = \frac{a}{2\sqrt{-b}} \text{ and } \tan\psi = \frac{1+C^2}{1-C^2}\tan\theta.$$

After expanding $\sin(k\theta + \psi)$, one needs the identities (ADAMS, 1922; JOLLEY, 1961):

$$\sum_{k=0}^{N-1} a^k \cos k\theta = \{(1-a\cos\theta)(1-a^N \cos N\theta) + a^{N+1}\sin\theta \sin N\theta\}/(1-2a\cos\theta+a^2)$$

$$\sum_{k=1}^{N-1} a^k \sin k\theta = \{a\sin\theta(1-a^N \cos n\theta) - (1-a\cos\theta)a^N \sin N\theta\}/(1-2a\cos\theta+a^2).$$

The summations give (REIHER and HUZZEN, 1967): $\phi = A + B$ with

$A = \cot\psi(1-2C\cos\theta + C^2)^{-1} \{NC\sin\theta - [2C^{N+2}\sin(N\theta) - C^{N+1}$
$\sin(N+1)\theta - C^{N+3}\sin(N-1)\theta + C(1-C^2)\sin\theta](1-2C\cos\theta + C^2)^{-1}\}$

$B = (1 - 2C\cos\theta + C^2)^{-2}[C^{N+3}\cos(N-1)\theta - 2C^{N+2}\cos N\theta$
$+ C^{N+1}\cos(N+1)\theta - C^3\cos\theta + NC\cos\theta - NC^2 - 2NC^2\cos^2\theta - NC^4$
$+ 3NC^3\cos\theta - C\cos\theta + 2C^2].$

REFERENCES

ADAMS, E.P. (1922). Smithsonian Mathematical Formulae. Washington.

ANIS, A.A. and LLOYD, E.H. (1953). On the range of partial sums of a finite number of independent random variable. Biometrika 40, 35-42.

BAYLEY, G.V. and HAMMERSLEY, J.M. (1946). The effective number of independent observations in an autocorrelated time series. J. Roy. Stat. Soc. B-8, 184-197.

BEARD, L.R. (1967). Simulation of daily streamflow. In Proceedings, The International Hydrology Symposium, September 6-8, Fort Collins, Colorado. Ed: Colorado State University. Fort Collins, Colorado, 624-632.

BROOKS, C.E.P. and CARRUTHERS, N. (1953). Handbook of statistical methods in meteorology, M.O. 538. Meteorological Office, London.

CLUIS, D. (1972). Relationship between stream water temperature and ambient air temperature. Nordic Hydrology 3, 67-71.

CLUIS, D. and BELLEMARE, R. (1983). Contenu en information et étendue des processus autoregressifs AR2. Rapport scientifique 149, INRS-Eau.

FELLER, W. (1951). The asymptotic distribution of the range of sums of independent variable. Ann. Math. Statistics 22, 427-432.

FISHER, R.A. (1921). The mathematical foundations of theoretical statistics. Phil. Trans. Roy. Soc. A-222, 309-368.

FISHER, R.A. (1925). Theory of statistical estimation. Proc. Cambridge Phil. Soc. 22, 700-725.

HIRTZEL, C.S., QUON, J.E. and COROTIS, R.B. (1982). Mean of autocorrelated air quality measurements. J. Env. Eng. EE3, 488-501.

HURST, H.E. (1951). Long term storage capacities of reservoirs. Am. Soc. Civil Engineers, Trans. 116, 776-808.

JOLLEY, L.B.W. (1961). Summation of series. Dover, New York.

KENDALL, M.G. (1945). Note on Mr Yule's paper. J. Roy. Statis. Soc. 108, 225-230.

KENDALL, M.G. (1951). The advanced theory of statistics, Vol. II. Griffin, London.

LETTENMAIER, D.P. (1967). Detection of trends in water quality data from record with dependent observations. Wat. Resour. Res. 12, 1037-1046.

LETTENMAIER, D.P. and BURGES, S.J. (1977). Design of trend monitoring net-works. J. Env. Eng. EE5, 785-802.

LETTENMAIER, D.P. (1978). Design considerations for ambient stream quality monitoring. Wat. Res. Bull. 14, 884-902.

LOFTIS, J.C. and WARD, R.C. (1980). Sampling frequency selection for regulatory water quality monitoring. Wat. Res. Bull. 16, 501-507.

MATALAS, N.C. and LANGBEIN, W.B. (1962). Information content of the mean. J. Geoph. Res. 67, 3441-3448.

O'CONNELL, P.E. (1974). ARIMA models in synthetic hydrology. In Mathematical models for surface water hydrology. Eds: Cirani, T.A., Maione, V. and Wallis, J.R., Wiley, New York, 51-68.

QUIMPO, R.G. (1967). Stochastic model of daily river flow sequences. Hydrol. Paper No 18, Colorado State University, Fort Collins, USA.

QUIMPO, R.G. (1969). Reduction of serially correlated hydrologic data. Bull. IASH 14, 111-118.

REIHER, B.S. and HUZZEN, C.S. (1967). Some comments of the effective sample size of second order Markov processes. Bull. IASH 12, 63-74.

SUTABUTRA, P. (1967). Reservoir storage capacity required when water inflow has a periodic and a stochastic component. PhD Dissertation, Colorado State University, Fort Collins, USA.

YEVJEVICH, V. (1967). Mean range of linearly dependent normal variables with application to storage problem. Wat. Resour. Res. 3, 663-671.

YEVJEVICH, V. (1972). Stochastic processes in hydrology. Water Resources Publications, Colorado State University, Fort Collins (Colorado), USA.

YULE, G.U. (1945). A method of studying time series based on their internal correlations. J. Roy Statis. Soc. 108, 208-225.

RAINFALL AT FORTALEZA IN BRAZIL REVISITED

Pedro A. Morettin and Afranio R. de Mesquita
University of São Paulo, C.P. 20570, São Paulo, Brazil

Jacira G.C. da Rocha
Federal University of Pernambuco, 50000, Recife, Brazil

The 131 years of annual rainfall data at Fortaleza is reanalyzed in order to determine if they support the existence of periodicities. The use of several tests (Fisher, Whittle, Hannan, Bartlett and Priestley) seems to suggest that thirteen and twenty-six year periodicities are present in the series. In the light of these results, a preliminary model is given for the series.

1. INTRODUCTION

The series of rainfalls at Fortaleza, Ceará, Brazil has been analysed by several authors recently, its importance depending on the fact that it is probably the longest series available for the study of the severe droughts that affect the Brazilian North-East.

The series consists of 131 years of annual data, from 1849 to 1979. It has been argued that this series is not appropriate for forecasting purposes, since the climate of Fortaleza is influenced by the sea and hence is not representative of the rest of the area. This question has been discussed by Girardi and Teixeira (1978) who have shown that there is a great similarity between the behavior of the rainfall at Fortaleza and that at other sites of the region.

Recent studies as in Markham (1974), using "seasonalized" annual totals, 1849-1970, concluded that there were thirteen and twenty-six year periodicities in the data, and provided speculation as to the causes of these apparent periodicities. See also Markham (1967).

Jones and Kearns (1976) reanalysed the same data and concluded that the hypothesis that the series consists of statistically independent observations could not be rejected at the 10% level. They based their conclusions on tests of serial correlation, estimated spectrum and cumulative periodogram.

Girardi and Teixeira (1978) used the same periodicities found by Markham to predict a severe drought in the area from 1978 to 1983. Further studies are those of Almeida et al. (1980) and Kantor (1982). This last author used the method of maximum entropy (Burg, 1975) to produce forecasts for the series.

In this paper we will provide a careful analysis of the rainfall series in order to find if the data supports the existence of periodicities. A mixed-spectrum analysis will be carried out and several tests will be used to detect the presence of harmonic terms.

In section 2 we present the data and some preliminary remarks. In section 3 we describe the statistical time series methodology we will use. The analysis of the series is performed in section 4 and a tentative model is discussed in section 5. We conclude the paper with some further comments.

2. THE DATA AND PRELIMINARY REMARKS

The observations are given in Appendix A and their plot is presented in Figure 1. From a visual inspection of both, it is not easy to detect noticeable trend and periodical patterns. The sample mean is 1425mm, the sample variance 230.52mm^2 while the minimum and maximum values are 468mm (1877) and 2512mm (1974), respectively.

The climate of the region may be classified as semi-arid, and the corresponding drought may be termed seasonal, occurring when the Inter-Tropical Convergence Zone (ITCZ) does not move up to the region in the period February-April.

During the period considered (1849-1979) major droughts occurred in 1877-1879, 1888-1889, 1898, 1900, 1903-1904, 1907-1908, 1915, 1919, 1932, 1936, 1951, 1953, 1958.

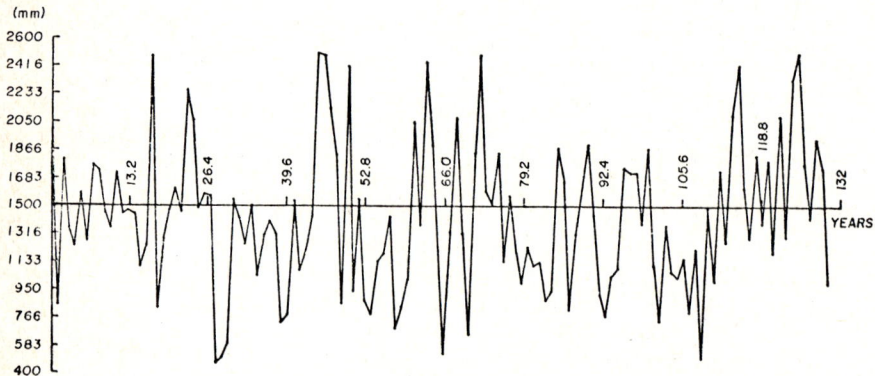

FIGURE 1: The Rainfall Series, 1849-1979

The least squares line for the raw data is

$x_t = 1390.3 + 0.537 (t-1849)$

which would indicate a very small positive trend; but this is not statistically significant, as can also be seen using other trend tests (Cox and Stuart test, for example; see Conover, 1971, p.130).

Figure 2 shows the plot of the autocorrelation function, and the periodogram of the data is given in Figure 3. The first sample autocorrelation is $r_1 = 0.24$ (significant at 5% level); this shows that there is a low year-to-year dependence. The remaining values tend to oscillate and do not show a definite periodical pattern.

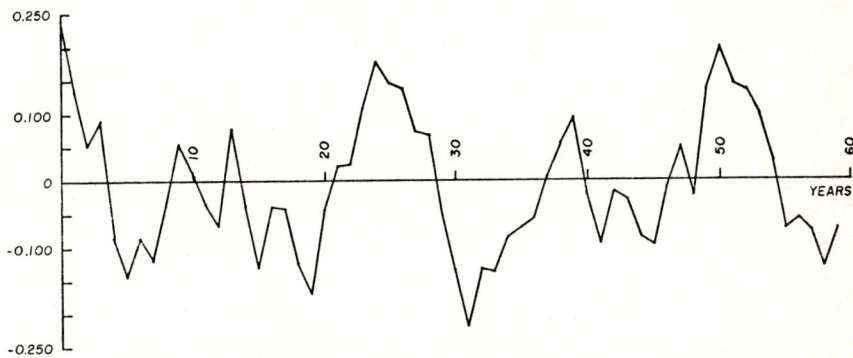

FIGURE 2: Autocorrelation Function of the Rainfall Series

The periodogram contains several peaks, the more prominent being those corresponding to 65, 26, 13, 4.8 and 3.6 years. The significance of these peaks will be discussed in section 4.

3. A MIXED SPECTRUM ANALYSIS FOR THE RAINFALL SERIES

The conclusions drawn by Jones and Kearns (1976), based on a shorter series, suggest that the (mean-corrected) rainfall at Fortaleza behaves like a white noise series, with a constant spectrum. If the series contains periodicities, its spectrum would have a mixed form; that is, peaks (corresponding to the periodical components) will emerge from a continuous spectrum. Then a model that seems adequate to describe the series may by developed as follows.

Denote by X_t, $t=1,2,\ldots,T$, the observations of a discrete parameter, zero mean, stationary process, with a mixed spectrum. This means that the spectral distribution function $F(\lambda)$ of the process X_t, $t=0,\pm1,\pm2,\ldots$ may be written

$$F(\lambda) = F_c(\lambda) + F_d(\lambda) \qquad (3.1)$$

where F_c (the continuous component) is absolutely continuous and $F'_c(\lambda)=f(\lambda)$ is the spectral density function, while F_d (the discrete component) is a step function with jumps $p_1,\ldots,p_J$ at frequencies $\lambda_1,\ldots,\lambda_J$. It follows that X_t can be written

$$X_t = Y_t + Z_t \qquad (3.2)$$

where Y_t and Z_t are uncorrelated processes, Y_t corresponding to the continuous part of the spectrum and Z_t to the discrete part. Further:

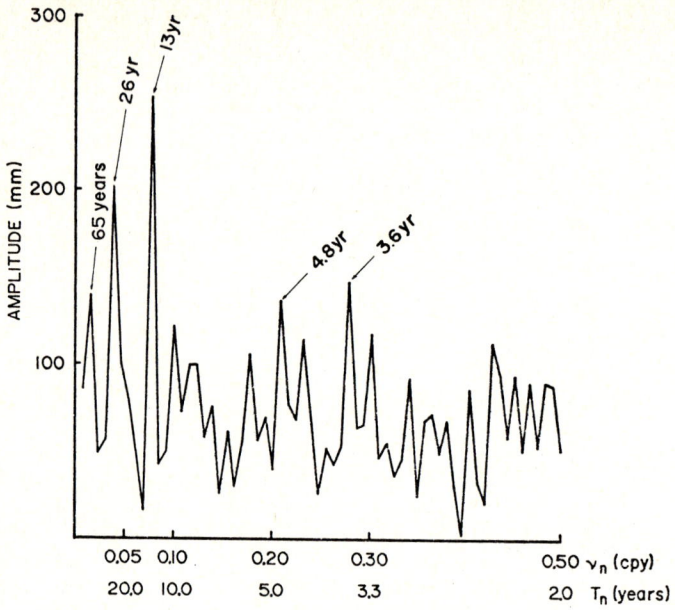

FIGURE 3: Periodogram of the Rainfall Series

i) Y_t is a linear process, that is,

$$Y_t = \sum_{k=0}^{\infty} \beta_k \varepsilon_{t-k} \qquad (3.3)$$

where ε_t is a white noise series, with mean zero, variance σ_ε^2, $E(\varepsilon_t \varepsilon_s) = 0$, $s \neq t$, and β_k are constants satisfying $\sum_{k=0}^{\infty} \beta_k^2 < \infty$ and $\sum_{k=0}^{\infty} k|\beta_k| < \infty$. Let $\gamma_Y(k)$ denote the autocovariance function of Y_t and

$$f_Y(\lambda) = \frac{1}{2\pi} \sum_{k=-\infty}^{\infty} \gamma_Y(k) e^{-i\lambda k} \qquad (3.4)$$

its spectral density function, assuming that $\sum_{k=-\infty}^{\infty} |\gamma_Y(k)| < \infty$;

ii) Z_t is a process of the form

$$Z_t = \sum_{j=1}^{J} A_j \cos(\lambda_j t + \phi_j) \qquad (3.5)$$

where A_j, λ_j are unknown constants, $j=1,\ldots,J$, and ϕ_j are independent, identically distributed, rectangular random variables on $[-\pi, \pi]$. If $\gamma_Z(k)$ is the autocovariance of Z_t, then

$$\gamma_Z(k) = \frac{1}{2} \sum_{j=1}^{J} A_j^2 \cos(\lambda_j k) \qquad (3.6)$$

and $\gamma_X(k) = \gamma_Y(k) + \gamma_Z(k)$. The $\{A_j, j=1,\ldots,J\}$ forms the discrete (or line) spectrum of X_t and $f_Y(\lambda)$ the continuous spectrum.

A mixed spectrum analysis of X_t consists of:

i) estimating the amplitudes A_j and the frequencies λ_j, in order to obtain the discrete spectrum of X_t;

ii) estimating the continuous spectrum $f_Y(\lambda)$ of X_t.

It follows that the first step in the analysis is to test for the presence of periodical components, i.e., to test the null hypothesis that $A_j=0$, for all j. If we find that there are J periodical components, we estimate the λ_j, A_j and remove the contribution of these components from X_t; after this is done, the spectrum $f_Y(\lambda)$ is estimated from the residuals $X_t - \hat{Z}_t$, using standard techniques of spectrum estimation.

Several tests are available for testing the existence of periodic components in a set of data. These tests were not considered by the previous authors who dealt with the rainfall series.

We decided to apply all the tests below since we are not aware of any study comparing their powers for finite samples.

a) An extension of Fisher's g-test (Fisher, 1929, Whittle, 1952)

b) Whittle's test (Whittle, 1952);

c) Hannan's test (Hannan, 1961);

d) Bartlett's test (Bartlett, 1955);

e) Priestley's $P(\lambda)$ test (Priestley, 1962a,b).

A comparision of the asymptotic powers of the Whittle's test, the grouped periodogram test and the $P(\lambda)$ test was given by Priestley (1962b). We also used a white noise test based on the cumulative periodogram (Jenkins and Watts, 1968).

In Appendix B we briefly describe these tests.

4. ANALYSIS OF THE RAINFALL SERIES

The tests mentioned in section 3 (and described in Appendix B) were applied to the data. Unless otherwise stated we shall assume for all tests an overall significance level $\alpha=0.05$.

(a) Extension of Fisher's Test

We used the suggestion of Whittle (1952) (see Appendix B). The value of $g^{(1)}$ is 0.136, and for m=65 we obtain the critical value 0.0961. Therefore, the peak corresponding to 13.1 years is significant.

Applying the test to the second largest ordinate, correcting the denominator of (B.2), we observe $g^{(2)}=0.1012$. Since, for m=64, the critical value is 0.0984, we accept the presence of a periodical component with period $131/5 = 26.2$ years. For the third largest ordinate we obtain $g^{(3)} = 0.0623$, which is not significant and we reject a periodicity of $131/36 = 3.64$ years.

(b) <u>Whittle's Test</u>

The largest periodogram ordinate occurs at $j=10$ (the frequency is $\lambda_{10} = 2\pi \times 10/131 \approx \pi/6$), with $g_W = 0.145$, which is significant (the critical value is 0.0964). Therefore, we accept a periodical component with period of 13.1 years.

For the second largest ordinate ($j=5$), $g_W = 0.092$, which is greater than the critical value 0.0699 and we conclude that a 26.2 year periodicity is present.

For the third largest ordinate, $g_W = 0.056$, which is not significant (0.061 is the critical value), and we reject the existence of a harmonic term with period 3.64 years.

(c) <u>Bartlett's Test</u>

Let us choose $k=4, 5, 7, 10, 20$. The corresponding values of $g_{k,\ell}^{(B)}$ are given in Table 1, for some values of ℓ. For the 5% significance level, the critical values are given by $0.05k/[T/2] = k(1-g)^{k-1}$ and these are also shown in Table 1. We see that the only significant value is that for $k=4$ and $\ell=3$, corresponding to the period of 13.1 years. We recall that taking k small is the only way to obtain a good approximation of $g_{k,\ell}$, using $g_{k,\ell}^{(B)}$. On the other hand, we do not expect to get significant results for large values of k, since this implies that we are assuming that the spectrum $f_Y(\lambda)$ is approximately constant over a broad band of frequencies.

TABLE 1: Values of $g_{k,\ell}^{(B)}$ and g

k \ ℓ	1	2	3	9	g
4	.5904	.6670	.9267*	.7417	.9084
5	.5478	.7625	.4276		.8335
7	.4421	.6748			.6973
10	.4115	.2528			.5492
20	.2983	.1724			.3143

(d) <u>Hannan's Test</u>

For the largest ordinate, we obtain $g_H = 0.145$, which is significant against the critical value 0.13135, and the harmonic with period 13.1 is accepted. Testing for the second largest ordinate, we get $g_H = 0.092$, which is greater than the tabled value 0.0699, and we accept as significant the peak corresponding to 26.2 years.

It is easy to see that we reject further periods. As expected, the test gives the same results as the Wittle test, due to the choice of the window $W(\theta)$ as that corresponding to the truncated periodogram.

(e) The $P(\lambda)$ Test

The calculations were done with $m=25$ and $n=55$; other values were tried, but we will only describe these results. The graph of $P(\lambda)$ (Figure 4) shows the presence of well defined peaks at $4\pi/131$, $10\pi/131$, $20\pi/131$, $30\pi/131$ and $40\pi/131$.

For $\lambda = 4\pi/131$, we have $J_2 = 1.225$ and for $\alpha=1\%$, we have $\alpha_0 = 2.33$, and since $J_2 < \alpha_0$, we see that the peak at this λ is not significant.

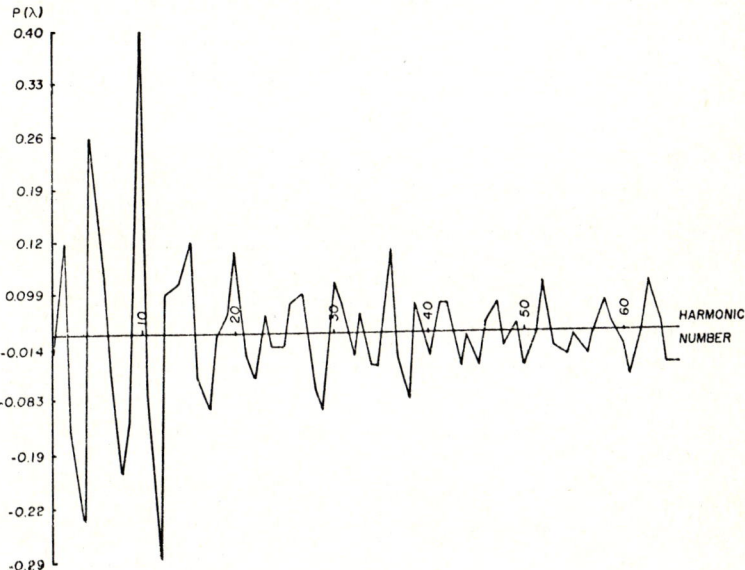

FIGURE 4: Graph of $P(\lambda)$ for the Rainfall Series

Next, in order of frequency, we test the peak at $\lambda = 10\pi/131$ and obtain $J_2 = 2.603$, which is significant ($\alpha_0=2.58$, corresponding to $\frac{\alpha}{2} = 0.005$) and we accept a harmonic component with period 26.2 years.

Removing the contribution of this significant harmonic term at the frequency $\hat{\lambda}_1' = 10\pi/131$, we obtain the new graph of $P(\lambda)$, which we call $P'(\lambda)$ (Figure 5). This has peaks at frequencies $20\pi/131$ and $72\pi/131$, and testing again in order of frequency we find that the peak at $\lambda = 20\pi/131$ is significant, since $J_2 = 3.026$, is greater than the critical value $\alpha_0=2.71$. Hence we accept a periodicity of 13.1 years.

Removing again the contribution of this component we get $P''(\lambda)$ (Figure 6) with

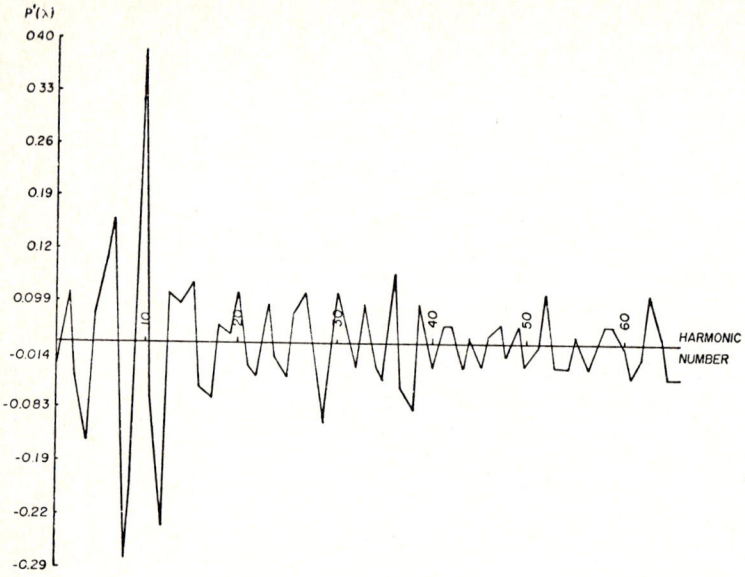

FIGURE 5: Graph of $P'(\lambda)$ for the Rainfall Series

peaks at the frequencies $46\pi/131$, $54\pi/131$, $66\pi/131$ and $72\pi/131$. For $\lambda = 46\pi/131$ we find $J_2 = 0.810$, which is not significant and therefore we reject a periodicity of 5.7 years. The remaining peaks are not significant.

(f) <u>The White Noise Test</u>

The use of the test described in section B.2 of Appendix B on the rainfall data gave the result shown in Figure 7. It differs from the one obtained by Jones and Kearns (1976), since in the present case the series cannot be taken as white noise. The only difference in the calculation is that the present series has 10 years of data points more than the series used by them. This suggests that even other periods could eventually be added to the set of significant periodicities, as more years of observations allow them to come up from the overall background continuous spectrum.

The results of the application of the several tests are summarized in Table 2, and seem to indicate that the 13.1 and 26.2 years periodicities are present in the data, since four out of the five tests detected them both, while all the tests detected the 13.1 year periodicity. Therefore, we shall proceed to establish a model for the series.

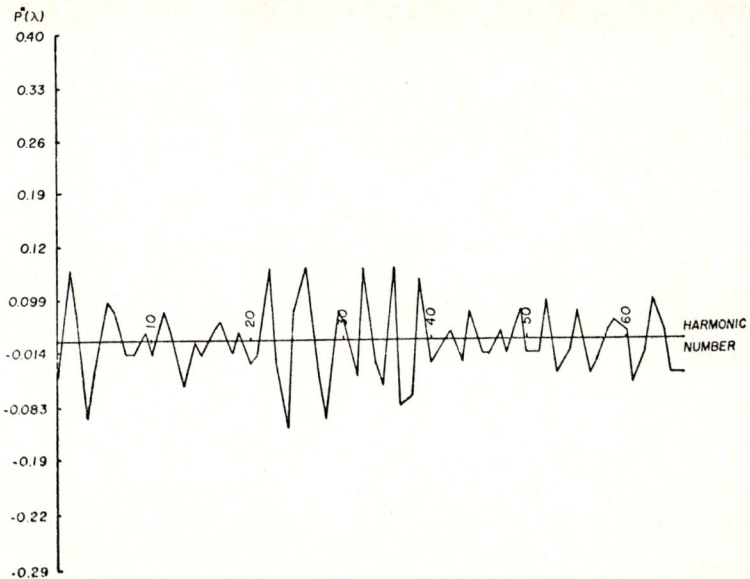

FIGURE 6: Graph of $P''(\lambda)$ for the Rainfall Series

TABLE 2: Summary of the application of the tests

Test	Periods or remarks
Ext. Fisher	13.1, 26.2 years
Whittle	13.1, 26.2 years
Bartlett	13.1 years
Hannan	13.1, 26.2 years
Priestley	13.1, 26.2 years
White Noise	series is not white woise

5. A MODEL FOR THE RAINFALL SERIES

From the considerations in section 4 we shall try two harmonic terms for the rainfall series: one with frequency $\hat{\lambda}_1 = 10\pi/131 = 0.2398$ and the other with frequency $\hat{\lambda}_2 = 20\pi/131 = 0.4796$, corresponding to the periods of 26.2 and 13.1 years respectively.

Therefore, we can write (3.5), with J=2, as

$$Z_t = \mu + A_1 \cos(\hat{\lambda}_1 t + \phi_1) + A_2 \cos(\hat{\lambda}_2 t + \phi_2) \tag{5.1}$$

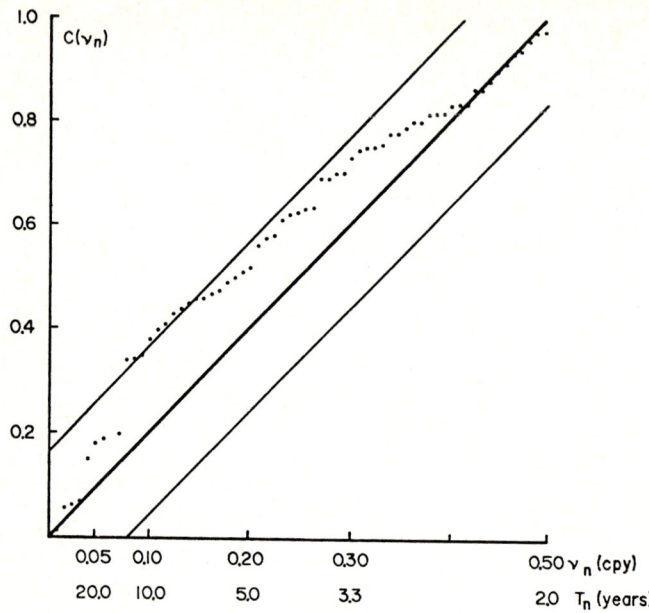

FIGURE 7: Normalized Accumulated Periodogram for the Rainfall Series, with Theoretical and Confidence Lines

and least squares estimates of A_i and ϕ_i (i=1,2) and μ are given by

$\hat{A}_1 = 202.5\text{mm}, \quad \hat{\phi}_1 = 114°$

$\hat{A}_2 = 255.6\text{mm}, \quad \hat{\phi}_2 = 126°$

$\hat{\mu} = \bar{X} = 1424.8\text{mm},$

respectively. Thus we can write the model for X_t as:

$$X_t = 1424.8 + 202.5 \cos(0.2398t + 1.99) + 255.6 \cos(0.4796t + 2.20) + Y_t \qquad (5.2)$$

where Y_t has a continuous spectrum.

Let $\hat{Z}_t$ be the estimated harmonic polynomial and $\hat{Y}_t = X_t - \hat{Z}_t$ be the residual series. Figure 8 shows the plots of the original and $\hat{Z}_t$ series.

It is readily seen that the sample autocorrelation of the $\hat{Y}_t$ series are all nonsignificant. Also, a white noise test applied to this residual series gives the results shown in Figure 9. Hence we may conclude that the Y_t series is white noise, with a constant spectrum.

6. FINAL REMARKS

In this paper several tests for detecting periodicities were applied to the rainfall series at Fortaleza in Brazil with the purpose of testing for the presence of unknown periodic terms. All the tests, except one, detected two significant harmonic terms with frequencies $2\pi/26.2$ and $2\pi/13.1$.

The evidence given in this paper and by other workers seems to suggest that the hypothesis that the series has a continuous (uniform) spectrum should be rejected and a mixed spectrum model is more appropriate for it.

But, as we have emphasized before, care should be taken in drawing definite conclusions. As we gather more observations, further analyses could be carried out and eventually other suspected periodicities, such as the one corresponding to 65 years, might be confirmed.

A suggested model for the series is given by (5.2), where Y_t is a white noise process. We also fitted an autoregressive process to the data, using both the Yule-Walker and Burg's method for estimating the coefficients and the FPE criterion of Akaike to determine the order of the model. In both cases a fourth-order model was obtained. The corresponding (maximum entropy or auto-regressive) spectral estimates did not detect any harmonic term. To resolve the periodicities of 13.1 and 26.2 years, a much higher order (of about T/2) was necessary.

The statistical model presented predicts a long period (1980-1985) of below average precipitation annual means. This, so far, is in fair agreement with the recent reports on the drought of the area.

The agreement draws attention to the periods of 13 and 26 years, pointed out in the analysis, and to their physical explanation. This has long been a matter of

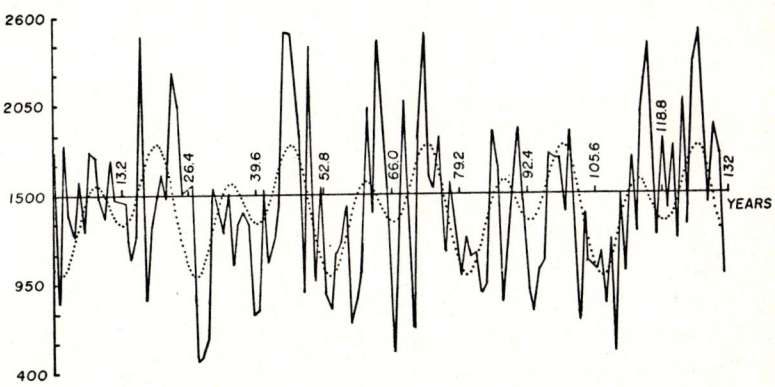

FIGURE 8: Original Series (Solid Curve) and Fitted Z_t (Dotted Curve)

concern, and many causes have been suggested to explain the occurrence of the
drought in the northeast area of Brazil. Some try to relate it to solar activity,
others try to find a possible connection with the occurrence of the "El Niño" on
the coast of Peru (Caviedes, 1973). The 13 years cycle may be a result of a
slightly out of phasing between the annual and the quasi-biannual period
oscillations of the trade winds of the South Atlantic (Reiter, 1979).

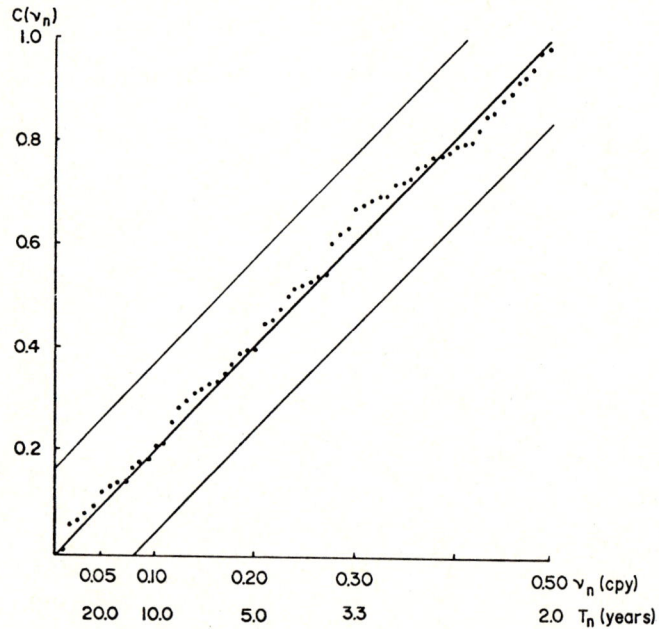

FIGURE 9: Normalized Accumulated Periodogram for the Residual
Series, with Theoretical and Confidence Lines

Recent preliminary analyses of the mean sea level, made by the authors at the port
of San Francisco, USA, show a periodicity of 12.4 years, among others. The
evidence so far gathered seems to indicate that there is a great deal of oceanic
causation in defining the 13 years periodicity of the Fortaleza rainfall. But
certainly large amounts of data, not yet available, must be collected on several
oceanographic, meteorological and hydrological time series, before definite
conclusions can be drawn.

AKNOWLEDGEMENT: The authors would like to thank two referees and the Editor for
comments which improved the presentation of the paper.

This work was partially supported by CNPq, FAPESP, CAPES and FINEP grants.

APPENDIX A: The Rainfall Series, 1849-1979

Year	Rainfall	Year	Rainfall	Year	Rainfall	Year	Rainfall
1849	2001	1882	1246	1915	530	1948	1384
1850	852	1883	1508	1916	1328	1949	1881
1851	1806	1884	1047	1917	2077	1950	1114
1852	1356	1885	1307	1918	1319	1951	747
1853	1233	1886	1399	1919	656	1952	1378
1854	1590	1887	1320	1920	1847	1953	1068
1855	1273	1888	736	1921	2496	1954	1032
1856	1770	1889	784	1922	1595	1955	1152
1857	1734	1890	1534	1923	1513	1956	806
1858	1457	1891	1077	1924	1847	1957	1225
1859	1357	1892	1211	1925	1137	1958	504
1860	1716	1893	1430	1926	1571	1959	1493
1861	1445	1894	2505	1927	1195	1960	1011
1862	1468	1895	2491	1928	995	1961	1737
1863	1452	1896	2144	1929	1230	1962	1258
1864	1098	1897	1839	1930	1107	1963	2102
1865	1238	1898	863	1931	1133	1964	2428
1866	2478	1899	2414	1932	879	1965	1630
1867	832	1900	940	1933	937	1966	1288
1868	1289	1901	1545	1934	1888	1967	1839
1869	1470	1902	878	1935	1661	1968	1385
1870	1628	1903	789	1936	820	1969	1805
1871	1459	1904	1136	1937	1313	1970	1192
1872	2256	1905	1189	1938	1586	1971	2093
1873	2058	1906	1430	1939	1911	1972	1299
1874	1487	1907	697	1940	1447	1973	2331
1875	1581	1908	834	1941	916	1974	2512
1876	1569	1909	1015	1942	780	1975	1778
1877	468	1910	2051	1943	1042	1976	1417
1878	503	1911	1373	1944	1090	1977	1941
1879	597	1912	2446	1945	1750	1978	1752
1880	1539	1913	1905	1946	1724	1979	996
1881	1423	1914	1512	1947	1726		

SOURCE: Kantor (1982). The figures are in mm of rainfall

APPENDIX B

B.1. Tests for Detecting Periodicities

Given T observations $X_1, \ldots, X_T$ of X_t, let

$$I_X^{(T)}(\lambda) = \frac{2}{T} \left| \sum_{t=1}^{T} X_t e^{-i\lambda t} \right|^2 \tag{B.1}$$

be the <u>periodogram</u> of these values. We shall evaluate (B.1) at the frequencies $\lambda_j = 2\pi j/T$, $j = 0, 1, \ldots, [T/2]$, called Fourier frequencies, and call $I_j^{(T)} = I_X^{(T)}(\lambda_j)$.

(a) <u>Fisher's Test and an Extension</u>

Fisher's test is used for testing the value of the largest peak in the periodogram. The model (3.2) is assumed for X_t, but Z_t is now assumed to be white noise. We test $H_0: A_j = 0$, for all j, under the condition that X_t is Gaussian. Writting $I_j = I_j^{(T)}$ for brevity, Fisher's g-statistic is given by

$$g = \max(I_j) / \sum_{j=1}^{m} I_j \tag{B.2}$$

where $m = |T/2|$. Fisher (1929) derived (under H_0) the exact distribution of g, which is given by

$$P(g > z) = m(1-z)^{m-1} - \binom{m}{2}(1-2z)^{m-1} + \ldots + (-1)^{k-1}(1-kz)^{m-1}, \tag{B.3}$$

where $k = [1/z]$. Tables of the distribution are given by Fisher (1929) and Shimshoni (1971).

If we reject H_0, we conclude that there is a periodicity in X_t at the frequency corresponding to $\max(I_j)$; if this occurs for $j = j'$, then this frequency is $\tilde{\lambda}_0 = 2\pi j'/T$.

Whittle (1952) suggests that we may test for the next largest ordinate by omitting the term I_j, from the denominator of (B.2) and adjust the value of m to m-1.

If we know that X_t has exactly r periodic components, when H_0 is false, then a test based on

$$g^{(r)} = I^{(r)} / \sum_{j=1}^{m} I_j \tag{B.4}$$

can be used, where $I^{(r)}$ is the r-th largest ordinate. See Grenander and Rosenblatt (1957) for the distribution of $g^{(r)}$, of which (B.3) is a special case.

(b) <u>Whittle's Test</u>

We now return to the general model (3.2). To test the null hypothesis H_0, we use the statistic

$$g_W = \max_j [I_j / 2\pi \hat{f}_Y(\lambda_j)] / \sum_{j=1}^{m} [I_j / 2\pi \hat{f}_Y(\lambda_j)] \tag{B.5}$$

and refer to Fisher's g-distribution with m degrees of freedom.

If $C_X(s) = T^{-1} \sum_{t=1}^{T-|s|} X_t X_{t+|s|}$ is the sample autocovariance, $\hat{f}_Y(\lambda)$ is the truncated periodogram estimate,

$$\hat{f}_Y(\lambda) = \frac{1}{2\pi} \sum_{s=-(\ell-1)}^{\ell-1} C_X(s)e^{-i\lambda s} \qquad (B.6)$$

$\ell<T$ being the truncation point.

(c) <u>Hannan's Test</u>

Hannan (1961) considers the problem of testing if a periodic component corresponds to a jump in the spectral distribution function $F(\lambda)$. The null hypothesis is that F is absolutely continuous, with derivative $f(\lambda)$, a smooth function, and the alternative hypothesis is that F has a jump.

The test statistic is essentially g_W, except that $f_Y(\lambda)$ is estimated through a windowed spectral estimate $f_Y^*(\lambda)$, of the form

$$f_Y^*(\lambda) = \int_{-\pi}^{\pi} I_X^{(T)}(\alpha) \, W_M(\lambda-\alpha) d\alpha. \qquad (B.7)$$

Here $W_M(\lambda)$ is the spectral window and M is the truncation point. Under the null hypothesis, the statistic

$$g_H = \max_j [I_j/2\pi f_Y^*(\lambda_j)] / \sum_{j=1}^m [I_j/2\pi f_Y^*(\lambda_j)] \qquad (B.8)$$

has approximately the Fisher distribution with m degrees of freedom.

(d) <u>Bartlett's</u> (<u>the grouped periodogram</u>) <u>Test</u>

Bartlett suggested a test for the purpose of separating spectral peaks with narrow bandwidths. Given a sample of T observations, it will be practically impossible to distinguish harmonic components from peaks in the continuous spectrum, with widths less than $2\pi/T$. See Bartlett (1957) and Priestley (1981).

Thus we will assume that $f_Y(\lambda)$ has bandwidth $B_f \geq 2\pi/T$. Let $k<B_f$ and divide the periodogram ordinates into $[T/2k]$ sets, each containing k ordinates. Let

$$g_{k,\ell} = I_{j'}/2\pi f_Y(\lambda_{j'}) / \sum_{j=(\ell-1)k+1}^{\ell k} I_j/2\pi f_Y(\lambda_j) \qquad (\ell=1,2,\ldots,[T/2]k^{-1}) \qquad (B.9)$$

where $I_{j'}/2\pi f_Y(\lambda_{j'}) = \max\{I_j/2\pi f_Y(\lambda_j): (\ell-1)k+1 \leq j \leq \ell k$.

Under H_0, $g_{k,\ell}$ has asymptotically the same distribution as the Fisher's g-statistic with k degrees of freedom. Then $g_{k,\ell}$ may be approximated by

$$g_{k,\ell}^{(B)} = I_{j'}/\sum_j I_j \qquad (B.10)$$

and even when $f_Y(\lambda)$ is unknown the test may be carried out considering $g_{k,\ell}^{(B)}$ as having a Fisher distribution with k degrees of freedom. It can be shown that $g_{k,\ell}^{(B)}$ can differ considerably from $g_{k,\ell}$ and there is not any systematic way of choosing k. A compromise must be achieved between making k smaller than B_f, and sufficiently large to retain sufficient degrees of freedom.

Assuming that we may use $g_{k,\ell}^{(B)}$, it remains to adjust the significance level of the test. If α is the level for the original Fisher's test applied to k ordinates, then the approximate significance level for a test based on $g_{k,\ell}^{(B)}$ is $\alpha'=\alpha k/[T/2]$. Then the procedure to follows is: i) choose some values of k; ii) for each k, test for the significance of the peaks, using $g_{k,\ell}^{(B)}$; usually there will be no need to test all values of ℓ (look at the periodogram); iii) the critical value for $g_{k,\ell}^{(B)}$ based on a significance level α is approximately given by $\alpha k/[T/2]=k(1-g)^{k-1}$, using only the first term in (B.3).

(e) <u>Priestley's $P(\lambda)$ Test</u>

The preceding tests have a number of disadvantages, which are summarized in Priestley (1981, p.625). The $P(\lambda)$ test, developed by Priestley (1962a,b) is not based on the periodogram but on the autocorrelation function of the observed series.

We saw in section 3 that the autocovariance function of the series X_t is given by $\gamma_X(s)=\gamma_Y(s)+\gamma_Z(s)$, where $\gamma_Y(s)\to 0$, as $|s|\to\infty$, since Y_t has a purely continuous spectrum, and $\gamma_Z(s)$ is a combination of cosine waves with the same frequencies as Z_t, and therefore $\gamma_Z(s)\not\to 0$, as $|s|\to\infty$.

Therefore, if some of the A_j are nonnull, the autocovariance of X_t does not wear off as $|s|\to\infty$, and its tail will behave like a linear combination of cosine waves with the same frequencies as those of the periodic component.

The $P(\lambda)$ test exploits this behavior of $\gamma_X(s)$, performing a Fourier analysis of the tail of $\gamma_X(s)$. Let m be such that $\gamma_Y(s)\approx 0$, for $|s|>m$. To test $H_0:A_j=0$, all j, proceed as follows:

(i) compute $\hat{f}_m(\lambda)$ and $\hat{f}_n(\lambda)$, estimates of $f_Y(\lambda)$ using some spectral window, with truncation points m and n, $n>2m$;

(ii) compute $P(\lambda)=\hat{f}_n(\lambda)-\hat{f}_m(\lambda)$, for $\lambda=2\pi j/T$, $j=0,1,\ldots,[T/2]$, and plot $P(\lambda)$ against λ. If $A_j\neq 0$, then $P(\lambda)$ will have well defined peaks;

(iii) test each peak in <u>order of frequency</u>; if the first peak appears at $\lambda_0=2\pi p/T$, subdivide the frequency range $(0,\pi)$ at intervals $2\pi/m$ on both sides of λ_0 and form

$$J_q = (\frac{T}{m}\Lambda_{n,m}^{-1})^{\frac{1}{2}} \sum_{s=0}^{q} P*(\frac{2\pi s}{m}+\delta) [\hat{g}(\pi)/2\pi]^{-\frac{1}{2}} \qquad (B.11)$$

for $q=0,1,\ldots,[T/2]$ and test whether $\max(J_q)<\alpha_0$, where α_0 is the upper 100α% point of the standard normal, for a given significance level α. In (B.11),

$P*(\lambda) = P(\lambda)/C_Y(0)$

$$\hat{g}(\pi) = \frac{1}{4\pi}\left\{2\sum_{s=-m+1}^{m-1} r_Y^2(s) - \sum_{s=-2m+1}^{2m-1} r_Y^2(s)\right\},$$

δ is chosen so that $\lambda_0 = 2\pi p/T = 2\pi s/m+\delta$, for some integer s, and

$$\Lambda_{n,m} = 2\pi \int_{-\pi}^{\pi} \{W^{(1)}(\alpha) - W^{(2)}(\alpha)\}^2 d\alpha. \tag{B.12}$$

In (B.12), $W^{(1)}(\lambda)$ and $W^{(2)}(\lambda)$ are the spectral windows corresponding to the truncation points n and m, respectively. If the Bartlett window is used for $W^{(1)}$ and $W^{(2)}$ we obtain

$$\Lambda_{n,m} = \frac{2}{3}n - \frac{4}{3}m + \frac{2m^2}{3n}.$$

In the formulae for $P^*(\lambda)$ and $\hat{g}(\pi)$, $C_Y(s)$ and $r_Y(s)$ denote the sample autocovariance and sample autocorrelation functions of Y_t, respectively;

(v) if $\max(J_q) > \alpha_0$, then the peak at λ_0 is judged significant and the amplitude of the harmonic term at λ_0 is then estimated by

$$\hat{A}_0^2 = 8\pi P(\lambda_0)/(n-m) \quad (n>2m); \tag{B.13}$$

(vi) the effect of this harmonic component is then removed, computing

$$C_Y^{(1)}(s) = C_Y(s) - \frac{1}{2}\hat{A}_0^2 \cos(s\hat{\lambda}_0), \tag{B.14}$$

and testing whether there are other harmonic terms, recomputing $P(\lambda)$ using $C_Y^{(1)}(s)$ and examining its peaks in order of frequency.

If α is the overall significance level, the significance level for testing the j-th peak in order of frequency, should be α/j. We continue until no further peaks of $P(\lambda)$ are significant.

B.2. A White Noise Test

This is not a periodicity peak detector, but it tests if an observed time series can be regarded as a realization of a white noise process. Let us denote frequency in cycles per unit time by ν and let $\nu_j = j/T$. If $\{Z_t, t=1,\ldots,T\}$ are observations of a stochastic process, denote by $F_Z(\nu)$ its spectrum and $I^{(T)}(\nu)$ the periodogram. If Z_t is white noise, then $f_Z(\nu) = 2\sigma_Z^2$, $0 \leqslant \nu \leqslant \frac{1}{2}$, and

$$F_Z(\nu) = \int_0^\nu f_Z(\alpha) d\alpha = \begin{cases} 0, & \nu < 0 \\ 2\sigma_Z^2 \nu, & 0 \leqslant \nu < \frac{1}{2} \\ \sigma_Z^2, & \nu \geqslant \frac{1}{2} \end{cases}$$

$F_Z(\nu)$ is the accumulated spectrum. Since $I^{(T)}(\nu)$ is an estimator of $f_Z(\nu)$, an estimator of $F_Z(\nu_j)$ is $T^{-1} \sum_{i=1}^{j} I_Z(\nu_i)$ and therefore

$$C(\nu_j) = \sum_{i=1}^{j} I_Z(\nu_i)/(T\hat{\sigma}_Z^2) \tag{B.15}$$

is an estimator of $F_Z(\nu_j)/\sigma_Z^2$, where $\hat{\sigma}_Z^2$ is an estimator of the variance of the process. $C(\nu_j)$ is the (normalized) accumulated periodogram. For a white noise process, the graph of $C(\nu_j) \times \nu_j$ will be scattered around the line passing through

(0,0) and (0.5,1).

To judge the deviations of $C(\nu_j)$ from this theoretical line, a test of significance of the Kolmogorov-Smirnov type is used. See Jenkins and Watts (1968) and Priestley (1981) for details.

REFERENCES

ALMEIDA, F.C., CALHEIROS, R.V., DIAS, P.L.S., XAVIER, T.M.B.S., KANTOR, I.J., KOUSKY, V.E., MEIRA, G.M., MOLION, L.C.B., PARADA, N.J., SRIVATSANGAM, S. and GRAY, W.M. (1980): Contribuição ao estudo da previsão de seca e modificação artificial do tempo e do clima do Nordeste brasileiro. INPE-1812-RPE/180, São José dos Campos, INPE.

BARTLETT, M.S. (1957): Discussion on "Symposium on spectral approach to time series". J. Roy. Statist. Soc., B, 19, 1-63.

BURG, J.P. (1975): Maximum entropy spectral estimates. PhD Dissertation, Stanford University.

CAVIEDES, C.N. (1973): Secas and El Niño: Two simultaneous climatic hazards in South America. Proc. Assoc. Geogr., 5, 44-49.

CONOVER, W.J. (1971): Practical Nonparametric Statistics, New York, Wiley.

FISHER, R.A. (1929): Tests of significance in harmonic analysis. Proc. Roy. Soc., A, 125, 54-59.

GIRARDI, C. and TEIXEIRA, L. (1978): Prognóstico do tempo a longo prazo. Relatório Técnico ECA-06/80, S. José dos Campos, CTA/IAE.

GRENANDER, U. and ROSENBLATT, M. (1957): Statistical analysis of stationary time series. New York, Wiley.

HANNAN, E.J. (1961): Testing for a jump in the spectral function. J. Roy. Statist. Soc., B, 23, 394-404.

JENKINS, G.M. and WATTS, D.G. (1968): Spectral analysis and its applications, San Francisco, Holden-Day.

JONES, R.H. and KEARNS, J.P. (1976): Fortaleza, Ceará, Brazil rainfall. Journal of Applied Meteorology, 15, 307-308.

KANTOR, I.J. (1982): Previsibilidade da série de precipitação de chuvas de Fortaleza, pelo método da máxima entropia de Burg. INPE-2546-RPE/420, São José dos Campos, INPE.

MARKHAM, C.G. (1967): Climatological aspects of drought in northeastern Brazil. PhD Dissertation, University of California, Berkeley.

MARKHAM, C.G. (1974): Apparent periodicities in Rainfall at Fortaleza, Ceará,

Brazil. *Journal of Applied Meteorology*, 13, 176-179.

PRIESTLEY, M.B. (1962a): Analysis of stationary processes with mixed spectra-I. *J. Roy. Statist. Soc.*, B, 24, 215-233.

PRIESTLEY, M.B. (1962b): Analysis of stationary processes with mixed spectra II. *J. Roy. Statist. Soc.*, B, 24, 511-529.

PRIESTLEY, M.B. (1981): *Spectral analysis and Time Series*, Vol. 1. London, Academic Press.

REITER, E. (1979): On the dynamic forcing of short term climate fluctuation by feedback mechanisms. *Environmental Research Papers*, Colorado State University, Fort Collins, USA.

SHIMSHONI, M. (1971): On Fisher's test of significance in harmonic analysis. *Geophys. J. R. Astron. Soc.*, 23, 373-377.

WHITTLE, P. (1952): The simultaneous estimation of a time series harmonic components and covariance structure. *Trabajos Estad.*, 3, 43-57.

POSTSCRIPT

After the paper was completed, we traced A.F. Siegel (1980): Testing for periodicity in a time series, *Journal Amer. Statist. Assoc.*, 75, 345-348 (thanks to a remark by J.K. Ord, during the ITSM in Toronto). We applied the test to our series and, at the 5% significance level, 3.6, 13 and 26 years periodicities were accepted; while, at the 1% level, 13 and 26 years periodicities were detected.

IDENTIFICATION AND PARAMETER ESTIMATION OF PHOSPHATE BALANCE MODELS IN DRINKING WATER RESERVOIRS

Herman Koppelman
Berninkholthoek 30, 7546 CA Enschede, Netherlands

Tom Schilperoort
Mathematics Branch, Delft Hydraulics Laboratory, P.O. Box 152,
8300 AD Emmeloord, Netherlands

In the storage reservoir "De Grote Rug" in the Netherlands, intake water with a high phosphate concentration must be pretreated. For a proper interpretation of the results of different treatments, a dynamic mathematical model for the phosphate balance is indispensable. To choose the most adequate model out of various possibilities, we applied identification techniques to data from the reservoir. For the parameter estimation, both the off-line prediction error method and the on-line Extended Kalman Filter were used. The on-line method, especially, is shown to be useful for model evaluation.

1. INTRODUCTION

In the drinking water storage reservoir "De Grote Rug" in The Netherlands, investigations are going on in order to find the most appropriate method to pretreat intake water with a high phosphorus concentration. This pretreatment is necessary to reduce the eutrophication (build up of nutrients) to an acceptable level, because otherwise algal growth in the reservoir may cause severe difficulties at various stages of the drinking water purification process. A possible pretreatment consists of adding phosphate precipitating iron and aluminium salts to the intake water, thereby reducing the phosphorus levels.

To interpret the results of different treatments, the use of a dynamic mathematical model, to describe the total phosphate mass balance, is indispensable. For this purpose, several models of different complexity levels are offered in the relevant literature, starting with simple phosphorus-algae models and ending with multi parameter ecological models that include almost all aquatic life forms and water quality parameters. From the practical point of view, a reasonably applicable model is needed, which should both use the available data base and focus on the main processes. Since various simple hypotheses exist to describe the processes affecting the mass balance, a number of simple mathematical models can be built, from which the most adequate one should be chosen. To investigate the feasibility of modern identification techniques to discriminate between these models, both an off-line and an on-line identification method were applied to data from the reservoir; and results are presented here. Details with respect to the eutrophication study, including the various pretreatment effects, can be found elsewhere (Bannink and Van der Vlugt 1978, Bannink et al 1980).

2. EXPERIMENTAL SET UP

To study the effect of different pretreatments, three experimental butylrubber reservoirs were installed in the storage reservoir. A detailed description of the experimental set-up can be found in Bannink and Van der Vlugt (1978). In this study, data from reservoir ## 2 were used. This reservoir was treated with the sewage water treatment agent AVR consisting of iron and aluminium salts. The data set included weekly observations of
- the waterdepth $D(t)$ [m]
- the total phosphate mass in the reservoir $m(t)$ [g]
- the phosphate mass inflow to the reservoir $m_{in}(t)$ [g/week]
- the phosphate mass outflow from the reservoir $m_{out}(t)$ [g/week]
- the chlorophyll mass in the reservoir $m_{chl}(t)$ [g]
- the suspended phosphate mass in the reservoir $m_{sus}(t)$ [g].

3. HYPOTHESES

Basically, the phosphate mass balance in the reservoir is determined by three types of biological and ecological processes: inflow and outflow of phosphates, sedimentation, and uptake from the bottom. Since in this specific investigation the uptake could be neglected, the basic model reads

$$\frac{dm(t)}{dt} = m_{in}(t) - m_{out}(t) - m_{sed}(t) \tag{1}$$

where $m_{sed}(t)$ denotes the sedimentated phosphate mass in g/week.
A number of different models were constructed by substituting in (1) different hypotheses for $m_{sed}(t)$. It was assumed that the reservoir could be considered as a continuous stirred tank reactor.

Organic sedimentation

Algae take up dissolved inorganic phosphate from the surface water. Part of this algal population dies at time t and sinks to the bottom, thereby removing (suspended organic) phosphate from the water. When it is assumed that
- the algal mass is proportional to the (measured) chlorophyll mass
- the algal mortality is proportional to the total algal mass
- the decrease of organic phosphate is proportional to the algal mortality then a first hypothesis for this process, called organic sedimentation, is

$$m_{os}(t) = \frac{k_{os} k_c}{D(t)} m_{chl}(t) = \frac{k_o}{D(t)} m_{chl}(t) \tag{2}$$

where $m_{os}(t)$ denotes the organic sedimentation [g/week], k_{os} the organic sedimentation rate [m/week], k_c the phosphate-chlorophyll ratio and

$k_o = k_{os} k_c$.

A second hypothesis suggests that the organic sedimentation is proportional to the decrease of chlorophyll

$$m_{os}(t) = \max\left(0, -\frac{k_o}{D(t)} \frac{dm_{chl}(t)}{dt}\right). \qquad (3)$$

This hypothesis corresponds more to the reality of alternating periods of algal bloom and mortality.

Inorganic sedimentation

Another part of the total phosphate mass settles after reacting with inorganic compounds. A first hypothesis is that this inorganic sedimentation is proportional to the inorganic particulate phosphate mass. This can be approximated by subtracting the organic particulate mass, which is supposed to be proportional to the chlorophyll mass, and the suspended phosphate mass from the total phosphate mass. Hence, this hypothesis yields

$$m_{as}(t) = \frac{k_{as}}{D(t)} \left[m(t) - m_{sus}(t) - k_c m_{chl}(t) \right] \qquad (4)$$

in which $m_{as}(t)$ denotes the inorganic sedimentation [g/week] and k_{as} the inorganic sedimentation rate [m/week]. Because the suspended phosphate measurements were rather unreliable, it could be useful to exclude the term $m_{sus}(t)$ from (4). This yielded the second hypothesis

$$m_{as}(t) = \frac{k_{as}}{D(t)} \left[m(t) - k_c m_{chl}(t) \right]. \qquad (5)$$

A very simple, alternative hypothesis states that the inorganic sedimentation is proportional to the total phosphate inflow

$$m_{as}(t) = k_i m_{in}(t) \qquad (6)$$

with k_i a proportionality constant. Although (6) is not very realistic from a physical point of view, it can provide in some cases a convenient description of the sedimentation.

4. CONSIDERED MODELS

From the models which can be constructed by combining the given hypotheses, only four will be considered here. In discrete time, the model equations read:

Model 1 (hypotheses 2 and 6)

$$\underline{m}(j+1) = \underline{m}(j) - m_{out}(j) + (1-k_i) m_{in}(j) - \frac{k_o}{D(j)} m_{chl}(j) + \underline{w}(j) \qquad (7)$$

Model 2 (hypotheses 3 and 6)

Equation (7), with $\phi(j)$ replacing $m_{chl}(j)$, and

$$\phi(j) = \max\{0, m_{chl}(j-1) - m_{chl}(j)\}. \qquad (8)$$

Model 3 (hypotheses 2 and 5)

$$\underline{m}(j+1) = a(j)\,\underline{m}(j) + \frac{D(j)}{k_{as}}\{1-a(j)\}\cdot\{m_{in}(j) - m_{out}(j) + \frac{b}{D(j)} m_{chl}(j)\} + \underline{w}(j) \qquad (9)$$

$$a(j) = \exp\{-\frac{k_{as}}{D(j)}\},\ b = (k_{as} - k_{os})k_c \qquad (10)$$

Model 4 (hypotheses 2 and 4)

Equation (9), with $\{1-a(j)\}\cdot m_{sus}(j)$ being added to the right hand side.

In these models, the values $m(j)$, $m_{in}(j)$, $m_{out}(j)$ etc. represent the averages of these signals over the time interval $(j-1)\Delta t$ to $j\Delta t$. Within the derivation of these discrete time versions the (physically realistic) assumption of a slowly changing waterdepth was used. Note that, in hypotheses (4) and (5), the inorganic sedimentation partly depends on the total phosphate mass in the reservoir. Substitution of these expressions in the basic model (1) therefore yields first order differential equations for $m(t)$, which explains the exponential coefficient $a(j)$ in the corresponding difference equations.

The models are necessarily stochastic because none of them describes the reality exactly; various chemical, physical and biological processes are modelled poorly, if modelled at all. To account for this uncertainty, a white noise process $\underline{w}(j)$ (system noise) was added to each model.

From (7) to (10), we see that each model has a state space representation

$$\underline{x}(j+1) = A(j)\,\underline{x}(j) + B(j)\,u(j) + \underline{w}(j). \qquad (11)$$

Here $x(j)$ denotes the state vector (in this case containing $m(j)$ only), $A(j)$ and $B(j)$ are matrices containing the model parameters, and $u(j)$ is the input vector containing $m_{in}(j)$, $m_{out}(j)$, $m_{chl}(j)$ and $m_{sus}(j)$. Since, normally, the input vector is assumed to be perfectly known, the measurement noise associated with these inputs was considered to be part of the system noise, $\underline{w}(j)$. To apply identification techniques, a measurement equation was added to (11), which describes the relation between the state vector and the measured output

$$\underline{y}(j) = H(j)\,\underline{x}(j) + \underline{v}(j). \qquad (12)$$

Here $y(j)$ is a scalar, and $H(j)$ equals one. To account for the measurement uncertainty, a white noise process $\underline{v}(j)$ (measurement noise) was added to (12). To discriminate between the four models, both the parameter matrices, $A(j)$ and $B(j)$, and the covariance matrices, $Q(j)$ and $R(j)$, of the noise processes, $\underline{w}(j)$ and $\underline{v}(j)$, have to be estimated. Then, using various validation criteria, the models can be compared and evaluated.

5. PARAMETER ESTIMATION AND MODEL VALIDATION

In this study, two parameter estimation methods and several validation criteria were used.

Off-line estimation of parameters

When at time j an optimal estimate of the state vector $x(j)$ is available, then the state, and hence the observation y, can be predicted at time $(j+1)$, using (11) and (12) with the noise terms put equal to zero. The difference between the predicted and the actual observations at time $(j+1)$ depends on the values of the unknown parameters. This difference, which is called the innovation or output prediction error, is given by

$$\underline{z}(j+1|\theta) = \underline{y}(j+1) - H(j+1)\,\hat{\underline{x}}(j+1|j). \tag{13}$$

Here θ denotes the unknown parameter vector, and $\hat{\underline{x}}(j+1|j)$ the one-step-ahead prediction of $x(j+1)$, based on the measurements $\{y(1),\ldots,y(j)\}$. This one-step-ahead prediction can be computed from the Kalman filter equations given in the Appendix. Now the best estimate of the unknown parameter vector θ is that vector $\hat{\theta}$ which minimizes the criterion

$$q(\theta) = \frac{1}{M}\sum_{j=1}^{M}\{z(j,\theta)\}^2. \tag{14}$$

Hence, all available measurements $y(j)$, $j=1,\ldots,M$, are used at once, off-line, to estimate one overall value of θ. This way of estimating θ is called the output-prediction-error (OPE) method. It can be proved (Goodwin and Payne, 1977) that the estimate $\hat{\theta}$ which minimizes (14) is asymptotically normally distributed, with mean value equal to θ and a covariance matrix P which, asymptotically, is given by

$$\lim_{M\to\infty} P = \frac{2}{M}\sigma^2\,(q_{\theta\theta})^{-1}. \tag{15}$$

Here σ^2 denotes the variance of the innovations, to be estimated by (14), and $q_{\theta\theta}$ the Hessian of $q(\theta)$ with respect to θ. In the case of a finite number of measurements, (15) provides a lower bound for the parameter uncertainty.

On-line estimation of parameters

An on-line estimation method can be constructed by transforming the parameter estimation problem into a state estimation problem. A suitable state representation for the parameters is

$$\underline{\theta}(j+1) = \underline{\theta}(j) + \underline{w}_\theta(j) \tag{16}$$

in which $\underline{w}_\theta(j)$ is a white noise process. The state vector $x(j)$ then can be augmented with $\theta(j)$, after which the Kalman filter can be applied to provide estimates of both $x(j)$ and $\theta(j)$. In general the extended system will be non-linear, which requires the use of the Extended Kalman Filter. See Appendix.

Model validation

When the parameters are estimated, there is no guarantee that the chosen model

structure with the estimated parameters is actually a good model. Therefore objective validation criteria should be applied to the identified model.
In this study the following criteria were used:
- **Whiteness of innovations** : when the model stucture and the parameter estimates are correct, then the prediction errors (or innovations) should be a zero-mean white noise sequence. This can be checked by calculating the mean value and the autocorrelation-function or the (cumulative) variance density spectrum of the innovations.
- **Correlation between input and innovations** : when the input of the system has been modelled correctly, then no systematic coherence should exist between the input and the innovations. This can be checked by calculating their cross correlation function.
- **Physical consistency** : a structural inadequate model does not contain explicit representations of all the significant processes associated with the system. If, in the on-line case, the Kalman filter "recognizes" a persistent error between the one step ahead predictions and the observations, it will respond by correcting the parameter estimates in its attempt to correct the inadequate model structure (Beck and Young, 1976). An evolution in time of these on-line estimates, which is not consistent with the underlying physics, therefore provides evidence of a model inadequacy. Obviously also, off-line parameter estimates should be physically realistic.
- **Model simulations** : the output of an adequate model, provided with the right initial conditions and input signals, should correspond closely to the observations. A visual comparison between model simulations and observations therefore is an evident criterion for model validation.
- **Forward prediction error (FPE)** : for an adequate model the one-step-ahead prediction $\hat{y}(j|j-1)$ of the output should correspond closely to the actual observation $y(j)$. Therefore, the value of the forward prediction error

$$FPE = \frac{1}{M} \sum_{j=1}^{M} \{y(j) - \hat{y}(j|j-1)\}^2$$

can be used as a validation criterion. Note that this corresponds to the variance of the innovations.

Estimation of noise parameters

To apply the Kalman filter equations, the noise covariance matrices, Q and R, must be known a priori. Mostly, however, this requirement is not met, in which case they have to be estimated from the data. In this study, three methods were used to estimate Q and R in the off-line case. The first method uses the property that the degree of whiteness of the innovation sequence $z(j)$ depends on the values of the variances. It can be proved that the (steady state) innovations depend only on the ratio, QR^{-1}, of the variances if the system model is time invariant (experimental results showed that this was also true for our models). The optimal value of QR^{-1}, then, is the value (found by trial and error) for which the innovations correspond as much as possible to white noise. The second approach is to consider QR^{-1} as an unknown parameter, and to estimate it simultaneously with the system parameter vector θ. The third, more systematic procedure, is based on adaptive filtering (Jazwinski, 1970). The idea is to let the innovations themselves determine appropiate noise levels in real time. That value of Q is computed which produces the most probable predicted innovation. It can be shown (Jazwinksi, 1970) that this Q is determined by

$$E\{z^2(j)\} = z^2(j).$$

It is much more difficult to estimate Q and R for on-line case. This is because Q is augmented with the variances of the noise associated with the parameter equations (16). The problem was solved by putting the values of the original Q and R equal to the off-line estimates; whereas Q_θ, being the part of Q corresponding to the parameter equations, was chosen such that the parameters showed a gradual change in time, with moderate variations. The consequences of this rather subjective choice will be examined in the next section.

6. IDENTIFICATION RESULTS

Two years of weekly observations of the variables, indicated in Section 2, were used in this study: data from 1977 for the parameter estimation, and data from 1978 for model validation. The results will be discussed below.

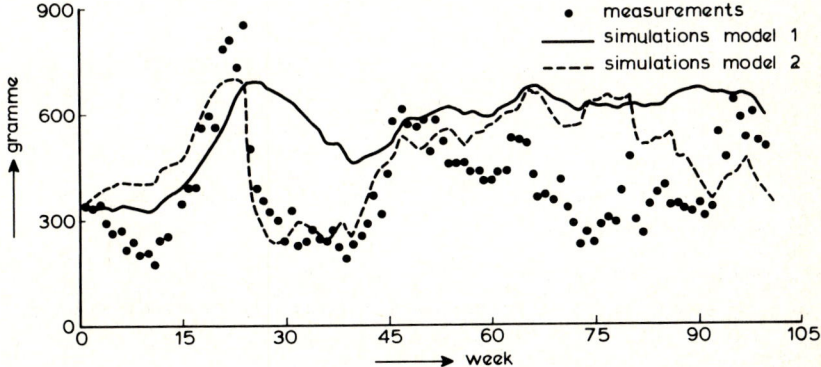

Figure 1. Measurements and simulations of the total phosphate mass (Models 1 and 2).

Model 1

The off-line parameter estimation resulted in the parameter values: $k_i = 0.89$, $k_o = -0.02$ m/week. This value of k_o could not have any physical meaning: it was known that the organic sedimentation was much more important than this value indicated; and anyhow k_o should be positive.

Also, since the model simulations were bad (Figure 1), it was concluded that this model did not provide an adequate representation of reality. Still, it was worthwile to investigate the behaviour of the on-line estimates. Augmenting (7) with the parameter equations yielded the following linear system with $\theta_2 = 1-k_i$ and $\theta_1 = k_o$:

$$m(j+1) = m(j) - m_{out}(j) + \theta_2(j) \, m_{in}(j) - \frac{\theta_1(j)}{D(j)} m_{chl}(j) + w(j)$$
$$\theta_1(j+1) = \theta_1(j) + w_1(j) \quad (17)$$
$$\theta_2(j+1) = \theta_2(j) + w_2(j).$$

Application of the linear Kalman filter to this system yielded the results given in Figure 2. It clearly shows the inadequacy of the model. Especially, the behaviour of $\theta_1(j)$ is not consistent with the underlying physics. Because the largest fluctuations in $\theta_1(j)$ occured during periods of algal growth and mortality, it was concluded that the hypothesis (2) was not satisfactory. In model 2, therefore, this hypothesis was replaced by Equation (3).

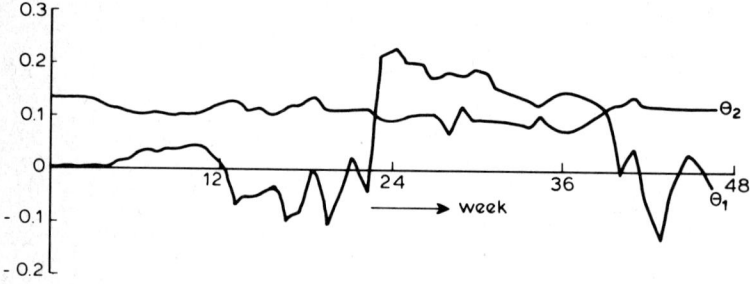

Figure 2. Time evolution of model 1 parameters.

Model 2

Here the off-line estimation resulted in $k_i = 0.81$, $k_o = 0.19$ m/week. The model simulations (Figure 1), although improved, are still not satisfactory. Because of the small number of measurements with $\dot{m}_{chl}(t) < 0$, only k_i could be estimated on-line, using the following augmented system with $\theta = 1-k_i$:

$$m(j+1) = m(j) - m_{out}(j) + \theta(j) \, m_{in}(j) - \frac{k_o}{D(j)} \phi(j) + w(j)$$
$$\theta(j+1) = \theta(j) + w_1(j) \quad (18)$$

which is also linear. The estimated θ is shown in Figure 3. The fluctuations in θ

suggested that the hypothesis (6), for the inorganic sedimentation, was not very realistic. Therefore, in models 3 and 4, this was replaced by hypotheses (5) and (6), respectively.

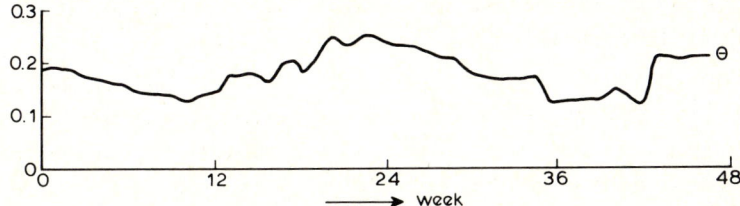

Figure 3. Time evolution of the model 2 parameter.

Models 3 and 4

The off-line estimation resulted in: k_{as}= 3.6 m/week, and b= 0.8 m/week, for model 3; and k_{as}= 8.0 m/week, and b= 1.9 m/week, for model 4. The model simulations, shown in Figure 4, improved considerably. On-line estimation of the parameters required the use of the Extended Kalman Filter, because the augmented system is nonlinear, being

$$m(j+1) = \exp\left\{\frac{-\theta_1(j)}{D(j)}\right\} m(j) + \frac{D(j)}{\theta_1(j)} \left\{1 - \exp\left(\frac{-\theta_1(j)}{D(j)}\right)\right\} [m_{in}(j) - m_{out}(j)$$
$$+ \frac{\theta_2(j)}{D(j)} m_{ch1}(j) + \frac{\theta_1(j)}{D(j)} m_{sus}(j)] + w(j) \quad (19)$$

$\theta_1(j+1) = \theta_1(j) + w_1(j)$

$\theta_2(j+1) = \theta_2(j) + w_2(j)$

with $\theta_1 = k_{as}$ and $\theta_2 = b$.

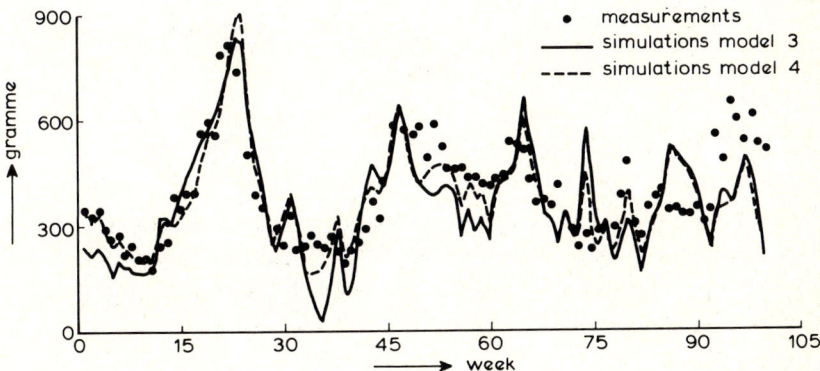

Figure 4. Measurements and simulations of the total phosphate mass (Models 3 and 4).

For model 3 the term $m_{sus}(j)$ was not included. The estimation results are shown in Figure 5. A comparison of these results with Figure 6 reveals a striking resemblance between the k_{as} evolution and the input of water. This could be attributed to the fact that the dosing of aluminium was affected by adding the AVR agent to the input water. Hence, for low input levels, the number of aluminium-particles in the reservoir decreased, which caused a lower inorganic sedimentation. The parameter k_{as} apparently did depend on the AVR concentration of the input.

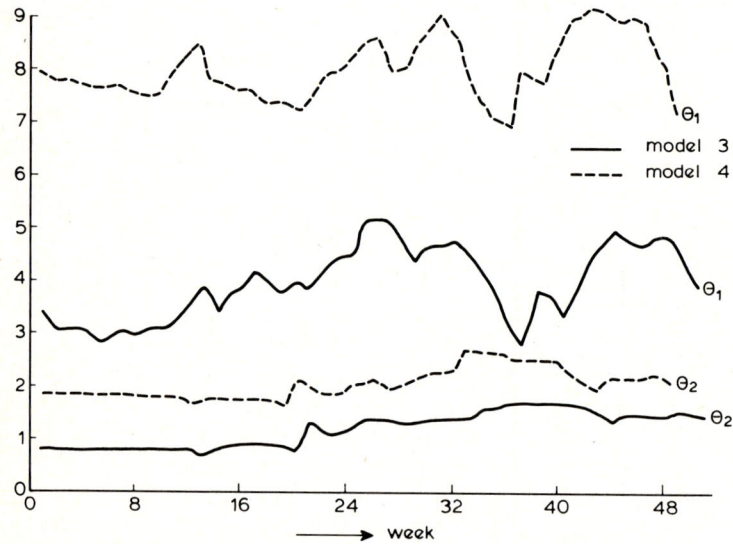

Figure 5. Time evolution of the parameters of Models 3 and 4.

This fact, which was not included in the hypotheses (4) and (5), was confirmed by the behaviour of the cross correlation functions between the innovations and the phosphate input. For models 3 and 4, significant negative values were observed

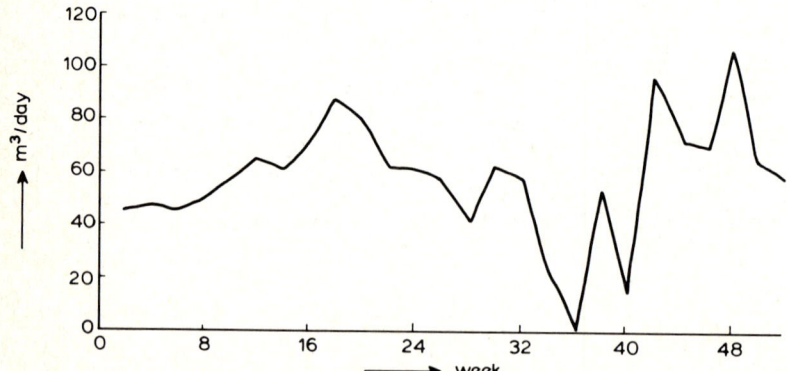

Figure 6. Time evolution of water intake in 1977.

(Figure 7), which indicate that high input levels were coupled to negative innovations. Hence, for high input levels, the model predictions were too high, due to an underestimated sedimentation. For low input levels the sedimentation was, correspondingly, overestimated.

Models 3A and 4A

For model 2, in which the inorganic sedimentation was coupled completely to the input, no significant cross correlation between innovations and input was

model	k_{as}	b	FPE * 10^4	
			1977	1978
3	3.6	0.8	1.25	1.40
3A	1.8	0.48	1.04	0.60
4	8.0	1.9	0.69	1.34
4A	4.4	1.1	0.75	0.76

Table I. Parameter values and FPE-values for different models.

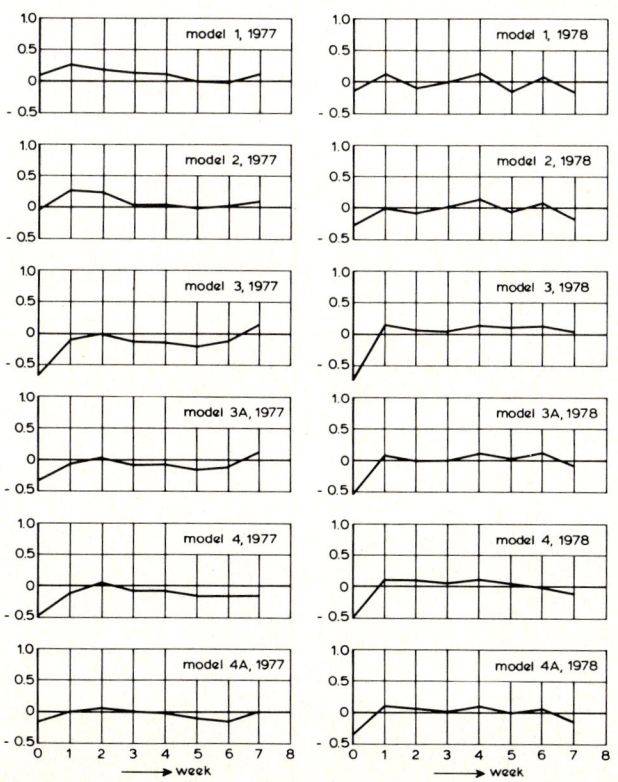

Figure 7. Cross correlations between innovations and phosphate input for different models.

observed (Figure 7). This suggested that the models 3 and 4 could be improved by replacing $m_{in}(t)$ with the suspended part of the total phosphate input. In this way a partial dependency between the input and the sedimentation was introduced, which was motivated by the reasoning that the particulate part of the input settles down as soon as it flows into stagnant water. Results of the models 3A and 4A thus obtained are shown in Figure 8.

Both the behaviour of the cross correlation between input and innovations and the model simulations are indeed improved. This was also reflected in the FPE-values for the different models (Table I).

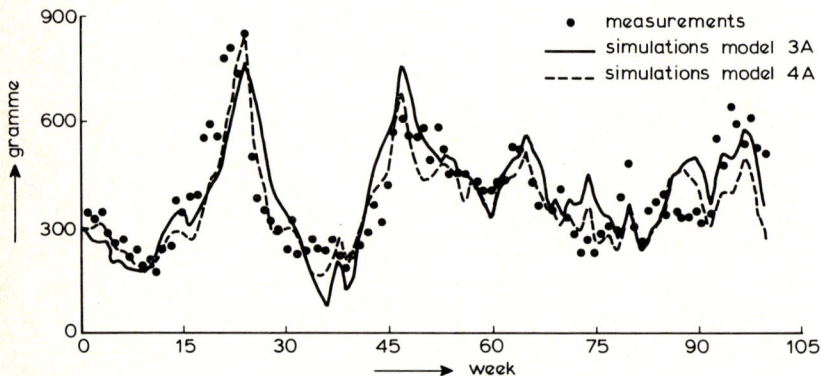

Figure 8. Measurements and simulations of the total phosphate mass (Models 3A and 4A).

Parameter investigation

For model 4A the estimated parameters were investigated in further detail. For this purpose the whole dataset (1977 and 1978) was used for the parameter estimation. The OPE-method resulted in k_{as} = 3.6 m/week ($\pm$ 0.5), and b = 0.9 m/week ($\pm$ 0.2). The values, indicating the precision, were obtained from (15); as were the (approximate) joint σ- and 2σ-confidence intervals for k_{as} and b, shown in Figure 9, which clearly indicate the linear dependency between k_{as} and b. Because this dependency was not modelled explicitly, it provides an independent confirmation of the relation b= k_c ($k_{as}-k_{os}$).

Moreover, the parameter b was estimated for various fixed values of k_{as}. This also resulted in a linear relation between b and k_{as}, from which k_{os} and k_c could be calculated. This yielded k_c= 0.37 and k_{os}= 1.2 m/week.

On physical grounds, the values of k_{as} and k_c were expected to be approximately 3.5 m/week and 0.35 respectively. Hence, the estimated and physical values showed a good correspondence.

Estimation of noise parameters

For the estimation of the noise parameters, Q and R, three methods were used in the off-line case (cf. Section 5). Because each method gave similar results, and because the parameter values turned out to be rather insensitive to the chosen Q and R values, it could be concluded that, in the off-line case, the noise

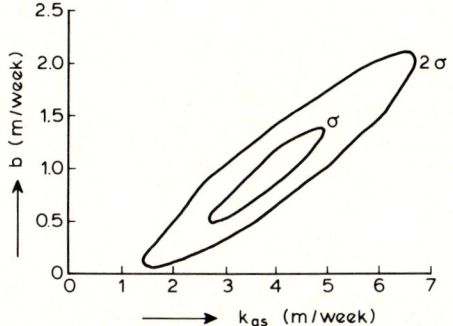

Figure 9. Joint confidence intervals for k_{as} and b of Model 4A.

parameters could be estimated accurately enough. More details can be found in Koppelman (1981). In the on-line case, however, the precise parameter evolution did depend heavily on the choice of the Q_θ values. For model 3 this is illustrated in Figure 10. The overall behaviour, however, does not change very much. Therefore, it can be concluded that the on-line estimation method provides

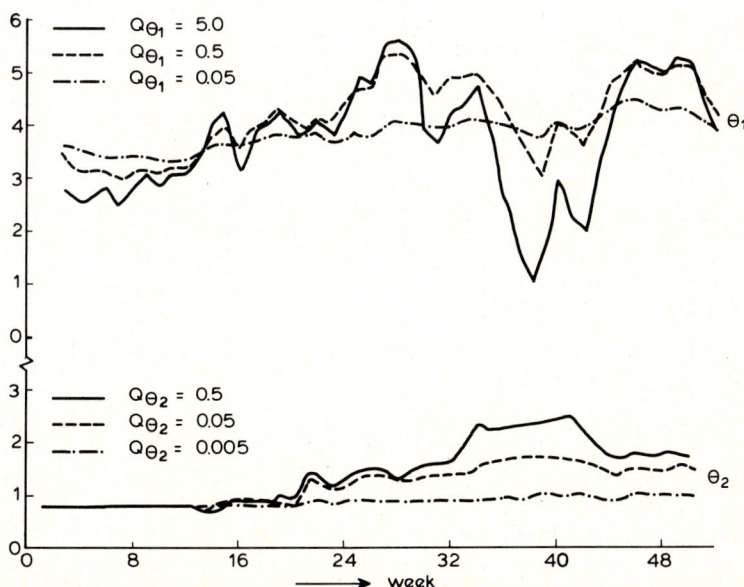

Figure 10. Time evolution of model 3 parameters for different Q_θ-values.

a very useful tool for model validation, even in the case where the noise variances are not accurately known.

7. CONCLUSIONS

An off-line and an on-line parameter estimation method have been applied to different models describing the same system. Most information about the validity of the models was provided by the on-line method. Especially, its ability to show any time variation in the parameter estimates which is not consistent with the underlying physics makes it very suitable for model structure evaluation. The behaviour of the parameters may suggest additional or alternative hypotheses and confirm relations between parameters and measured variables. So, the on-line technique is a useful tool in the process of developing an adequate model. Compared to the off-line method, the on-line method has some drawbacks. It is more sensitive to the choice of initial conditions and noise variances. Moreover, these variances are much more difficult to estimate. Also, the estimated parameters and their variances are possibly inaccurate, and divergence of the on-line algorithm is not an exception. This last drawback can be reduced, however, by using the results of the off-line method as initial conditions for the on-line method. So, it can be concluded that the off-line method is more suitable to estimate the parameter values, whereas the on-line method provides more information about the model structure.

From a mathematical, as well as a physical, point of view, model 4A proves to be the most adequate. Still, this model gives only a very rough description of reality. Further improvements may be possible by applying the investigated identification techniques on more complex models.

APPENDIX.   THE DISCRETE KALMAN FILTER EQUATIONS

Suppose the system under study can be represented by the following linear system model

$$\underline{x}(j+1) = A(j) \underline{x}(j) + B(j) u(j) + \underline{w}(j) \tag{A.1}$$

with $\underline{w}(j)$ an $N(0, Q(j))$ white noise sequence. Furthermore, assume the following linear observation model

$$\underline{y}(j) = H(j) \underline{x}(j) + \underline{v}(j) \tag{A.2}$$

with $\underline{v}(j)$ an $N(0, R(j))$ white noise sequence. Let the initial conditions be

$$E\{\underline{x}(o)\} = \hat{x}_o \tag{A.3}$$

$$\text{var}\{\underline{x}(o)\} = P_o$$

and assume the processes $\underline{v}$ and $\underline{w}$ to be independent, i.e.

$$E\{\underline{w}(k)\,\underline{v}(j)^T\} = 0 \quad \text{for all } j,k. \tag{A.4}$$

Then the optimal one-step-ahead prediction $\underline{\hat{x}}(j+1|j)$ of $x(j+1)$, given the measurements $\{y(1),\ldots,y(j)\}$, equals

$$\underline{\hat{x}}(j+1|j) = A(j)\,\underline{\hat{x}}(j|j) + B(j)\,u(j) \tag{A.5}$$

with covariance matrix

$$P(j+1|j) = A(j)\,P(j|j)\,A^T(j) + Q(j). \tag{A.6}$$

When at time $(j+1)$ the observation $y(j+1)$ is available, the one-step-ahead prediction can be updated, yielding

$$\underline{\hat{x}}(j+1|j+1) = \underline{\hat{x}}(j+1|j) + K(j+1)\left[\underline{y}(j+1) - H(j+1)\,\underline{\hat{x}}(j+1|j)\right] \tag{A.7}$$

with covariance matrix

$$P(j+1|j+1) = [I - K(j+1)\,H(j+1)]\,P(j+1|j). \tag{A.8}$$

The Kalman gain matrix is given by

$$K(j+1) = P(j+1|j)\,H^T(j+1)\left[H(j+1)\,P(j+1|j)\,H^T(j+1) + R(j+1)\right]^{-1}. \tag{A.9}$$

From the initial conditions (A.3), (A.5) to (A.9) can be solved for each time step recursively.

In the non-linear case, with the system model and observation model given by

$$\underline{x}(j+1) = f(j,x(j),u(j)) + \underline{w}(j) \tag{A.10}$$

$$\underline{y}(j) = h(j,\underline{x}(j)) + \underline{v}(j) \tag{A.11}$$

the Extended Kalman Filter can be used. Then, the following equations hold

$$\underline{\hat{x}}(j+1|j) = f(j,\underline{\hat{x}}(j|j),u(j)) \tag{A.12}$$

$$P(j+1|j) = F(j,\underline{\hat{x}}(j|j),u(j))\,P(j|j)\,F^T(j,\underline{\hat{x}}(j|j),u(j)) + Q(j) \tag{A.13}$$

$$\underline{\hat{x}}(j+1|j+1) = \underline{\hat{x}}(j+1|j) + K(j+1)\left[y(j+1) - h(j+1,\underline{\hat{x}}(j+1|j))\right] \tag{A.14}$$

$$P(j+1|j+1) = [I - K(j+1)\,H(j+1,\underline{\hat{x}}(j+1|j))]\,P(j+1|j) \tag{A.15}$$

$$K(j+1) = P(j+1|j)\,H^T(j+1,\underline{\hat{x}}(j+1|j))\,[H(j+1,\underline{\hat{x}}(j+1|j))\,P(j+1|j)$$
$$H^T(j+1,\underline{\hat{x}}(j+1|j)) + R(j+1)]^{-1} \tag{A.16}$$

where

$$F(j,\underline{\hat{x}}(j|j),u(j)) = \frac{\partial}{\partial x}\,f(j,x,u) \tag{A.17}$$

$$H(j+1,\underline{\hat{x}}(j+1|j)) = \frac{\partial}{\partial x}\,h(j,x) \tag{A.18}$$

with the partial derivatives being evaluated at $x = \underline{\hat{x}}(j|j)$ and $x = \underline{\hat{x}}(j+1|j)$ respectively.

REFERENCES

BANNINK, B.A. and Van der VLUGT, J.C. (1978). Hydrobiological and chemical response to the addition of iron and aluminium salts, studied in three Lund-type butylrubber reservoirs. Verh. Internat. Verein. Limnol. 20, 1816-1821.

BANNINK, B.A., Van der MEULEN, J.H.M., PEETERS, J.C.N. and Van der VLUGT, J.C. (1980). Hydrobiological consequences of the addition of phosphate precipitants to inlet water of lakes. Hydrobiol. Bull. 14, 73-89.

BECK, M.B. and YOUNG, P.G. (1976). Systematic identification of DO-BOD model structure. ASCE, Env. Eng., 102, 909-924.

GOODWIN, G.C. and PAYNE, R.L. (1977). Dynamic System Analysis: Experimental Design and Data Analysis. Academic Press, New York.

JAZWINSKI, A.H. (1970). Stochastic Processes and Filtering Theory. Academic Press, New York.

KOPPELMAN, H. (1981). Application of parameter estimation methods to a waterquality problem (in Dutch). Report S456- S459, Delft Hydraulics Laboratory, Delft.

Part 2
Geophysical Time Series

THE RIGHT-HALF AUTOCORRELATION THEOREM

Enders A. Robinson
Geosciences Dept., The University of Tulsa, Tulsa, Oklahoma, USA 74104

Sven Treitel
Amoco Production Co., Tulsa, Oklahoma, USA 74102

From a given signal, the autocorrelation function and the spectral density may be computed. The autocorrelation and spectral density are Fourier transform pairs. The positive-definite property of the autocorrelation is equivalent to the property that the spectral density is positive for all values of frequency. The impedance and admittance functions of a physical system have the same mathematical properties, so the coined word "immittance" can designate either one. By physical reasoning, an immittance function is necessarily stable, causal, and invertible. Equivalently, we may say that an immittance function is necessarily minimum-delay. The frequency spectrum of an immittance function has both a real and an imaginary part. A well-known theorem of Brune is that the real part of the frequency spectrum of an immittance is positive for all frequencies. Thus, an immittance is called positive-real. We establish the following theorem: A minimum-delay signal (a_0, a_1, a_2, ...) has a positive-real spectrum if and only if the signal is the weighted right-hand side (r_0, $2r_1$, $2r_2$, ...) of a positive-definite function r_n. In other words, the theorem states that an immittance is the weighted right-half of an autocorrelation, and conversely, a weighted right half of an autocorrelation is an immittance. We call this principle the Right-Half Autocorrelation Theorem. Thus the weighted right-hand side of a positive-definite function (or equivalently, the right-hand side of an autocorrelation) is necessarily minimum-delay. However, a minimum-delay signal may or may not have a positive-real frequency spectrum. Various procedures are given for the numerical checking of the positive-real as well as of the minimum-delay properties of given signals.

1. INTRODUCTION

Autocorrelations are positive-definite functions, while impedances and admittances (i.e., "immittances") are positive-real functions. There has been a general tendency to treat the positive-definite concept and the positive-real concept as separate subjects, even though they are closely related. Both concepts are of importance in seismology because of the bearing they have on the proper design of deconvolution operators for the marine seismogram (Robinson and Treitel, 1977). An abbreviated discussion has recently been given by Robinson, Loewenthal, and Treitel (1978).

We begin by reviewing the well-known relationships between the autocorrelation function and the spectral density function. After consideration of the positive-definite property of the autocorrelation and the positive-real property of the immittance, we give the main result of this paper, namely the Right-Half Autocorrelation Theorem. The theorem states that an immittance is the weighted right-half of an autocorrelation, and conversely that a weight right-half of an autocorrelation is an immittance. Finally, we describe several numerical tests to establish the positive-definite and the positive-real properties of signals.

2. THE AUTOCORRELATION AND THE SPECTRAL DENSITY

A _discrete signal_ a_n and its _frequency spectrum_ $A(f)$ are Fourier transform pairs:

$$a_n = \int_{-0.5}^{0.5} A(f) \, e^{2\pi i f n} \, df$$

$$A(f) = \sum_{n=-\infty}^{\infty} a_n \, e^{-2\pi i f n} \, .$$

Here the sample interval is defined as one time unit, so frequency f is in units of cycles per time unit. Because the shortest physical wavelength that can be observed as such in a discrete waveform is 2 time units, the highest non-aliased frequency is 1/2 or 0.5. The frequency 0.5 cycles per time unit is called the Nyquist frequency, after Harry Nyquist (1928), who contributed to telegraph transmission in the 1920's. Note that 1 sample per time unit is the sampling rate, so the Nyquist frequency is one-half the sampling rate.

The _energy_ of a discrete signal is defined as the sum of the squared magnitudes of the wave; any signal whose energy is finite is called an _energy signal_. It can be shown (see e.g., Robinson, 1967a, p. 216) that the energy is equal to the integral of the squared magnitude of the frequency spectrum. This result, called Bessel's equality, may be written as:

$$\text{Energy} = \sum_{n=-\infty}^{\infty} |a_n|^2 = \int_{-\infty}^{\infty} |A(f)|^2 \, df \quad .$$

The <u>energy spectral density</u> $R(f)$ of the energy signal a_n is defined as the squared magnitude of its frequency spectrum $A(f)$; that is, the energy spectral density is defined as

$$R(f) = |A(f)|^2 \geq 0 \quad .$$

By definition the energy spectral density $R(f)$ is a positive function of frequency. More precisely, $R(f)$ is a non-negative function of frequency, but in this paper we will deal with the strictly positive case.

The z-transform $A(z)$ of the signal a_n is defined as the formal power series

$$A(z) = \sum_{n=-\infty}^{\infty} a_n z^n \quad .$$

We note that if we let $z = e^{-2\pi i f}$ in the z-transform $A(z)$, we obtain the frequency spectrum

$$A(e^{-2\pi i f}) = \sum_{n=-\infty}^{\infty} a_n e^{-2\pi i f n} = A(f) \quad .$$

Note that we use the same symbol A to denote both the z-transform $A(z)$ and the frequency spectrum $A(f)$. More precisely, if $A(z)$ is the z-transform, then $A(e^{-2\pi i f})$ is the frequency spectrum, but for simplicity we write $A(e^{-2\pi i f})$ as $A(f)$.

Now let us define the function $\overline{A(z)}$ as

$$\overline{A(z)} = \sum_{n=-\infty}^{\infty} a_n^* z^{-n}$$

where the superscript asterisk denotes the complex-conjugate of the quantity to which it is attached. That is, if we take A(z) and change each coefficient to its complex-conjugate and each power of z to its reciprocal, we obtain $\overline{A(z)}$. Thus $\overline{A(z)}$ is the z-transform of the complex-conjugate time-reverse of the signal. We know that in order to autocorrelate a signal we convolve the signal with its complex-conjugate time-reverse. Since convolution corresponds to multiplication of z-transforms, it follows that R(z) defined by the equation

$$R(z) = A(z)\overline{A(z)}$$

is the z-transform of the <u>autocorrelation</u> r_n; that is,

$$R(z) = \sum_{n=-\infty}^{\infty} r_n z^n .$$

If we let $z = e^{-2\pi i f}$ in R(z), we obtain the energy spectral density $R(e^{-2\pi i f}) = A(f) A^*(f) = R(f)$, which as we have already noted is a positive function. In other words, the autocorrelation r_n and the energy spectral density R(f) are Fourier transform pairs:

$$r_n = \int_{-0.5}^{0.5} R(f) e^{2\pi i f n} df$$

$$R(f) = \sum_{t=-\infty}^{\infty} r_n e^{-2\pi i f n} .$$

This correspondence is one of the main results of harmonic analysis. As we have derived it, the result is for signals with finite energy. Generalized harmonic analysis introduced by Wiener (1930) deals with waveforms with infinite energy, but finite power (where power = average energy per unit time). Any signal whose power is finite is called a <u>power signal</u>. The same Fourier correspondence between r_n and R(f) holds if we now interpret r_n as the autocorrelation of a power signal and R(f) as the power spectral density.

3. THE POSITIVE-DEFINITE PROPERTY OF THE AUTOCORRELATION

The results which we will give are valid in either the case of energy signals or the case of power signals. The outstanding property of a spectral density $R(f)$ is that it is a positive function for all values of f. The corresponding property of the autocorrelation is that it is a positive-definite function. The connection between these two properties is found by means of the Toeplitz form. Extensive background material on Toeplitz forms can be found in Grenander and Szegö (1958). The <u>Toeplitz form</u> T is the quadratic (Hermitian) form formed from the autocorrelation coefficients as follows:

$$T = \sum_{k=0}^{N} \sum_{n=0}^{N} x_k \, r_{k-n} \, x_n^*$$

where $x_0, x_1, \ldots, x_N$ is an arbitrary complex signal (and where N can be any non-negative integer). If we substitute our Fourier expression for r_n in this equation we obtain:

$$T = \sum_{k=0}^{N} \sum_{n=0}^{N} x_k \left[\int_{-0.5}^{0.5} R(f) \, e^{2\pi i f(k-n)} \, df \right] x_n^*$$

$$= \int_{-0.5}^{0.5} R(f) \left[\sum_{k=0}^{N} x_k \, e^{2\pi i f k} \right] \left[\sum_{n=0}^{N} x_n \, e^{2\pi i f n} \right]^* df$$

$$= \int_{-0.5}^{0.5} R(f) \, |X(f)|^2 \, df \; .$$

Because the signal $x_0, x_1, \ldots, x_N$ is arbitrary, this equation shows that the Toeplitz form T and the spectral density $R(f)$ are both positive together, i.e., the positiveness of one implies the positiveness of the other. Any function r_n whose Toeplitz form is positive for any signal $x_0, x_1, \ldots, x_N$ is called <u>positive definite</u>. Thus we have shown that the positive-definite property of the autocorrelation is equivalent to the positive property of the spectral density. (As a matter of interest, in probability theory and mathematical statistics the same result holds; namely, the positive-definite property of the characteristic function is equivalent to the positive property of the probability density.) In summary, by its very definition the spectral density $R(f)$ is a positive function of frequency; equivalently, the autocorrelation r_n is a positive-definite function of the time index n.

4. IMPEDANCE, ADMITTANCE AND IMMITTANCE

Let us now consider a physical system which admits an input signal and produces an output signal in response. If the input signal is an impulse δ_n (where $\delta_n = 1$ for $n = 0$ and $\delta_n = 0$ for all other n), then the resulting output a_n is called the <u>impulse response</u> of the system. The system is said to be <u>stable</u> if the impulse response is an energy signal, and the system is said to be <u>causal</u> if the impulse response is one-sided (that is, if $a_n = 0$ for all negative n). In other words, a system is stable if an impulsive stimulus produces a finite-energy response and a system is causal if the response cannot occur in time before its stimulus.

An important class of physical systems is characterized by appropriate <u>impedance</u> and <u>admittance</u> functions. Impedance for a mechanical system is defined as the ratio of force exerted to velocity. In electrical engineering, impedance is defined as the ratio of voltage to current. In acoustics and seismic work, impedance is the ratio of pressure to velocity. For these systems the admittance is physically defined as the inverse of the impedance.

Let the signal x_n represent the force (or voltage or pressure as the case may be) and let the signal y_n represent the velocity (or current or acoustic or seismic velocity, as the case may be).

Either signal x_n or y_n can be considered as the input, and then the other represents the output. In the case when the force x_n is the input and the velocity y_n is the output, the impulse response a_n represents the admittance function. Because we are considering time-invariant linear systems, the velocity y_n is given by the convolution of the force x_n with the admittance a_n, that is,

$$y_n = \sum_{k=0}^{\infty} a_k x_{n-k} .$$

In the case when velocity y_n is the input and the force x_n is the output, the impulse response b_n represents the impedance function. In this case we can write the force x_n as the convolution of the velocity y_n with the impedance b_n, that is,

$$x_n = \sum_{k=0}^{\infty} b_n y_{n-k} .$$

Let $X(f)$ be the frequency spectrum of the force signal x_n, and let $Y(f)$ be the frequency spectrum of the velocity signal y_n. We let $A(f)$ denote the frequency spectrum of the admittance and $B(f)$ the frequency spectrum of the impedance. Since convolution in the time domain corresponds to multiplication in the frequency domain, the above two equations give

$$Y(f) = A(f) X(f)$$

and

$$X(f) = B(f) Y(f)$$

so

$$B(f) = \frac{1}{A(f)} .$$

Because admittance and impedance represent a physical phenomenon in real time, they each must be stable and causal. Also by definition each is the inverse of the other. Only a minimum-delay system can be stable and causal and have a stable and causal inverse. Therefore it follows that the impedance is minimum-delay, and that the admittance is also minimum-delay. A further discussion of the <u>minimum-delay</u> concept can be found in Treitel and Robinson (1964), where it is also shown that the concept of minimum-delay is identical to the older concept of minimum-phase. In summary, we have shown that both admittance and impedance are minimum-delay. Because both <u>imp</u>edance and ad<u>mittance</u> have the same mathematical structure, Bode (1945) coined the word <u>immittance</u> to designate either one. In summary, we have shown up to this point that an immittance function is always minimum-delay.

5. THE POSITIVE-REAL PROPERTY OF AN IMMITTANCE

A <u>positive-real</u> function is defined as a causal stable signal which has the additional property that the real part of its frequency spectrum is positive for all values of frequency f. The concept of positive-real was introduced by Brune (1931) in his classic work on network synthesis. A recent discussion of positive-real functions and their use in network synthesis and in stability analysis, optimality, and sensitivity of dynamic systems is given by Jury (1977).

We now want to show that the admittance spectrum A(f) and the impedance spectrum B(f) each have positive real parts; that is, we want to show that both admittance and impedance are positive-real functions. For convenience let us assume that force and velocity are energy signals; a similar argument can be carried out for power signals by making use of generalized harmonic analysis. Energy is the sum over time of force times velocity. Following a method commonly used, force and velocity are represented by complex waveforms x_n and y_n, so the expression for energy E becomes

$$E = \text{Re} \sum_{n=-\infty}^{\infty} y_n x_n^* .$$

Here, as usual, Re denotes "the real part of" and the asterisk indicates the complex conjugate. We now make use of Parseval's equality which states that

$$\sum_{n=-\infty}^{\infty} y_n x_n^* = \int_{-\infty}^{\infty} Y(f) X^*(f) df .$$

Because $Y = AX$ we see the energy is given by

$$E = \text{Re} \int_{-\infty}^{\infty} A(f) X(f) X^*(f) df$$

which is

$$E = \int_{-\infty}^{\infty} \text{Re } A(f) \left| X(f) \right|^2 df .$$

Because the signal x_n is arbitrary, this equation shows that the energy E and the real part Re $A(f)$ of the admittance spectrum are both positive together. Because energy is always positive, it follows that Re $A(f)$ is always positive. Thus we have shown that the positive property of energy demands the positive-real property of the admittance. The same argument also shows that the positive property of energy demands the positive-real property of the impedance. We have thus obtained the celebrated theorem of Brune (1931) which states that both admittance a_n and impedance b_n are positive-real signals. More concisely, Brune's theorem is: An immitance signal is a positive-real function; that is, the frequency spectrum of an immitance signal has a positive real part.

In the previous section we have shown that an immittance is always a minimum-delay signal. In this section we have given Brune's theorem which states that an immittance is always a positive-real signal. We now want to establish a theorem relating the concepts of minimum-delay and positive-real.

As we have stated earlier, the concept of minimum-delay is mathematically identical to the concept of minimum phase. The magnitude spectrum A(f) and the phase-lag spectrum $\phi(f)$ are defined as the magnitude and negative phase, respectively, of the frequency spectrum A(f); that is,

$$A(f) = |A(f)| e^{-i\phi(f)} .$$

A stable causal signal a_n is said to be minimum-phase (or more properly minimum-phase lag) provided that its net phase-lag displacement $\phi(0.5)-\phi(-0.5)$ is a minimum in the class of all causal stable signals with the same gain. In fact, for all minimum-phase signals encountered in physical applications, this minimum will actually be zero. Thus we can describe a minimum-phase (or minimum-delay) signal as a causal stable signal with phase-lag at the positive Nyquist frequency 0.5 equal to the phase-lag at the negative Nyquist frequency -0.5, that is,

$$\phi(0.5) = \phi(-0.5) .$$

In words, we can say that a causal stable signal is minimum-delay provided that its phase-lag curve at the end returns to its initial value.

In terms of z-transforms, the equation $w = A(z)$ represents a mapping from the complex z-plane to the complex w-plane. The frequency spectrum $A(f) = A(e^{-2\pi i f})$ is the curve in the w-plane corresponding to the unit circle $z = e^{-2\pi i f}$ in the z-plane. A well-known characterization is that a causal stable signal is minimum-delay if and only if its z-transform has no zeros within the unit circle (Treitel and Robinson, 1964). As a result, the curve representing the frequency spectrum A(f) of a minimum-delay signal will not enclose the origin in the w-plane. The phase-lag is the negative of the angle of the vector in the w-plane from the origin to this curve. Since for a minimum-delay system the curve does not enclose the origin, the swings (if any) that the vector makes around the origin in one direction must be offset by the same number of swings in the other direction as f goes from -0.5 to 0.5. Thus the phase-lag at 0.5 is the same as the phase-lag at -0.5, and for a minimum-delay signal the vector ends at the same place as it began. In the minimum-delay case, the vector can rotate forward as much as it wants to, as long as it rotates backward by the same amount, so that the net angle traced as f goes from -0.5 to 0.5 is zero.

Let us now show that the positive-real condition requires a more stringent phase-lag behavior than that of minimum-delay. For a strictly positive-real function, the real part of the frequency spectrum is greater than zero; that is,

$$\text{Re } A(f) > 0 \ .$$

We recall that the phase-lag spectrum is the negative of the phase of $A(f)$; that is, it is the negative of the angle of the vector from the origin to the $A(f)$ curve. The real component of this vector is Re $A(f)$ and the imaginary component is Im $A(f)$. Since Re $A(f)$ is positive, this vector can never leave the first and fourth quadrants in the w-plane; that is, the angle of this vector must always lie between $-\pi/2$ and $\pi/2$. Since the phase-lag is defined as the negative of this angle, that is,

$$\phi(f) = -\tan^{-1}\left[\frac{\text{Im } A(f)}{\text{Re } A(f)}\right]$$

it follows that the phase-lag spectrum must be bounded within the same limits, that is,

$$-\frac{\pi}{2} \leq \phi(f) \leq \frac{\pi}{2} \ .$$

In summary, the curve representing the frequency spectrum $A(f)$ of a positive-real signal will not enclose the origin in the w-plane, and in addition will lie completely in the first and fourth quadrants of the w-plane. As a result the $A(f)$ vector cannot swing outside these two quadrants as f goes from

-0.5 to 0.5. Thus for a positive real signal, the angle of this vector will always be in the range $-\pi/2$ to $\pi/2$, and the vector will return to the same place it started from. Accordingly, we have the following result: A causal stable signal is positive-real if and only if its phase-lag spectrum $\phi(f)$ satisfies

$$\phi(0.5) = \phi(-0.5)$$

and

$$-\frac{\pi}{2} \leq \phi(f) \leq \frac{\pi}{2}$$

for all f in the Nyquist range $-0.5 \leq f \leq 0.5$. This result can be expressed alternatively as follows: A causal stable signal is positive-real if and only if the signal is minimum-delay and $\phi(f) \leq \pi/2$ for all f in the Nyquist range. Thus minimum-delay signals fall into two categories, namely (1) those which are positive real, in which case their phase-lag spectra do not exceed $\pi/2$ in magnitude, and (2) those which are not positive real, in which case their phase-lag spectra exceed $\pi/2$ in magnitude for some frequencies. Immittance functions fall into the first category. The mapping $w = A(z)$ is illustrated in Figure 3 for a non-positive-real minimum-delay signal, and in Figure 6 for a positive-real (and hence minimum-delay) signal.

6. THE RIGHT-HALF AUTOCORRELATION THEOREM

We now want to establish our main result, namely, the theorem that an immittance function is the weighted right-hand side of an autocorrelation function, and conversely. We call this result the right-half autocorrelation theorem.

The autocorrelation r_n is a two-sided function of the time index n, so it can be written as $(\ldots, r_{-2}, r_{-1}, r_0, r_1, r_2, \ldots)$. As we have seen, an autocorrelation can be characterized by saying that its discrete Fourier transform, namely the spectral density, is positive. The center value r_0 is positive and the other values are conjugate-symmetric around this center value, that is,

$$r_{-n}^* = r_n \ .$$

In the case of a real autocorrelation function, we have, of course, $r_{-n} = r_n$. As we have seen earlier, the Toeplitz form associated with an autocorrelation function is positive for all signals $x_0, x_1, \ldots, x_n$. The matrix of this Toeplitz form is

$$\begin{bmatrix} r_0 & r_1 & \cdots & r_N \\ r_{-1} & r_0 & \cdots & r_{N-1} \\ r_{-N} & r_{-N+1} & \cdots & r_0 \end{bmatrix}$$

which is called a Toeplitz matrix. We see that the elements on the main diagonal, as well as the elements on any parallel diagonal, are the same. Because the Toeplitz form is positive for any signal $x_0, x_1, \ldots, x_N$, the Toeplitz matrix for any N must be positive-definite. Thus the Toeplitz matrix for N = 0, 1, 2, ... must have a positive determinant as well as positive eigenvalues.

An immittance a_n is a one-sided function of the time index n, so it can be written as $(a_0, a_1, a_2, \ldots)$. As we have seen, an immittance can be characterized by saying that the real part of its frequency function is positive. Values of the signal a_n are zero for any negative time indices, i.e., $a_n = 0$ for $n < 0$. We will require that the leading value a_0 be positive, i.e., $a_0 > 0$. An immittance signal, as we know, is a minimum-delay signal whose phase-lag spectrum is bounded in magnitude by $\pi/2$. In brief, an immittance is a positive-real function.

According to our notation, the autocorrelation is r_n and the immittance is a_n. Because an autocorrelation is two-sided, we want to convert it into an equivalent one-sided signal. This one-sided signal is

$(r_0, 2r_1, 2r_2, \ldots)$

which we call the weighted right-half autocorrelation. We see that in the weighted right-half, each autocorrelation coefficient except the leading one has a weighting factor of 2. The reason why the r_0 term does not have a factor 2 is that when we split the autocorrelation in two halves, the center term must be equally divided between left and right. Thus the term r_0 only has one-half the weighting as the other terms. The weighted right-half of the autocorrelation is equivalent to the autocorrelation in the sense that we can reconstruct the full autocorrelation from the weighted right-half by making use of the conjugate symmetry $r_{-n} = r_n^*$.

The theorem that we wish to establish can be stated as follows: There is a one-to-one correspondence between the class of autocorrelations r_n and the class of immittances a_n given by

$$(a_0, a_1, a_2, \ldots) = (r_0, 2r_1, 2r_2, \ldots) .$$

In other words, the theorem states that an immittance and a weighted right-half autocorrelation are the same thing. We call this theorem the <u>right-half autocorrelation theorem</u>.

We will establish our theorem by showing that the real part of the frequency spectrum of the immittance is the same thing as the spectral density derived from the autocorrelation. As we know, the spectral density $R(f)$ can be computed from the autocorrelation r_n by the discrete Fourier transform

$$R(f) = \sum_{n=-\infty}^{\infty} r_n e^{-2\pi i f n} .$$

We can write this equation as

$$R(f) = \sum_{n=-\infty}^{-1} r_n e^{-2\pi i f n} + \frac{r_0}{2} + \frac{r_0}{2} + \sum_{n=1}^{\infty} r_n e^{-2\pi i f n} .$$

The right-hand side is made up of two components. If we recognize the first component as the complex conjugate of the second component, we obtain

$$R(f) = 2 \, \text{Re} \left[\frac{r_0}{2} + \sum_{n=1}^{\infty} r_n e^{-2\pi i f n} \right]$$

which is

$$R(f) = \text{Re} \left[r_0 + \sum_{n=1}^{\infty} 2 r_n e^{-2\pi i f n} \right] \,.$$

This equation states that the spectral density is the real part of the spectrum of the weighted right-half autocorrelation. Because the spectral density is positive, the weighted right-half autocorrelation must be a positive-real signal. That is, the weighted right-half autocorrelation is an immittance (a_0, a_1, a_2, ...). If we carry out the above argument in the reverse direction, we see that an immittance is a weighted right-half autocorrelation (r_0, $2r_1$, $2r_2$, ...). Q.E.D.

Thus, for each autocorrelation there is an immittance such that

$$R(f) = \text{Re}\left[A(f)\right]$$

or, in words, the spectral density is equal to the real-part of the frequency spectrum of the admittance. We can summarize as follows. Given a positive function of frequency, we may consider it either as a spectral density, in which case the corresponding time function is an autocorrelation, or as the real part of a frequency spectrum, in which case the corresponding time function is an immittance signal.

Our theorem can also be stated in the following way. A signal is minimum-delay with a positive-real frequency spectrum if and only if the signal is the weighted right-half of a positive definite function. The implication is that the weighted right-half of an autocorrelation is positive real, and thus necessarily minimum-delay. However, a minimum-delay signal may or may not represent the weighted right-half of an autocorrelation. At this point, let us note that the right-half autocorrelation (without weighting), namely r_0, r_1, r_2, ...), is always positive real. This result follows immediately from the positive-realness of the weighted right-half.

7. NUMERICAL TESTS FOR THE POSITIVE-DEFINITE AND POSITIVE-REAL PROPERTIES

The right-half autocorrelation theorem states that a signal is positive real if and only if the signal is the weighted right-hand side of a positive definite function. That is, if the positive-real signal is $(a_0, a_1, a_2, ...)$ and the positive-definite function is r_n, then their relationship is

$$(a_0, a_1, a_2, ...) = (r_0, 2r_1, 2r_2, ...) .$$

For the purposes of this section we assume that all signals are real-valued. However, the numerical tests that we give can be readily adapted to the case of complex signals.

Let us consider two finite-length signals of the form (a_0, a_1, a_2), which we call A and B:

A: (1, -1.45, 0.525)

B: (1, -0.55, 0.075) .

First, let us see whether signal A is minimum-delay. A direct method is to factor its z-transform

$$1 - 1.45z + 0.52z^2 = (1 - 0.75z)(1 - 0.70z) .$$

Since the zeros 1/0.75 and 1/0.70 lie outside the unit circle, signal A is minimum-delay. Likewise, it can be shown that signal B is minimum-delay.

Factoring high-degree polynomials can be difficult. A computationally more efficient algorithm for testing the minimum-delay condition is based on the Toeplitz recursion method. A sequence $c_1, c_2, \ldots, c_M$ can be regarded as a sequence of reflection coefficients provided that the magnitude of each coefficient is less than one. The Topelitz recursion (Levinson, 1947) generates a sequence of prediction error operators from a sequence of reflection coefficients. Each of these prediction error operators is minimum-delay. If $A_k(z)$ denotes the z-transform of the kth operator, then the z-transform of the time-reverse of the kth operator is defined as

$$A_k^R(z) = z^k A_k(z^{-1}) \ .$$

With the initial z-transform given by $A_0(z) = 1$, the Toeplitz recursion is (Robinson, 1967b, equation [3], p. 125)

$$A_k(z) = A_{k-1}(z) - c_k z A_{k-1}^R(z) \ .$$

From this equation we see that the coefficient of z^k is $A_k(z)$ is $-c_k$. Solving this equation for both $A_{k-1}(z)$ and $A_{k-1}(z^{-1})$ and combining the resulting expression, we obtain the backward Toeplitz recursion

$$A_{k-1}(z) = \frac{1}{1 - c_k} [A_k(z) + c_k A_k^R(z)] \ .$$

In order to apply the backward Toeplitz recursion, we divide the original signal by its leading coefficient so as to obtain a scaled signal with leading coefficient unity. The last coefficient is then the negative of the reflection coefficient, so we can apply the backward Toeplitz recursion to find the next lower order signal. This signal has one less term than the original signal. The leading term on this signal is one, so the last term is the negative of the next lower reflection coefficient. We can then repeat the process until we obtain all the reflection coefficients. If all these reflection coefficients are less than unity in magnitude, then they are indeed legitimate reflection coefficients, so the original signal is (strictly) minimum-delay. This backward Toeplitz recursion test is mathematically identical to the Schur-Cohn test (Loewenthal, 1977).

For signal A, the reflection coefficient is $c_2 = -0.525$.

$$\frac{1}{1 - (0.525)^2} \left[1 - 1.45z + 0.525z^2 - 0.525 (0.525 - 1.45z + z^2) \right]$$

$$= 1 - \frac{1.45}{1 + 0.525} z = 1 - 0.9508z$$

so $c_1 = 0.9508$. Because c_1 and c_2 are each less than one in magnitude, they are legitimate reflection coefficients, hence the test shows that signal A is minimum-delay. Likewise, the test shows that signals B and C are minimum-delay.

Now let us consider whether signal A is positive real. Because the signal is real-valued, the real part of its frequency spectrum is its cosine transform

$$1 - 1.45 \cos \omega + 0.525 \cos 2\omega$$

where $\omega = 2\pi f$. When $f = 0$, its cosine transform is positive. In order for the signal to be positive-real, its cosine transform must be positive for the entire Nyquist frequency range $-0.5 \leq f \leq 0.5$. We can check this condition by letting $u = \cos \omega$ in the expression for the cosine transform. Then u lies in the range $-1 \leq u \leq 1$. Since

$$\cos 2\omega = 2\cos^2 \omega - 1 = 2u^2 - 1$$

the cosine transform is

$$(1 - 0.525) - 1.45u + 2(0.525)u^2 .$$

This polynomial has real roots at 0.53435 and 0.84660, which correspond to the frequencies ±0.09 and ±0.16. Thus the cosine transform becomes negative in the range -0.16 < f < -0.09 and in the range 0.09 < f < 0.16. As a result wavelet A is not positive real. In Figure 1 we plot the cosine transform. In Figure 2 we plot the phase-lag spectrum, which shows that the phase-lag exceeds $\pi/2$ in magnitude in the same ranges.

In Figure 3, we show the mapping $w = A(z)$ for the non-positive-real wavelet A. Note that the w-plane curve does not include nor touch the origin $w = 0$, which indicates that A is minimum-delay, and that it swings beyond the first and fourth quadrants, which indicates that A is non-positive-real.

If we apply the same test to signal B we obtain the polynomial

$(1 - 0.75) - 0.55u + 2(0.075)u^2$.

This polynomial has no real zeros in the range $-1 \leq u \leq 1$, which means that signal B is positive-real. In Figure 4 we plot its cosine transform, which we see is indeed positive. In Figure 5 we plot its phase-lag spectrum, which we see never exceeds $\pi/2$ in magnitude.

In Figure 6, we show the mapping $w = A(z)$ for the positive-real wavelet B. Note that the w-plane curve once again neither includes nor touches the origin $w = 0$, which indicates that B is also minimum-delay. However, the curve does NOT swing beyond the first and fourth quadrants, and this indicates that B is positive-real.

The cosine transform test may also be described as follows. The cosine transform of a real-valued finite-length signal is

$a_0 + a_1 \cos\omega + a_2 \cos2\omega + \ldots a_M \cos M\omega$.

Now we make use of the Chebyshev polynomials

$T_n(u) = \cos[n \cos^{-1} u]$

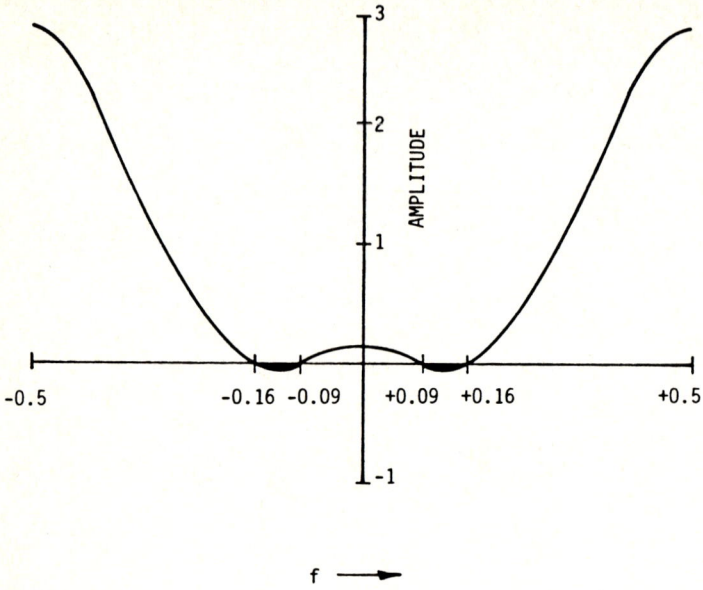

FIG. 1 THE COSINE TRANSFORM OF THE NON POSITIVE-REAL WAVELET A. THE SHADING INDICATES THE ZONES IN WHICH THE COSINE TRANSFORM IS NEGATIVE.

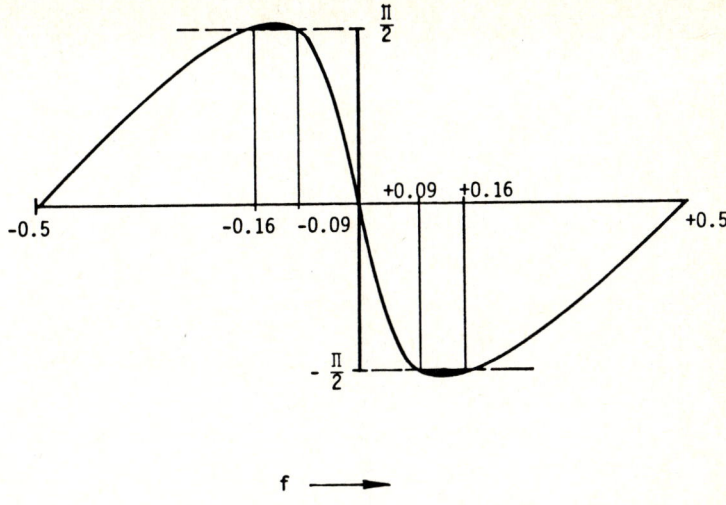

FIG. 2 THE PHASE-LAG SPECTRUM OF THE NON POSITIVE-REAL WAVELET A. THE SHADING INDICATES THE ZONES IN WHICH THE PHASE-LAG SPECTRUM EXCEEDS $\pi/2$ IN MAGNITUDE.

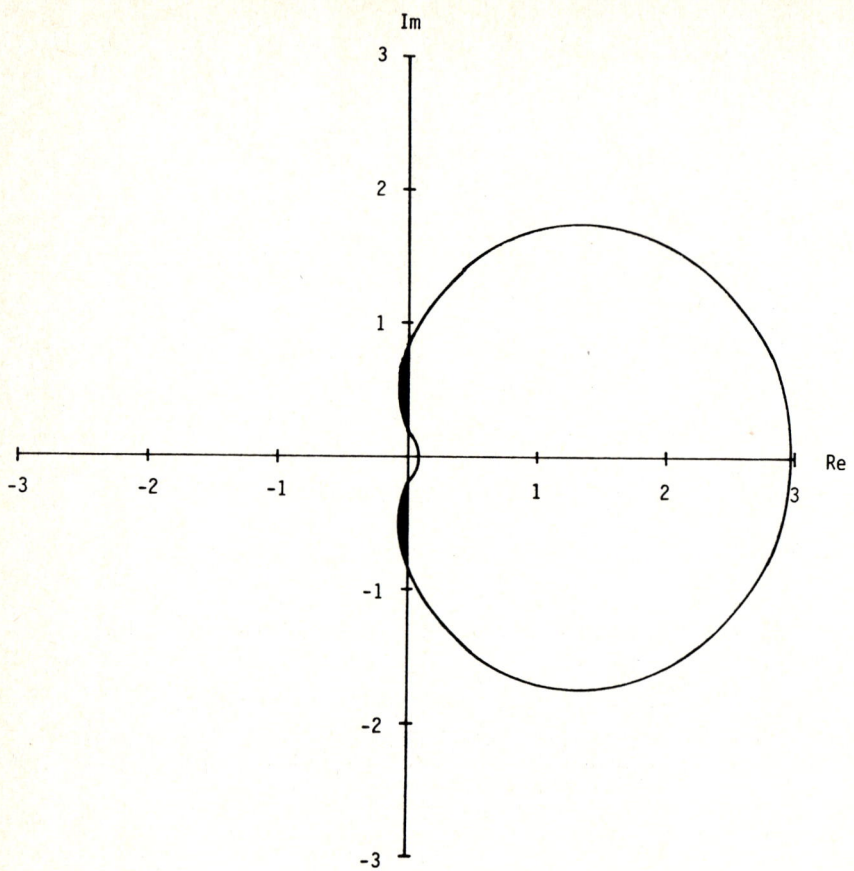

FIG. 3 THE MAPPING w = A(z) FOR THE MINIMUM-DELAY, NON-POSITIVE-REAL WAVELET A. THE SHADING INDICATES THE ZONES IN WHICH THE w-PLANE CURVE SWINGS OUTSIDE THE FIRST AND FOURTH QUADRANTS.

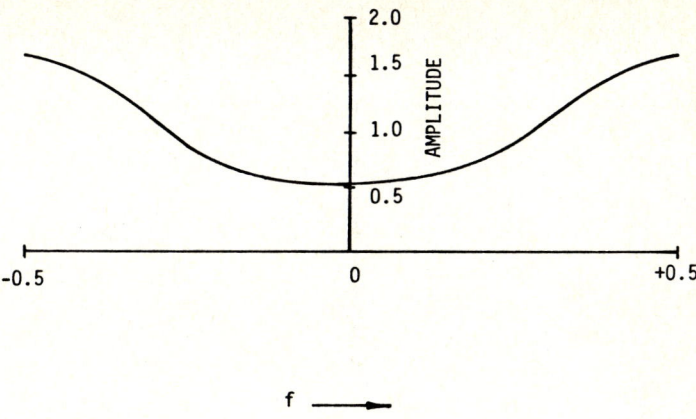

FIG. 4   THE COSINE TRANSFORM OF THE POSITIVE-REAL WAVELET B.

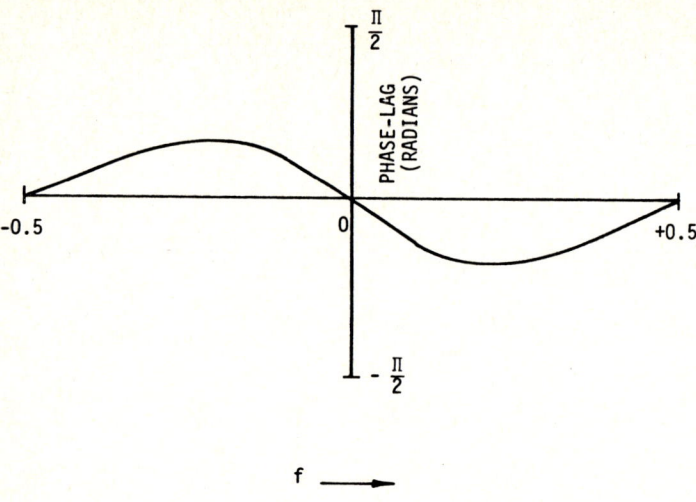

FIG. 5   THE PHASE-LAG SPECTRUM OF THE POSITIVE-REAL WAVELET B.

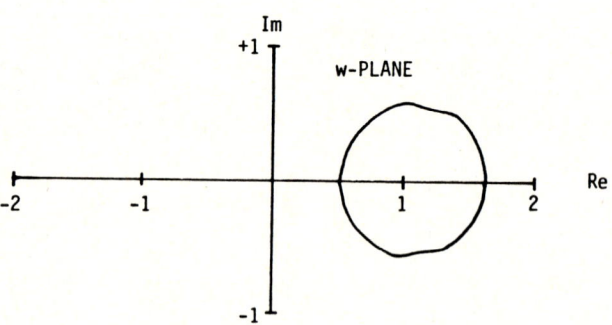

FIG. 6   THE MAPPING w=A(z) FOR THE MINIMUM DELAY, POSITIVE-REAL WAVELET B.

(see e.g., Hamming, 1977, p. 197 ff.). If we let $u = \cos\omega$, the cosine transform becomes the polynomial in u given by

$$P(u) = a_0 + a_1 T_1(u) + a_2 T_2(u) + \ldots + a_M T_M(u) \; .$$

The Chebyshev polynomials can be found by the recursion

$$T_{n+1}(u) - 2u\, T_n(u) + T_{n-1}(u) = 0 \; .$$

Since $T_0(u) = 1$ and $T_1(u) = u$, we have in particular that

$$T_2(u) = 2u^2 - 1$$

$$T_3(u) = 4u^3 - 3u$$

$$T_4(u) = 8u^4 - 8u^2 + 1$$

$$\vdots$$

Thus the coefficients of the polynomial P(u) can be readily obtained. The <u>cosine transform test</u> can now be stated as follows: A signal $a_0, a_1, \ldots, a_M$) is strictly positive-real if $a_0 + a_1 + \ldots + a_M > 0$ and if the polynomial P(u) has no zeros in the range $-1 \leq u \leq 1$. The cosine transform test is mathematically similar to the theorem of Wold (1938) on the autocorrelation of a process of moving averages.

Instead of testing a signal whether it is positive-real, we can set the signal equal to a weighted right-half autocorrelation function, and then test the function whether it is positive-definite. Thus we convert the signal $(a_0, a_1, \ldots, a_M)$ into the two-sided function

$$r_0 = a_0 = 1$$

$$r_k = r_{-k} = 0.5 a_k \qquad \text{for } k = 1, 2, \ldots, M$$

$$r_k = r_{-k} = 0 \qquad \text{for } k > M \; .$$

Given the autocorrelation r_n, we can form Toeplitz matrices of any order. If any of the eigenvalues of any of these matrices are not positive, then the function r_n is not positive definite, and as a result the signal a_n is not positive real. For signal A the two-side function is $r_0 = 1$, $r_1 = r_{-1} = -0.725$, $r_{-2} = r_2 = 0.2625$, $r_k = 0$ for $k > 2$. If we not compute the eigenvalues of the associated 11 x 11 Toeplitz matrix we find that all the eigenvalues are positive; the smallest is 0.00039 and the largest is 2.36205. The Sturm separation theorem (see e.g., Wilkinson, 1965, p. 103) guarantees that the eigenvalues of all the lower order Toeplitz matrices fall within these limits, so we know that all the Toeplitz matrices through order 11 have positive determinants. We might not be tempted to stop there and say that signal A is positive-real. However, if we compute the eigenvalue of the corresponding 12 x 12 Toeplitz matrix, we find that one eigenvalue, namely -0.00398 is negative. Thus the function r_n is not positive definite and hence the signal A is not positive-real. The eigenvalue test can accordingly be used only as follows. If a negative eigenvalue is encountered, the two-sided function is not positive definite and the corresponding signal is not positive-real. If only positive eigenvalues are encountered, no statement can be made.

The right-half (r_0, r_1, r_2, ...) of a positive-real function is in fact a marine seismogram (Robinson and Treitel, 1977). The reflection coefficients determined from (r_0, r_1, r_2, ...) can also serve as the basis of a test. In the case of signal A, we have the two-sided function $r_0 = 1$, $r_{-1} = r_1 = -0.725$, $r_{-2} = r_2 = 0.2625$, $r_k = 0$ for $k > 2$. If we compute the corresponding reflection coefficients, we find that each c_k for $k = 1, 2, ..., 10$ is less than one in magnitude, but that $c_{11} = -13.0467$ greatly exceeds one in magnitude. Thus r_n is not a positive definite function, and signal A is not positive-real. Let the determinant of the k x k Toeplitz matrix be denoted by $\Delta(k-1)$. Then it can be shown (Robinson, 1967b, p. 144) that

$\Delta(0) = r_0$

.
.
.

$\Delta(k) = r_0(1 - c_1^2)(1 - c_2^2) \ldots (1 - c_k^2) \Delta(k-1)$.

As we have previously seen, the $\Delta(k)$ for signal A are positive for k = 0, 1, 2, ..., 10 (i.e., the 11 x 11 Toeplitz matrix and all Toeplitz matrices of lower order are positive). This result agrees with the fact that c_k < 1 for k = 1, 2, ..., 10. However, as we have seen $\Delta(11)$ is negative, which agrees with the fact that c_{11} has magnitude greater than one. Hence the <u>reflection coefficient test</u> can be stated as follows. If a reflection coefficient greater than one in magnitude is encountered, then the two-sided function is not positive definite and the corresponding signal is not positive-real. If only reflection coefficients less than one in magnitude are encountered, no statement can be made.

8. CONCLUSIONS

The Right-Half Autocorrelation Theorem provides the required link between the weighted right-half of a positive-definite autocorrelation and the corresponding positive-real immittance. While the weighted right-hand side of an autocorrelation is necessarily minimum-delay and positive-real, a minimum-delay signal is NOT necessarily positive-real. We have described three numerical tests that can be used to establish whether a given minimum-delay signal is or is not positive-real. They are:

(1) The cosine transform test

(2) The eigenvalue test

(3) The reflection coefficient test.

The cosine transform test provides both the necessary as well as the sufficient criteria for a given signal to the positive-real. On the other hand, the eigenvalue and the reflection coefficient tests provide merely the necessary, but NOT the sufficient criteria for the given signal to be positive-real.

REFERENCES

BODE, W. H., 1945, Network analysis and feedback amplifier design: New York, Van Nostrand.

BRUNE, O., 1931, Synthesis of a finite two-terminal network whose driving-point impedance is a prescribed function of frequency: J. Math. Phys., v. 10, 191-236.

GRENANDER, U. and G. SZEGO, 1958, Toeplitz forms and their applications, Berkeley, Univ. of Calif. press.

HAMMING, R. W., 1977, Digital filters: Englewood Cliffs, N.J., Prentice-Hall.

JURY, E. I., 1977, Inners and stability of dynamic systems: New York, John Wiley & Sons.

LEVINSON, N., 1947, The Wiener RMS (root mean square) error criterion in filter design and prediction: J. Math. Phys., v. 25, 261-278.

LOEWENTHAL, D., 1977, Numerical computation of the roots of polynomials by spectral factorization: Topics in Numerical Analysis, v. 3, 237-254, London, Academic Press Inc.

NYQUIST, H., 1928, Certain topics in telegraph transmission theory: Trans. AIEE, 617-644.

ROBINSON, E. A., 1967a, Statistical communication and detection: London, Charles Griffin & Co., Ltd., New York, Hafner Publishing Co..

ROBINSON, E. A., 1967b, Multichannel time series analysis with digital computer programs: San Francisco, Holden-Day Inc..

ROBINSON, E. A. and S. TREITEL, 1977, The spectral function of a layered system and the determination of the waveforms at depth: Geoph. Prosp., v. 25, 434-459.

ROBINSON, E. A., D. LOEWENTHAL, and S. TREITEL, 1978, Numerical testing of minimum-delay, positive-real, and positive-definite digital filters: J. Comp. Phys., v. 29, 421-430.

TREITEL, S. and E. A. ROBINSON, 1964, The stability of digital filters: IEEE Trans. on Geoscience Electr., v. GE-2, 6-18.

WIENER, N., 1930, Generalized harmonic analysis: Acta Math., v. 55, pp. 117-258 (also reprinted as an MIT Press paperback edition, Cambridge, Mass., 1966).

WILKINSON, J. H., 1964, The algebraic eigenvalue problem: Oxford, Clarendon Press.

WOLD, H., 1954, A study in the analysis of stationary time series (2. edition): Uppsala, Almqvist & Wiksells.

A REVIEW OF NONLINEAR REGRESSION AND ITS APPLICATIONS TO GEOPHYSICAL
INVERSE PROBLEMS

Larry R. Lines and Sven Treitel
Amoco Production Company, Tulsa, Oklahoma, USA

Geophysical inversion involves the estimation of an earth model from a set
of observations. Since our models can be nonlinear functions of estimated
parameters, nonlinear regression proves to be a useful method for performing the inversion. A common type of nonlinear regression applies
iterative damped linear least squares through use of the Marquardt-
Levenberg method. Iterative least squares modeling can be used in a wide
variety of geophysical problems. Three examples in time and spatial
sequence analysis are given: (1) seismic wavelet deconvolution, (2) the
location of a buried wedge from surface gravity data, and (3) the traveltime inversion of vertical seismic profiles. In general, least squares
inversion can be used to estimate earth models for any set of observed
geophysical time or spatial sequences for which an appropriate mathematical description is available.

This paper is essentially an abbreviated version of a similar paper on
least squares inversion by the authors (to be published in _Geophysical
Prospecting_), and it includes an additional section on traveltime inversion of vertical seismic profiles. In this summary of these two papers we
used the term "nonlinear regression" to refer to the _iterative_ use of the
damped "least squares inversion." For the context of this discussion the
two terms will be used interchangeably.

1.0 Geophysical Inverse Problems

Geophysical inversion may be viewed as a problem in which one seeks to estimate
a set of subsurface earth model parameters by use of a finite set of observations. Although there is usually no guarantee of uniqueness in our answers,
inverse theory can provide information about subsurface geology, and consequently there has been a multitude of studies on inversion during the last two
decades. These include the inversion of global earth data by use of free oscillations (Backus and Gilbert, 1967, 1968; Wiggins, 1972), the inversion of traveltime data (Crosson, 1976; Gjoystdal and Ursin, 1981), the inversion of electromagnetic data (Inman, 1975; Jupp and Vozoff, 1975; Oristaglio and
Worthington, 1980), and the inversion of gravity data (Oldenburg, 1974;
Vigneresse, 1977). A comprehensive summary of inverse methods for remote
sensing has been given by Twomey (1977).

Some of our geophysical observations are nonlinear functions of the model parameters. A powerful and flexible procedure to estimate such parameters is the "Marquardt-Levenberg" nonlinear regression approach.

2.0 Nonlinear Regression

For the geophysical inverse problem, the use of nonlinear regression involves the iterative computation of constrained least squares solutions for our data sets. A comprehensive discussion of the relation between nonlinear and linear regression methods is given by Draper and Smith (1981).

2.1 Linear Least Squares Inversion

Before proceeding to nonlinear problems, we examine the linear geophysical inverse problem as discussed by Jackson (1972) and by Wiggins (1972), and by Aki and Richards (1980).

The basic strategy is to minimize the sum of squares of the errors between the model response and the observations. Let the n observations for a set of earth data be represented by the vector

$$y = \text{col}(y_1, y_2, \ldots, y_n) \tag{1}$$

and let the model response be the vector

$$f = \text{col}(f_1, f_2, \ldots, f_n) . \tag{2}$$

The model is a function of p parameters which are elements of a vector θ,

$$\theta = \text{col}(\theta_1, \theta_2, \ldots, \theta_p) . \tag{3}$$

Let θ_j^o be an initial estimate of the parameter θ_j, (j = 1, ..., p), and let f^o be the initial model. If the model f is a linear function of the parameters, a perturbation of the model about θ^o can be represented by the first order Taylor expansion

$$f = f^o + \sum_{j=1}^{p} \frac{\partial f}{\partial \theta_j}\bigg|_{\theta=\theta^o} (\theta_j - \theta_j^o) \tag{4}$$

or in matrix notation,

$$f = f^o + Z\delta \tag{5}$$

where Z is the n x p Jacobian matrix of partial derivatives whose values are given by

$$Z_{ij} = \frac{\partial f_i}{\partial \theta_j} \tag{6}$$

and $\delta = \theta - \theta^o$ is the parameter change vector containing the changes, or perturbations in the parameter vector θ.

Our choice of perturbations in θ will be made so as to minimize the sum of squares of the errors between our model response and the data. Let e represent the error vector containing the differences between the model response and the data,

$$y - f = e . \tag{7}$$

Combining (5) and (7) yields

$$y - f^o = Z\delta + e . \tag{8}$$

The vector $y - f^o$, which contains the differences between the initial model response and the observations, is also termed the discrepancy vector, g, so that

$$Z\delta = g - e . \tag{9}$$

Before proceeding to a means of solving (8) for the parameter change vector, δ, we note that such geophysical problems are generally not well posed. That is, the n x p Jacobian matrix Z is usually not square and of full rank. Many geophysical problems are overdetermined, that is, the number of data points exceeds the number of model parameters, namely n > p. In such cases, the Z matrix may or may not be of full rank. In the case of rank deficiency, techniques exist (such as the Marquardt-Levenberg method) to alleviate the resulting problem of an ill-conditioned Z matrix. The cases for which $p > n$ imply that our problem is underdetermined. However, experience indicates that underdetermined problems occur less frequently for geophysical inverse problems. A review of the various types of least squares problems has been given by Lawson and Hanson (1974).

Although an exact solution of (8) for the ill conditioned case cannot be obtained by conventional matrix inversion, Lanczos (1961) has shown how a least squares solution can be found nevertheless.

In the least squares or Gauss-Newton approach, the L2 norm of the error vector e^2, namely S, is minimized. The parameter change vector δ is chosen to minimize

$$S = e^T e$$

or

$$S = (Z\delta - g)^T (Z\delta - g) \tag{10}$$

Minimization requires that

$$\frac{\partial S}{\partial \delta} = 0 . \tag{11}$$

(Differentiation with respect to a vector is described by Graybill (1969) and implies that $\partial S / \partial \delta_i = 0$ for all i).

Substituting (10) into (11) gives

$$\frac{\partial}{\partial \delta} (\delta^T Z^T Z \delta - g^T Z \delta - \delta^T Z^T g + g^T g) = 0 \tag{12}$$

Carrying out the differentiation with respect to δ, we obtain the normal equations

$$Z^T Z \delta = Z^T g . \tag{13}$$

The least squares solution for the parameter change vector δ is then

$$\delta = (Z^TZ)^{-1}Z^Tg \ . \tag{14}$$

This very useful set of equations has had many important applications in inverse theory and digital filtering. In the present context, however, this so-called "unconstrained" linear least squares method may have some undesirable properties. These we now describe.

An obvious difficulty arises when the inverse of Z^TZ does not exist. This occurs whenever the matrix Z^TZ has zero eigenvalues (Wiggins, 1972). Furthermore, the unconstrained least squares method may encounter the numerical problems of either solution divergence or of slow convergence when the initial estimate of the model, f^o, is poor (Draper and Smith, 1981). Whenever $(Z^TZ)^{-1}$ becomes nearly singular, the elements of the solution vector δ tend to grow without bound. As pointed out by Smith and Shanno (1971), this tends to catapult the updated parameter vector $\theta = \theta^o + \delta$ to some region of parameter space which is far removed from an acceptable solution.

2.2 The Marquardt-Levenberg Method

In order to reduce the difficulties which arise when the matrix Z^TZ is nearly singular, one may solve a damped least squares problem subject to the condition that the sum of the squares, or energy of the elements of the parameter change vector δ be bounded by a finite quantity, say δ_o^2. This approach was introduced by Levenberg (1944) and later described in detail by Marquardt (1963).

The constrained least squares solution arises by solving a Lagrange multiplier problem in which e^Te is minimized subject to the constraint that $\delta^T\delta = \delta_o^2$. Thus, we choose δ to minimize a cost function

$$\hat{S}(\delta,\beta) = e^Te + \beta(\delta^T\delta - \delta_o^2) \tag{15}$$

where β is a Lagrange multiplier.

Differentiation with respect to the vector δ yields a modified form of the normal equations

$$(Z^T Z + \beta I)\delta = Z^T g \ . \tag{16A}$$

so that

$$\delta = (Z^T Z + \beta I)^{-1} Z^T g \tag{16B}$$

Comparison of equation (14) with equation (16B) shows that the constraint has produced a method for avoiding singularities or near singularities in the $Z^T Z$ matrix. By adding a constant, β, to the main diagonal of $Z^T Z$, we have effectively added a DC level to the eigenvalues of the $Z^T Z$ matrix so that none of the eigenvalues can vanish. Levenberg (1944) terms the Lagrange multiplier β a "damping factor", since it effectively damps out changes in the parameter vector θ by limiting the energy in the parameter discrepancy vector δ.

The solution (16B) has several other interesting characteristics. It is hybrid, because it combines the method of steepest descent with the method of least squares. The steepest descent vector is perpendicular to a contour for which $S(\theta) = e^T e =$ constant or $dS(\theta) = 0$. This condition is satisfied by a column vector

$$\delta_g = -\nabla S(\theta)$$

where ∇ is the gradient operator. We note that $+\nabla S(\theta)$ is then the steepest ascent vector.

Since $S(\theta) = (y - f)^T(y - f)$, the elements of ∇S are

$$\frac{\partial S}{\partial \theta_j} = -2\Sigma(y_i - f_i)\frac{\partial f_i}{\partial \theta_j} \tag{17}$$

and the steepest descent vector is

$$\delta_g = 2Z^T(y - f) = 2Z^T g \ . \tag{18}$$

Since δ_g is in the direction of decreasing S, convergence tends to occur, but at a usually slow rate. Further, computational problems can occur when the step size becomes infinitesimally small.

Generally speaking, the steepest descent method is optimal when $S(\theta)$ is large, while the least squares method becomes effective when $S(\theta)$ is small. The Marquardt-Levenberg approach thus exploits the useful properties of both methods. The technique determines the parameter change vector δ at each step in the iteration. Marquardt (1963) proved that the vectors δ and δ_g lie within 90° of each other, and that this is a necessary property for convergence. Comparison of (16B) with (18) shows that as β becomes large $\delta \sim \beta^{-1} Z^T g$, i.e., δ becomes proportional to the gradient vector δ. However, as β becomes larger and larger, the elements of the parameter change vector δ become smaller and smaller. Marquardt also showed that the angle between δ and δ_g is a monotonically decreasing function of β.

A particular choice of β in equation (16B) allows either the linear least squares method or the steepest descent method to dominate the parameter search. Setting $\beta=0$ implies that the linear least squares method predominates, while allowing $\beta \to \infty$ moves the technique towards the method of steepest descent. Initially β is set to a large positive value, so that the good initial convergence properties of the steepest descent method can come into play. Then, as this happens, β is reduced by multiplying it by a constant factor <1 so that the linear least squares method may take over in the region closer to a solution. If divergence occurs during a given iteration, β is divided once more by this factor until the error drops and convergence resumes.

A further means to speed the convergence of (16B) involves scaling. Marquardt (1963) showed how to determine a diagonal matrix D such that ZD^{-1} places ones along the main diagonal of the scaled matrix $(ZD^{-1})^T(ZD^{-1})$. The element d_i of D is equal to the root mean sum of squares (RMS) value of the elements in the i-th column of the unscaled Jacobian matrix Z. The resulting solution $\delta^{(D)}$ must then be rescaled in the form

$$\delta = D^{-1} \delta^{(D)} .$$

Smith and Shanno (1971) give the details of this very useful scaling procedure.

One might conclude that the solution (16B) minimizes the cost function $S(\delta,\beta)$ of equation (15) merely because $(Z^TZ + \beta I)$ is always positive definite (Marquardt, 1963). However, Dennis (1977) shows that this is so only if the errors are linear functions of the model parameters. In particular, we note that $\partial S/\partial \delta = 0$ is a necessary but not sufficient condition for the minimization of S. To establish sufficiency, we require knowledge of the second derivative or Hessian matrix H, whose elements are

$$H_{jk} = \frac{\partial^2 S}{\partial \theta_j \partial \theta_k} \qquad (19)$$

which must be positive definite (Fletcher (1980), p. 10-11).

The method of constrained least squares has also been used in the design of stabilized Wiener deconvolution filters. The addition of a positive constant to the main diagonal of the Z^TZ matrix is known as "prewhitening" in digital filtering terminology, since the addition of white noise to a signal will add power to the main diagonal of its associated autocorrelation matrix. In the case of filter design, the constraint is placed on the power of the filtered noise. A discussion of constrained least squares in this context has been given by Treitel and Lines (1982).

One of the main advantages of the nonlinear regression method is that inversion may be attempted for any problem for which a direct (or forward) solution exists. It is generally easier to solve the forward problem that transforms a set of model parameters into a synthetic data set, than to proceed in the opposite direction and solve the inverse problem. Having found a method of finding the model response f from the parameters θ, we must compute the Jacobian matrix of partial derivatives, $Z_{ij} = \partial f_i/\partial \theta_j$. These derivatives can be determined by formal differentiation if the model is simple enough. In other cases, the partial derivatives are replaced by finite differences. This can be computationally expensive since we need to determine two model responses for every value of θ_j even if we use the simplest forward difference formula which approximates $\partial f_i/\partial \theta_j$ by

$$\frac{f_i(\theta_j + \Delta\theta_j) - f_i(\theta_j)}{\Delta\theta_j} .$$

Clearly one should avoid using more parameters θ_j ($j = 1, \ldots, p$) than is absolutely necessary (the principle of parsimony!). Alternatively, one may have recourse to some special methods for finding partial derivatives, see e.g. Oristaglio and Worthington (1980). Otherwise, iterative least squares modeling is very versatile and can be adapted to a wide range of geophysical inverse problems, as evidenced by the number of references previously given.

2.3 Iterative Applications of Least Squares

In the preceding discussions we have outlined the Marquardt-Levenberg method for the constrained least squares solution. Our nonlinear regression approach uses constrained least squares iteratively to update the parameter vector for a given geophysical model. Although our model is often a nonlinear function of some or all of the parameters θ_i, we can often estimate the parameter vector θ by a sequence of constrained linear least squares estimates.

An outline of the nonlinear regressive approach to geophysical inversion is described in the flow diagram of Figure 1. The parameter changes from the initial response estimates are determined by use of the fundamental relation (16B). We obtain an updated set of parameters, which are then used to compute a new model estimate. At each stage, the sum of squares of the errors between the model response and the model estimate response is monitored. The iterative search for the parameter estimates terminates whenever either the squared error or a percentage change in the squared error become less than a prespecified value. After these convergence criteria have been satisfied, we may say that the estimated geophysical parameters have produced a model which has matched the data within our specifications. However, as will be seen in a later discussion

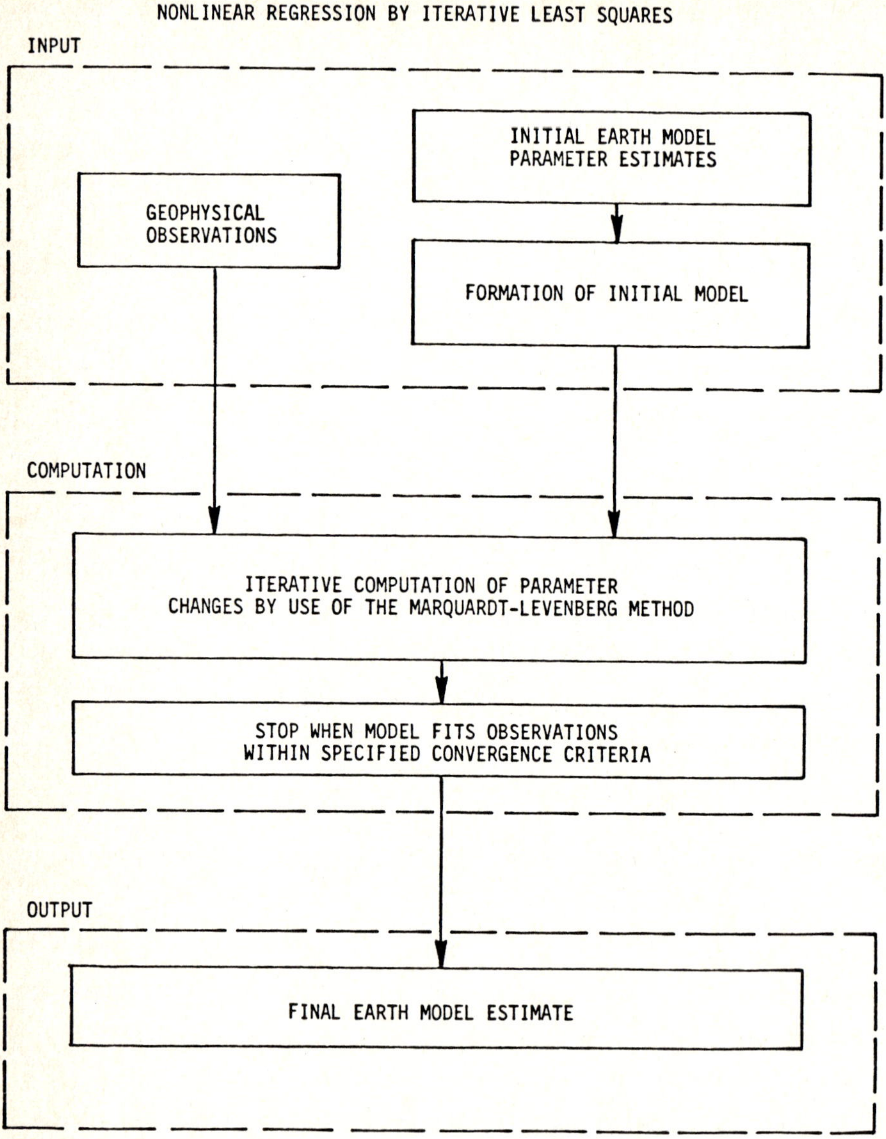

FIG. 1 FLOW DIAGRAM OF NONLINEAR REGRESSIVE SOLUTIONS TO GEOPHYSICAL INVERSION PROBLEMS

on nonuniqueness, a good match between the model and the observations does not necessarily guarantee that the correct model has been found.

3.0 Singular Value Decomposition
3.1 Computational Aspects

We have shown that the linear least squares solution can be found by solving the normal equations (13)

$$Z^T Z \delta = Z^T g \ . \tag{13}$$

This solution requires that $Z^T Z$ and $Z^T g$ be formed by matrix multiplication. $Z^T Z$ is a nonnegative definite square symmetric matrix which can always be made positive definite by use of Marquardt's damping factor. We could therefore solve (13) by use of the Cholesky decomposition, so that no matrix inversions need be calculated explicitly (Lawson and Hanson, 1974).

However, the formation of $Z^T Z$ and $Z^T g$ involves numerical inaccuracies. Such inaccuracies can be troublesome for large values of n and p. Golub and Reinsch (1970) recognized this problem and proposed that instead of dealing with (13), it is better to attempt a solution of the rectangular system

$$Z\delta = g \tag{20}$$

where we recall that the Jacobian matrix Z is generally n x p. The formal solution of (21) is

$$\delta = Z^{-1} g$$

but the conventional inverse Z^{-1} exists only if Z is square (n = p) and non-singular. For most geophysical problems n >> p, and in that case the inverse Z^{-1} must be identified with the so-called natural, or generalized inverse; see e.g. Lanczos (1961, Chapter III).

Golub and Reinsch (1970) developed an efficient method and a computer algorithm for solving (21) called Singular Value Decomposition (SVD). This procedure factors Z into a product of three matrices,

$$Z = U\Lambda V^T \tag{21}$$

where U is an n x p matrix whose columns contain the p orthonormal "observation" eigenvectors u_i, which are <u>not</u> associated with the at least n-p null eigenvalues of ZZ^T. These u_i's satisfy

$$ZZ^T u_i = \lambda_i^2 u_i \qquad i = 1, 2, \ldots, n \; . \tag{22}$$

Here V is a p x p matrix whose columns contain the p orthonormal "parameter" eigenvectors v_i which satisfy

$$Z^T Z v_i = \lambda_i^2 v_i \; . \tag{23}$$

Finally, Λ is a p x p diagonal matrix containing the at most p non-null singular values $+\lambda_i$, or positive square roots of the eigenvalues, λ_i^2 of $Z^T Z$. A more detailed description of the SVD method is given in Appendix I. For the following SVD manipulations, we note that $U^T U = V^T V = VV^T = I$.

Wiggins et al. (1976) have pointed out that the factorization of Z in terms of U and V is analogous to ordinary spectral decomposition in terms of sines and cosines. In this case, however, the sines and cosines are replaced by the observation and parameter eigenvectors u_i and v_i.

Writing the equation $Z\delta = g$ in terms of U and V gives

$$U\Lambda V^T \delta = g$$

or
$$\Lambda V^T \delta = U^T g \; . \tag{24}$$

Following Wiggins et al. (1976), these expressions may be recast in terms of the individual non-null eigenvalues λ_i and the eigenvectors u_i and v_i. Thus,

$$\lambda_i v_i^T \delta = u_i^T g \qquad \text{for } i = 1, 2, \ldots, p \tag{25}$$

Since $v_i^T \delta$ and $u_i^T g$ are the projections of the i-th orthonormal eigenvectors in the parameter change and observation spaces respectively, u_i and v_i can be viewed as spectral components of the discrepancy vector g and of the parameter change vector δ, respectively. Furthermore, the singular values λ_i determine how a spectral component in parameter space is weighted to provide a corresponding spectral component in observation space.

For the geophysical inverse problems we examine here, we usually have n>p, that is, in our model responses the number of data points exceeds the number of parameters. Further, experience suggests that the singular values λ_i may become small, but rarely do they vanish completely.

For the cases n>p, an inverse for Z will not exist in the usual sense. However, the decomposition $Z = U \Lambda V^T$ suggests use of the Lanczos or least squares inverse

$$Z_L^{-1} = V \Lambda^{-1} U^T \qquad (26)$$

to solve $Z\delta = g$ (Jackson, 1972). We obtain

$$\delta = V \Lambda^{-1} U^T g = Z_L^{-1} g . \qquad (27)$$

This solution is in fact the least squares solution (14), but written in terms of the singular value decomposition (22). We show this by substituting (22) into (14),

$$\begin{aligned}
\delta &= (Z^T Z)^{-1} Z^T g \\
&= (V \Lambda U^T U \Lambda V^T)^{-1} V \Lambda U^T g \\
&= (V \Lambda^2 V^T)^{-1} V \Lambda U^T g \\
&= V \Lambda^{-2} V^T V \Lambda U^T g \\
&= V \Lambda^{-1} U^T g \\
&= Z_L^{-1} g
\end{aligned}$$

which is equation (27).

3.2 Single Value Decomposition and Marquardt's Factor

Having discussed the advantages of Marquardt's method, we now describe a method of including it in the SVD formulation. This prescription for avoiding singularities has been treated by Lawson and Hanson (1974) and by Jupp and Vozoff (1975), among others.

We recall that the solution to the modified normal equations (16A) is

$$\delta = (Z^T Z + \beta I)^{-1} Z^T g \ . \tag{28}$$

We write $(Z^T Z)$ in terms of U, Λ, and V

$$Z^T Z = V \Lambda^2 V^T$$

so that

$$(Z^T Z)^{-1} = V \Lambda^{-2} V^T \ . \tag{29}$$

The matrix $(Z^T Z + \beta I)$ becomes

$$(Z^T Z + \beta I) = V \Lambda^2 V^T + \beta I$$
$$= V(\Lambda^2 + \beta I) V^T$$

Hence,

$$(Z^T Z + \beta I)^{-1} = V(\Lambda^2 + \beta I)^{-1} V^T$$

$$= V \, \mathrm{diag}\left(\frac{1}{\lambda_j^2 + \beta}\right) V^T \tag{30}$$

where $(\Lambda^2 + \beta I)^{-1}$ is a diagonal matrix of the form

$$(\Lambda^2 + \beta I)^{-1} = \begin{bmatrix} \frac{1}{\lambda_1^2 + \beta} & 0 & \cdots & 0 \\ 0 & \frac{1}{\lambda_2^2 + \beta} & & \vdots \\ \vdots & & \ddots & \\ 0 & \cdots & & \frac{1}{\lambda_p^2 + \beta} \end{bmatrix}$$

We then substitute (31) and (22) into (16B) to obtain

$$\delta = V \text{ diag}(\frac{1}{\lambda_j^2 + \beta})V^T V \Lambda U^T g$$

$$\delta = V \text{ diag}(\frac{\lambda_j}{\lambda_j^2 + \beta})U^T g \quad . \tag{31}$$

Comparing (32) with (28), we note that SVD can be combined with Marquardt's method by replacing the element $1/\lambda_j$ in the Λ^{-1} matrix by the element

$$\frac{\lambda_j}{\lambda_j^2 + \beta}$$

where β is Marquardt's damping factor. It now becomes clear how β can obviate the problem of matrix singularities: in the event that $\lambda_j \to 0$, division by zero can no longer occur.

4.0 Solutions to Geophysical Inverse Problems

As remarked earlier, iterative least squares methods have been used for many kinds of geophysical inverse problems. Let us describe three examples.

4.1 Deconvolution of Normal Incidence Seismograms

The reflected seismic signal, x_t, for a vertically incident P (compressional) wave may be expressed as a convolution of a source generated seismic wavelet, w_t, with the impulse response of a layered earth, i_t (Robinson and Treitel, 1980). A digitized version of this signal may be written

$$x_t = \sum_\tau w_\tau i_{t-\tau} \quad . \tag{32}$$

The process of seismic deconvolution is an attempt to remove the source wavelet, w_t, from the trace, x_t, and subsequently estimate normal-incidence reflection coefficients (or acoustical impedance contrasts) for interfaces within the earth. Deconvolution is an example of a nonunique inverse problem, in which we seek to determine w_t and i_t from knowledge of x_t. This nonuniqueness can best be appreciated by considering the Fourier transform of the trace, $X(\omega)$,

$$X(\omega) = W(\omega)I(\omega) \qquad (33)$$

where $W(\omega)$ and $I(\omega)$ are the Fourier transforms of the wavelet and of the impulse response. Having knowledge of only $X(\omega)$ leaves us with one equation and two unknowns. Hence, deconvolution may be viewed more as a digital processing "art" in which we attempt to separate w_t and i_t by using the different statistical properties of the two time sequences. Some of these statistical methods have been discussed by Lines and Ulrych (1977). Nonlinear regression gives us a method of perturbing classical solutions for w_t and i_t in order to provide a closer match between model responses and seismic trace data. Similar approaches have been used by Bilgeri and Carlini (1981), and by Cooke and Schneider (1981).

To justify the application of a constrained linear least squares method for seismic inversion, we recall the seismic trace model of equation (33). Because the trace x_t is the convolution of a source wavelet w_t with an impulse response i_t, the trace x_t is a <u>linear</u> function of the wavelet coefficients w_t. On the other hand, and as we show below, the impulse response i_t is itself generally a <u>nonlinear</u> function of the reflection coefficients c_t. Nevertheless, if the reflection coefficients c_t are sufficiently small in magnitude, then the impulse response i_t becomes a <u>quasilinear</u> function of the coefficients c_t, and for such cases the Taylor series expansion of the nonlinear model (4) becomes applicable.

Our layered earth model has been described by Goupillaud (1961), and is shown in Figure 2. It consists of n layers embedded between two halfspaces. The normal incidence response of this model has been treated in several papers, including Kunetz and d'Erceville (1962) and Treitel and Robinson (1966). Although all waves travel vertically in the present one-dimensional model, the wave paths in Figure 2 are drawn at nonvertical incidence to simulate the change of the wave's vertical position with time. Each layer is constructed to have unit two-way travel time, so that the unit delay operator z represents two-way travel time in each layer.

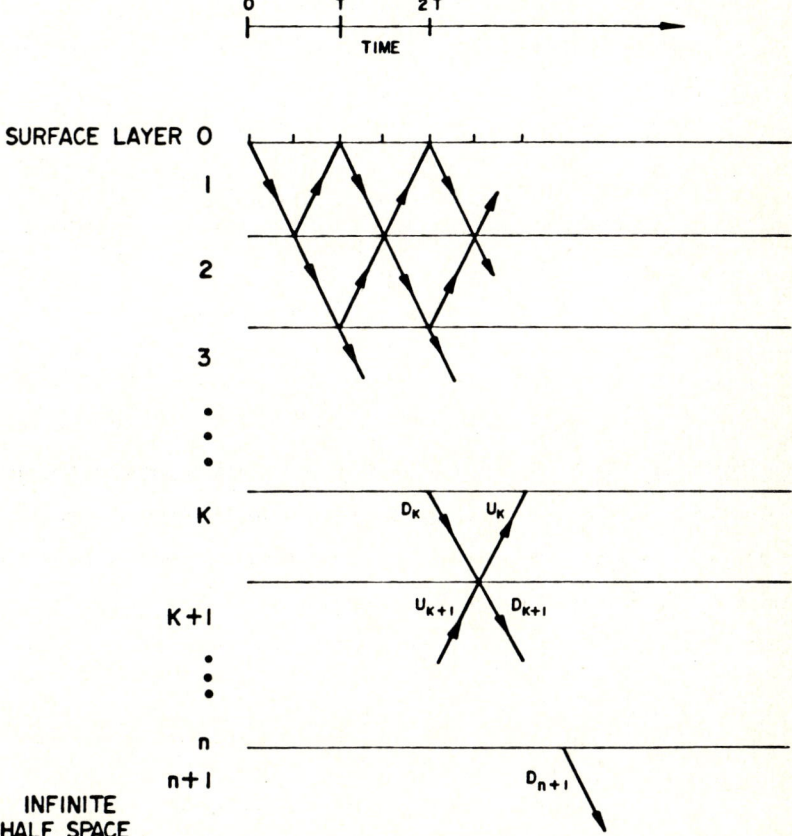

FIG. 2 GOUPILLAUD LAYERED EARTH MODEL

The pressure reflection coefficient at the boundary between the k-th and the (k+1)-st layers is defined as the ratio of the pressure of the reflected wave to the pressure of the incident wave. If the incident wave is in the k-th layer, this reflection coefficient is

$$c_k = \frac{\rho_{k+1} V_{k+1} - \rho_k V_k}{\rho_{k+1} V_{k+1} + \rho_k V_k} \tag{34}$$

where ρ_k and V_k are the density and compressional wave velocity in the k-th layer (Robinson and Treitel, 1980, p. 325-328). The transmission coefficient t_k for pressure waves is

$$t_k = \frac{2\rho_{k+1} V_{k+1}}{\rho_{k+1} V_{k+1} + \rho_k V_k} \tag{35}$$

By use of these reflection and transmission coefficients, the upgoing and downgoing waves in each layer can be related to the surface excitation. If $U_k(z)$ and $D_k(z)$ are the z-transforms of the upgoing and downgoing waves at the top of the k-th layer, the relationship between the waves in each layer is

$$\begin{bmatrix} D_o(z) \\ U_o(z) \end{bmatrix} = \prod_{i=0}^{k} M_i \begin{bmatrix} D_{k+1}(z) \\ U_{k+1}(z) \end{bmatrix} \tag{36}$$

where

$$M_i = \frac{1}{t_i} \begin{bmatrix} z^{-\frac{1}{2}} & c_i z^{-\frac{1}{2}} \\ c_i z^{\frac{1}{2}} & z^{\frac{1}{2}} \end{bmatrix} \tag{37}$$

is the communication, or layer matrix.

For the n-layered model shown in Figure 2, it turns out that $U_{n+1}(z) = 0$ and $D_o(z) = 1 - c_o U_o(z)$ (Robinson, 1967, Chapter 3). Here $U_o(z)$ is the z transform of the impulse response of the layered earth. $U_o(z)$ contains all primary and multiple reflections from the layered medium as they are recorded just above the surface (k = 0). Now $U_o(z)$ contains products of reflection coefficients, and is therefore a nonlinear function of these coefficients. However, under certain conditions this impulse response can be approximated as a linear function of the reflection coefficients. Consider the case of an isolated layer, say the k-th. Then U_k is

$$U_k(z) = \frac{c_k + c_{k+1}z}{1 + c_k c_{k+1} z} \qquad (38)$$

The wave recorded just above the k-th interface is a nonlinear function of the reflection coefficients but is approximately linear if

$$c_k c_{k+1} \ll 1 \; .$$

In such cases,

$$U_k(z) \sim c_k + c_{k+1} z \; . \qquad (39)$$

Thus $U_k(z)$ is the impulse response for a pressure geophone located just above the k-th interface. Physically speaking, (40) says that the impulse response is dominated by primary reflections whenever the absolute value of the reflection coefficients is much less than unity. Since reflection coefficients are rarely larger than 0.3, the linearity assumption is usually a good one.

We illustrate our nonlinear regression approach to deconvolution with a simple model.

Figure 3(a) shows an example in which a 7 point wavelet is convolved with the impulse response due to the reflection coefficient sequence (Figure 3(a))

$$C(z) = 0.200z^{10} - 0.300z^{15} + 0.100z^{16}$$

The initial wavelet guess, shown in Figure 3, was obtained with Kolmogorov's minimum phase estimation procedure (Lines and Ulrych, 1977). The w_t estimates

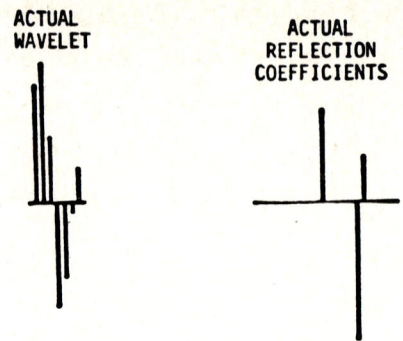

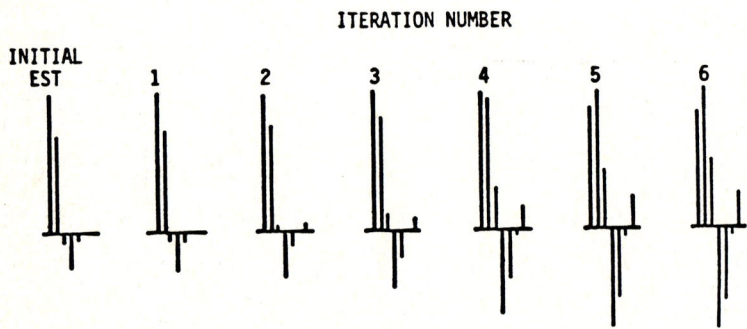

FIG. 3(a) EXAMPLE SHOWING THE EVOLUTION OF WAVELET ESTIMATES AS A FUNCTION OF ITERATION NUMBER

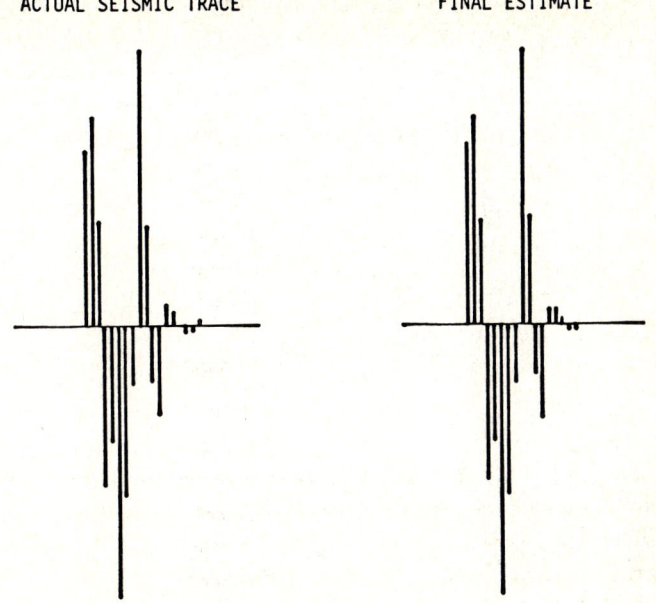

FIG. 3(b)   COMPARISON OF SEISMIC TRACE AND FINAL ESTIMATE.

converge within a specified error criterion after six iterations. Figure 3(a) shows the evolution of the wavelet estimates as a function of the number of iterations. The accurate reflection coefficients estimates remained essentially unchanged in these iterations. Figure 3(b) shows the similarity between the final model response and the actual trace.

It turns out that a suboptimal choice of β was used in the example of Figure 3. In this case the reflection coefficient estimates are accurate, and the problem essentially reduces to simple linear regression, so $\beta = 0$ is an appropriate choice. Thus a single iteration is needed for convergence to occur. With an inaccurate initial choice of reflection coefficients or in the presence of additive noise, more than one iteration would have been required. (An initial choice of $C(z) = 0.10z^{10} + 0.10z^{15} + 0.10z^{16}$ requires two iterations.) Our experience has usually shown that β controls the number of iterations for convergence, not the final solution.

Although there appear to be no a priori "foolproof" methods for choosing the damping factor in nonlinear problems, it is encouraging to see that our inverse problem solutions are reasonably robust with respect to the choice of β. This choice determines the rate of convergence, but in general does not adversely affect the final answer.

A physically more realistic wavelet estimation problem is pictured in Figure 4. The seismic wavelet was convolved with the reflectivity function from a sonic log in order to produce the shown seismic trace. Random noise of different variances was added to this trace to produce synthetic traces with different signal-to-noise ratios. Nonlinear regression was then used to produce wavelet estimates for signal-to-noise ratios of infinity (no noise), 5, 3, and 1. Estimates generally degrade in quality as noise increases, but do so gradually. Unlike other inverse techniques, nonlinear regression produces robust parameter estimates in noisy environments. This characteristic is never to be underrated when dealing with geophysical data!

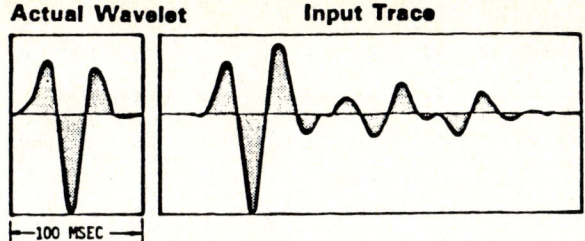

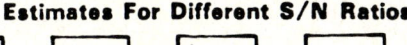

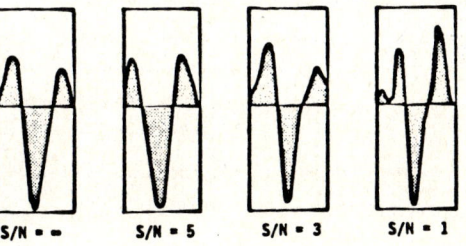

FIG. 4 WAVELET ESTIMATION FOR DIFFERENT LEVELS OF ADDITIVE NOISE. WAVELET ESTIMATES ARE SHOWN FOR SIGNAL-TO-NOISE RATIOS OF ∞, 5, 3, AND 1.

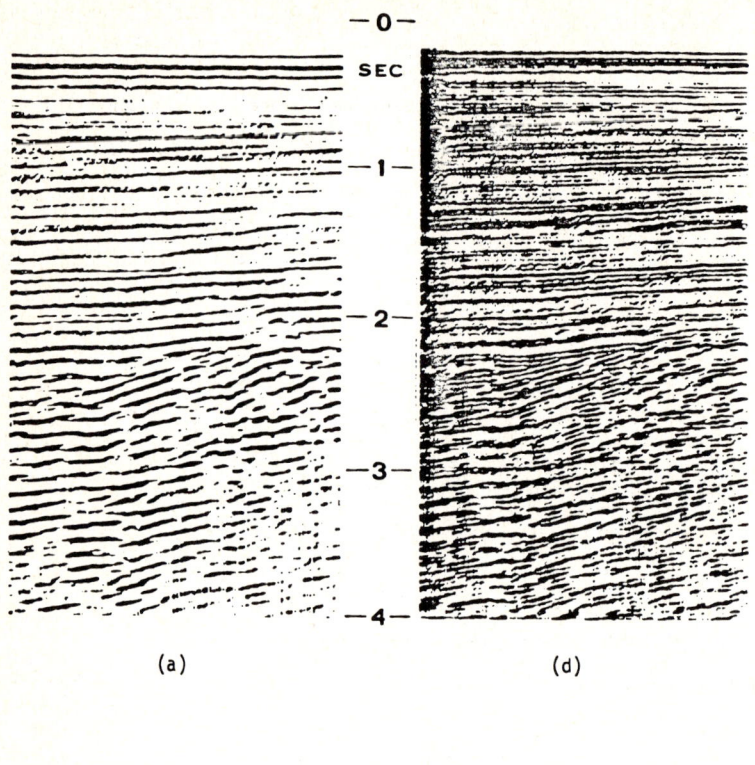

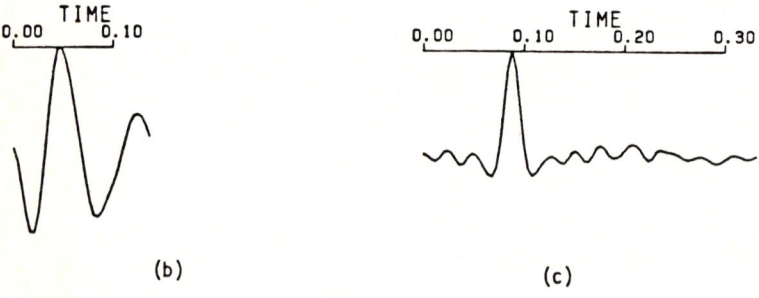

FIG. 5. (a) INPUT DATA, (b) WAVELET ESTIMATE, (c) WIENER FILTER OUTPUT, (d) WAVELET DECONVOLUTION

For a real data application of nonlinear regressive wavelet estimation, we use the seismic section from Treitel, Gutowski, and Wagner (1982). Figure 5(a) shows the input data set which has had plane wave processing applied. This processing decomposes a point-source seismic recording into a set of plane wave seismograms at various angles of incidence (Treitel, Gutowski, and Wagner (1982)).

Figure 5(b) displays the wavelet estimate obtained by nonlinear regression, and Figure 5(c) shows the "resolving kernel" or output obtained by convolving the wavelet with a Wiener spiking filter. Figure 5(d) shows the section resulting from the application of the Wiener filter to the seismic section in Figure 5(a). It is apparent that the deconvolution has enhanced a wedge feature at about 2.200 sec in the middle of the section and has vastly improved resolution on the section.

4.2 Inversion of Gravity Data for a Buried Wedge

Another geophysical example which illustrates the use of nonlinear regression involves the determination of the position and density contrast of a buried wedge or slab of rock, see Figure 6(a). The change in the vertical component of gravitational acceleration due to this slab, g_z, is (see e.g. Dobrin, 1976),

$$g_z = 2\gamma\rho t(\frac{\pi}{2} - \tan^{-1}\frac{x}{Z}) \tag{40}$$

where ρ is the density contrast of the anomalous slab of rock, t is the slab's thickness, x is the horizontal displacement from the edge of the slab, Z is the depth to the middle of the slab, and γ is the universal gravitational constant.

A possible inverse problem would be the following. Given a gravity profile, determine the position and density contrast for the slab. A direct means of determining the horizontal location of the wedge edge requires that we simply find the value of x for which $\partial^2 g_z/\partial x^2 = 0$. The determination of the rock density and of the wedge's vertical position are both nonunique.

GRAVITY RESPONSE OF A BURIED WEDGE

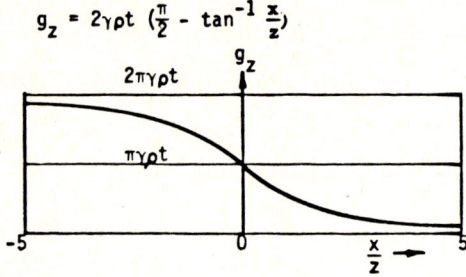

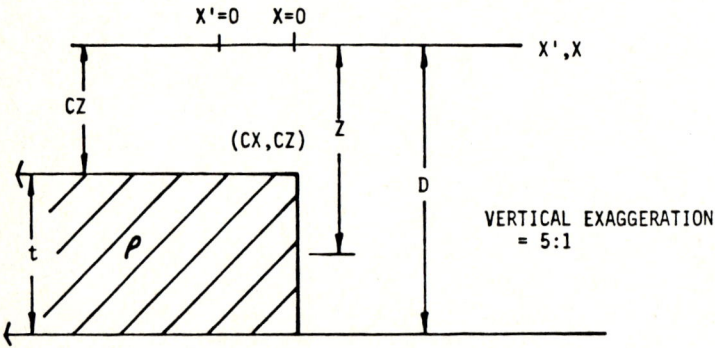

(CX,CZ) = LOCATION OF WEDGE CORNER

X = X' - CX = HORIZONTAL DISTANCE OF STATION FROM WEDGE EDGE

D = DEPTH TO BASEMENT

t = D-CZ = THICKNESS OF SLAB

$Z = \frac{D+CZ}{2}$

ρ = DENSITY CONTRAST

FIG. 6(a)   GRAVITY RESPONSE OF A BURIED WEDGE.

As in the case of seismic deconvolution, we recognize that the nonuniqueness involved in estimating the slab's density contrast and thickness arises because we only know g_z, which by (41) means that we may at best solve for the product of slab density contrast and thickness, ρt. This represents an ambiguity commonly found in gravity interpretation. The problem was recognized by Parker (1974), who placed bounds on both density contrast and depth.

Solutions for the wedge's location and density contrast can be obtained by nonlinear regression. Consider a problem in which the slab rests on a basement at depth D. The gravity model response g_z is a function of three parameters: the wedge corner location (CX, CZ) and the density contrast, ρ. Other variables may be defined as

$x = x' - CX$ where x' = gravity station location
$t = D - CZ$

and $Z = \dfrac{D + CZ}{2}$.

We can find analytic expressions for the Jacobian by computing

$\dfrac{\partial g_z}{\partial \rho}$, $\dfrac{\partial g_z}{\partial CX}$, and $\dfrac{\partial g_z}{\partial CZ}$.

The observations were created by placing our wedge corner at a depth of 3600 ft with a density contrast of 0.25 gm/cm^3. A profile of 19 stations spaced 360 ft (109.728 m) apart was run over the top of the wedge. The true wedge corner position is at position (50,50) on a 100 x 100 grid, with a grid spacing of 72 ft (21.946 m). The contour plot in Figure 6(b) illustrates the squared error S as a function of (CX, CZ) for the correct density contrast. We see that there is a range of (CX, CZ) values which produces low S values. This simply means that a shallow wedge with less total mass under the gravimeters gives almost the same response as a deeper wedge with more mass under the stations. Nevertheless, the parameters can converge to the correct answer if we examine some

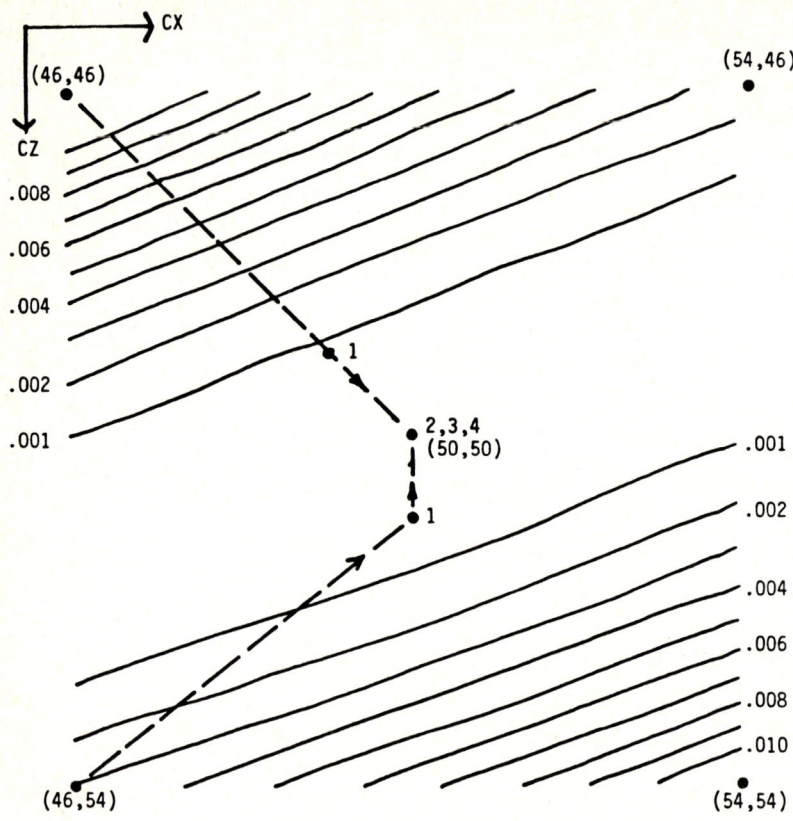

FIG. 6(b) GRAVITY ERROR CONTOURS FOR THE BURIED WEDGE PROBLEM. CONVERGENCE PATHS ARE SHOWN FOR EXAMPLES 2 AND 3 IN TABLE 4.

examples in terms of the initial values of ρ, CX, and CZ. This is illustrated in Table 1 which shows the convergence rates and solutions for various initial estimates of ρ and an initial guess of the corner location at (CX,CZ) = (46,46). The "normalized mean square error" (MSE) is given by the quotient, $e^T e$, divided by the sum of squares of the g_z values.

Table 1. Convergence for the Wedge Inversion Problem

Example 1: Density Contrast Estimate Low, Wedge Depth Estimate Shallow

Iteration No.	ρ	(CX,CZ)	Normalized MSE
0 (initial model)	0.200000	(46,46)	1.203279×10^{-2}
1	0.248564	(50,51)	6.712922×10^{-4}
2	0.249875	(50,50)	2.521883×10^{-7}
3	0.250007	(50,50)	8.723082×10^{-10}

Example 2: Density Contrast Estimate High, Wedge Depth Estimate Deep

Iteration No.	ρ	(CX,CZ)	Normalized MSE
0 (initial model)	0.300000	(46,54)	1.812316×10^{-2}
1	0.251544	(50,51)	2.031701×10^{-4}
2	0.249874	(50,50)	2.540213×10^{-7}
3	0.250008	(50,50)	9.821132×10^{-10}

Example 3: Density Contrast Estimate High, Wedge Depth Estimate Shallow

Iteration No.	ρ	(CX,CZ)	Normalized MSE
0	0.300000	(46,46)	1.216016×10^{-1}
1	0.248564	(49,49)	4.747731×10^{-4}
2	0.249904	(50,50)	1.492059×10^{-7}
3	0.250008	(50,50)	9.348184×10^{-10}

Example 4: Density Contrast Estimate Low, Wedge Depth Estimate Deep

Iteration No.	ρ	(CX,CZ)	Normalized MSE
0	0.200000	(46,54)	5.960400×10^{-2}
1	0.251539	(52,49)	1.562625×10^{-4}
2	0.250492	(50,50)	3.863768×10^{-6}
3	0.250008	(50,50)	9.348184×10^{-10}

All initial guesses have converged in a few iterations with the choice $\beta = 0$. The observed behavior suggests that our problem is quasilinear in the parameters CX and CZ, but linear in ρ. This is evident for the case of CX by inspection of the g_z versus x curve in Figure 6(a). We note that the profile has been run from $(x/z) = -1$ to $(x/z) = 1$, and g_z is approximately linear in x over this range. Since $CX = x'-x$, g_z is approximately linear in CX also. A choice of nonzero β in this specific problem has generally caused convergence to be much slower than for $\beta = 0$.

Aside from the inevitable nonuniqueness question, nonlinear regression provides a valid means to locate the wedge and to estimate its density contrast.

4.3 Traveltime Inversion of Offset Vertical Seismic Profiles

This inversion example is an abbreviation of a feasibility paper on traveltime inversion by Lines, Bourjeois and Covey (1983). The technique of traveltime inversion is certainly not new to geophysics and has been successful applied to earthquake data by Crosson (1976), and to exploration data by Gjoystdal and Ursin (1981). The advent of vertical seismic profiling (VSP), in which the seismic source is at the surface and the seismometers are in the wellbore (Figure 7a) has led to exploitation of this technology.

The data which we wish to fit are the traveltimes for direct arrivals in this "earth tomography" experiment. Primary reflections also prove to be useful in the procedure for deeper interfaces.

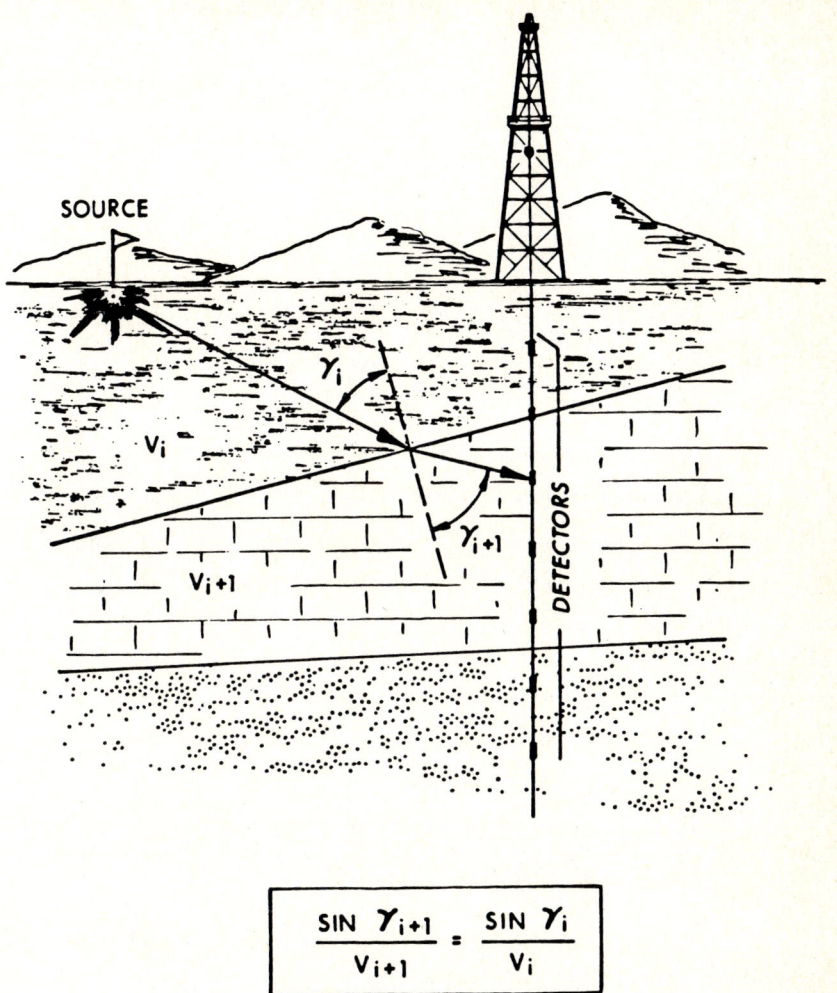

Figure 7(a) MODEL FOR VERTICAL SEISMIC PROFILE RAY TRACING

The ray tracing program used in our forward modeling, due to J. Covey, applies what is termed the "shooting method." In this iterative method rays are traced through the layered medium using Snell's law (i.e., $(\sin \gamma_i)/v_i$ = constant). For a given receiver, this process terminates when the ray arrives within some prescribed distance of the receiver. For each ray, the traveltime through each layer is computed and accumulated to give a total traveltime.

Our technique of estimating subsurface geology from traveltimes is essentially the same as proposed by Crosson (1976) and involves iterative use of the Marquardt-Levenberg method of damped least squares.

By this method we attempt to compute a model which fits a set of observed traveltimes for an array of receivers. If we denote the vector containing these observations by $\underline{t}$, and the vector containing the model traveltimes by $\underline{f}$, we seek to minimize

$$S = \|\underline{t} - \underline{f}\|^2 = \sum_{i=1}^{n} (t_i - f_i)^2 \tag{41}$$

for n observations.

The minimization is performed by adjustment of the model parameters which are a function of the local geology. If we assume that the velocities of these layers can be estimated from sonic logs or checkshot surveys, and that the earth has a layered structure, the parameters of interest are the dips of the rock layers. Setting x = 0 as the horizontal location of the well, we consider problems for which the local geology of each layer interface is described by:

$z_k = \theta_k x + b_k$

where z_k = depth of interface

θ_k = interface slope

b_k = depth at which interface intersects the well

x = offset distance from wellbore.

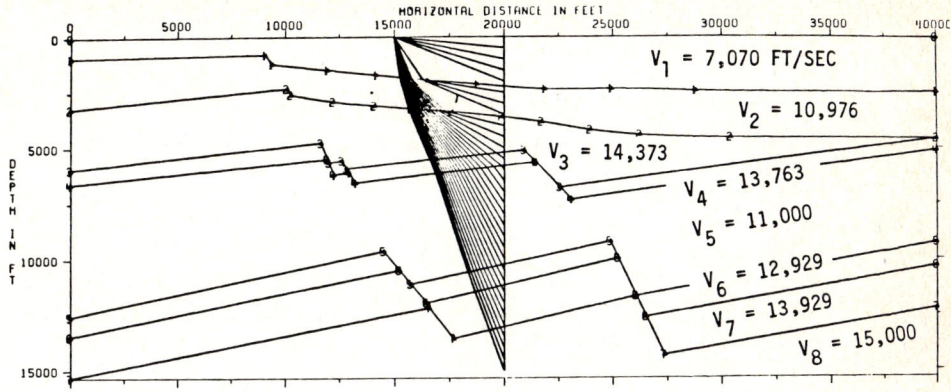

Figure 7(b) LAYERED MODEL USED IN CREATING TEST DATA SET (WITH DIRECT ARRIVAL RAYS).

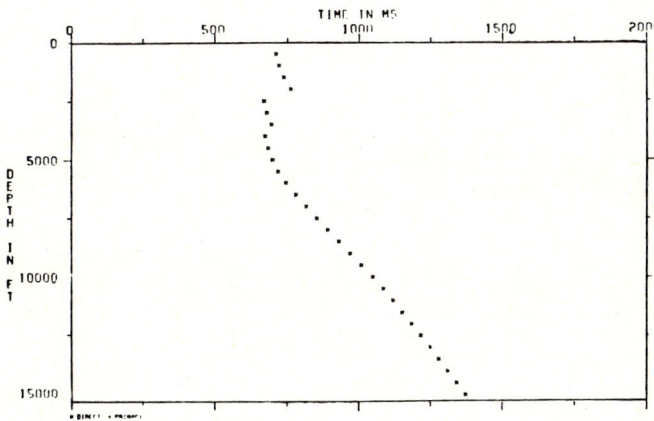

Figure 7(c) TRAVEL TIMES FOR DIRECT ARRIVALS.

If we assume that the intercepts, b_k, are known from well information, then the parameters vector θ contains the slopes of p layers $\theta = (\theta_1, \theta_2, \ldots, \theta_p)$. Our model vector of traveltimes can be written as $f_i = f_i(\theta)$.

In our least squares approach, we consider model perturbations to be linear functions of the parameter changes, δ_j, in a neighborhood of some initial guess, θ_0. This initial guess is usually the flat layer case where $\underline{\theta} = 0$.

The technique has been applied to a synthetic case and a real data case. The Gulf of Suez synthetic geologic model is shown in Figure 7b. The data traveltimes for this model are shown in Figure 7c). For the purposes of this study, we confine our aperture of rays to the area away from the faults since our inversion does not resolve discontinuities in slope and is similar to the traveltime inversion approach of Gjoystdal and Ursin (1981).

The general procedure has involved solving for the slopes of the upper layers first and then freezing these quantities in order to solve for slopes of the lower layers. This approach is often termed a "layer stripping" method.

Accurate solutions for the dips of the four upper interfaces are obtained in a few iterations. These are shown in Figure 8a along with starting estimates for layers 5, 6, 7. The inversion of layers 5, 6, 7 by use of direct arrivals was not successful due to the fact that the direct arrival traveltimes were insensitive to changes in slopes of deep layers. This becomes apparent when we compare the traveltimes for the case where layers 5, 6, and 7 are horizontal (as in Figure 8b) to the traveltimes for the true model of Figure 7b. The small difference in traveltimes between the two models indicates that our model responses are essentially unaffected by slope changes for these layers rendering the inversion ineffective. Mathematically,

$$\frac{\partial f_i}{\partial \theta_j} \to 0$$

whenever we are dealing with interfaces for which the ratio of

FOUR LAYER INVERSION

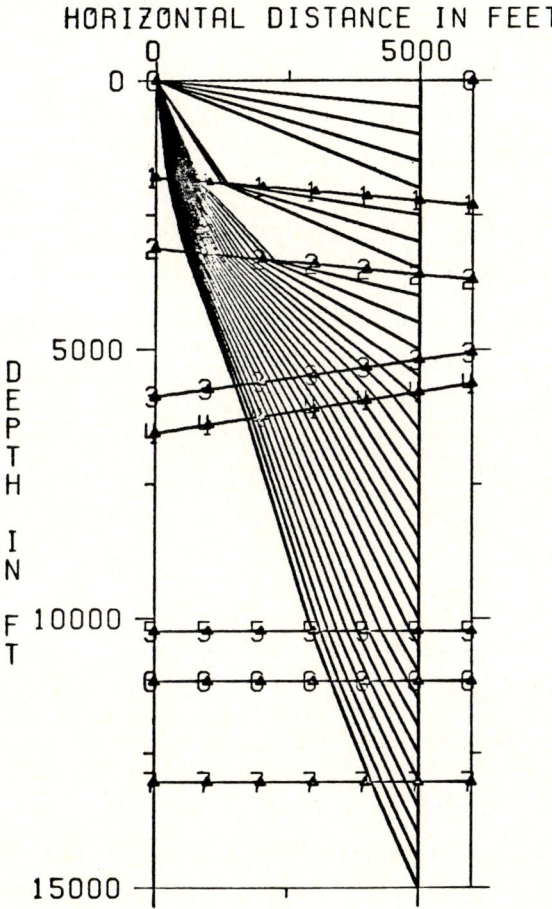

Figure 8(a) SOLUTION FOR LAYERS 1-4 WITH INITIAL ESTIMATE FOR LAYERS 5-7.

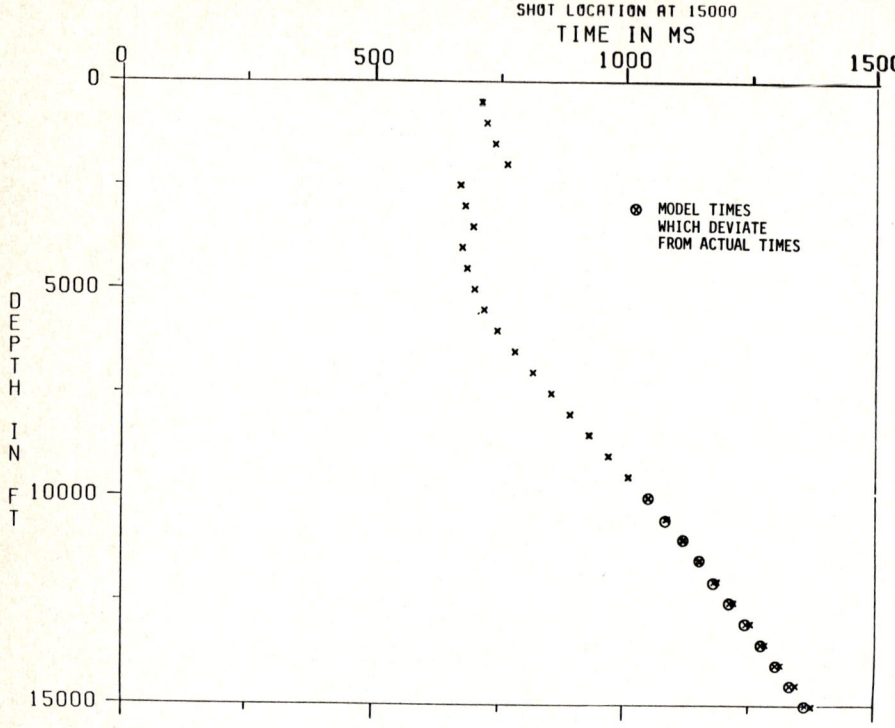

Figure 8(b)   TRAVEL TIMES FOR MODEL IN FIG. 5(a) COMPARED TO ACTUAL TRAVEL TIMES.

$$\frac{\text{source offset}}{\text{receiver depth}}$$

is small.

This unfortunate situation causes large errors in $\underline{\delta}$ for small errors in traveltime.

This problem of estimating slopes for deeper layers can possibly be solved by using the reflections from wide offset sources.

In order to invert for layer 5, we use a wide offset source as in Figure 9a. The direct arrival and reflection traveltimes are shown in Figure 9b. A good inversion is obtained by using the reflections for the receivers between interfaces 4 and 5. Figure 9c shows the reflections for a horizontal layer 5, and Figure 9d displays the difference in reflection traveltimes between the case of a horizontal layer 5 and a sloping layer. Clearly, the Jacobian values will not vanish for this problem, and, in fact, the slope of layer 5 is determined in two iterations to yield the model in Figure 9e.

The traveltimes for primary arrivals are more difficult to estimate from a seismogram than those for direct arrivals. In order to enhance these primary events and deeper events, a dip reject filter in the frequency wave number domain may be applied to the VSP data. The downtraveling waves in the VSP include direct arrivals, while the upgoing waves include the primary reflections. Since the wave numbers, k, of the upgoing and downgoing waves have opposite sign, the separation of the waves can be accomplished by setting half of the f-k plane to zero. This is demonstrated in the paper by Lines, Bourgeois, and Covey (1983).

The traveltime inversion technique was applied to a real data case in another geologic setting where an offset VSP and sonic well log data were available. The results were encouraging but again there was the ubiquitous nonuniqueness problem with our model. As shown by Figures 10 and 11, two earth models gave

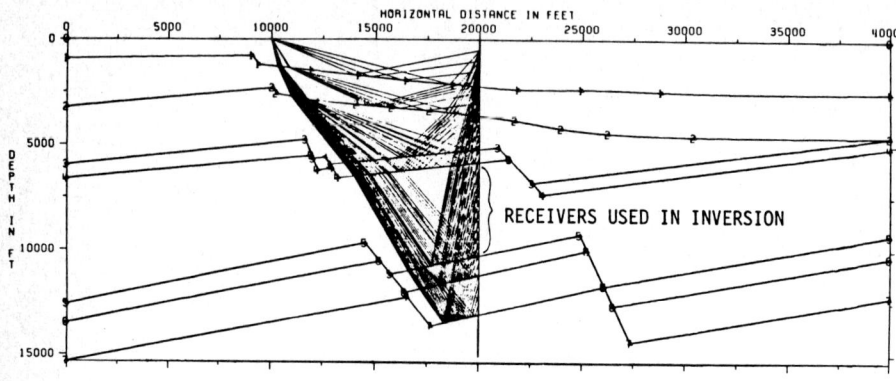

Figure 9(a) WIDE OFFSET REFLECTIONS FOR SOURCE AT 10,000 FT. OFFSET.

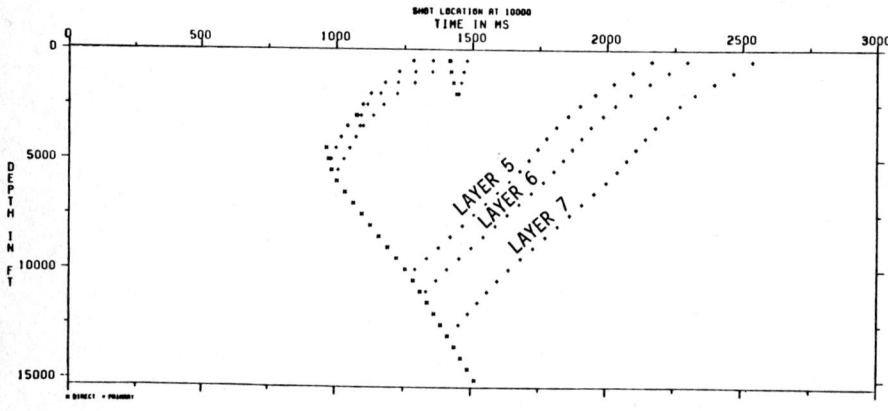

Figure 9(b) PRIMARY REFLECTIONS AND DIRECT ARRIVALS FOR MODEL.

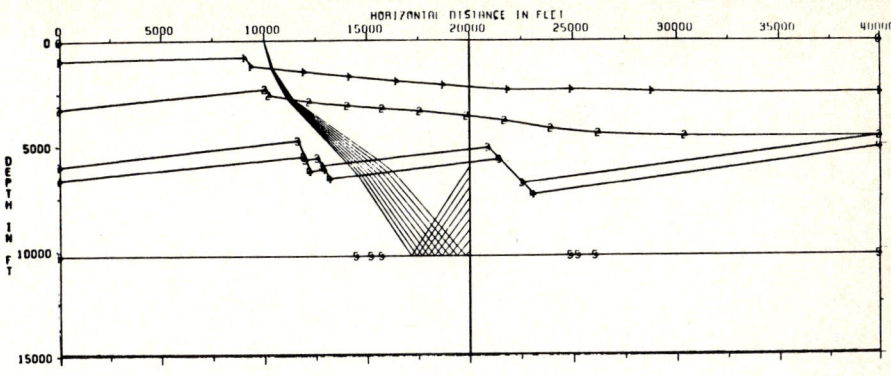

Figure 9(c)   INITIAL MODEL FOR LAYER 5.

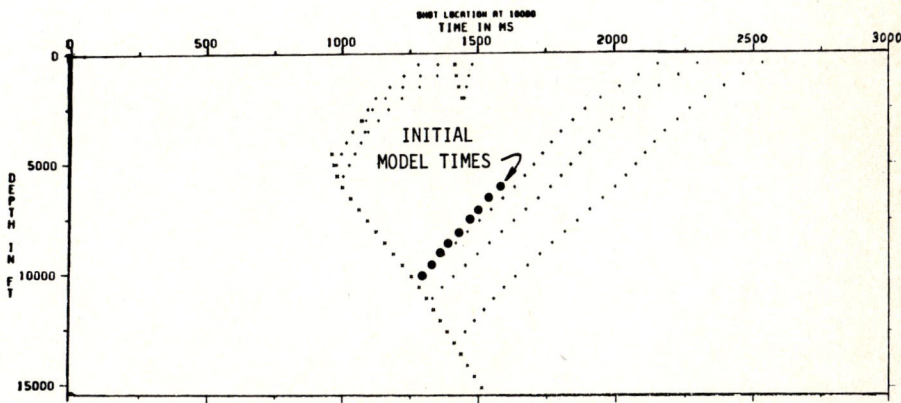

Figure 9(d)   COMPARISON OF TRAVEL TIMES FOR RAYS IN FIGURE 9(b) TO ACTUAL TRAVEL TIMES FOR MODEL.

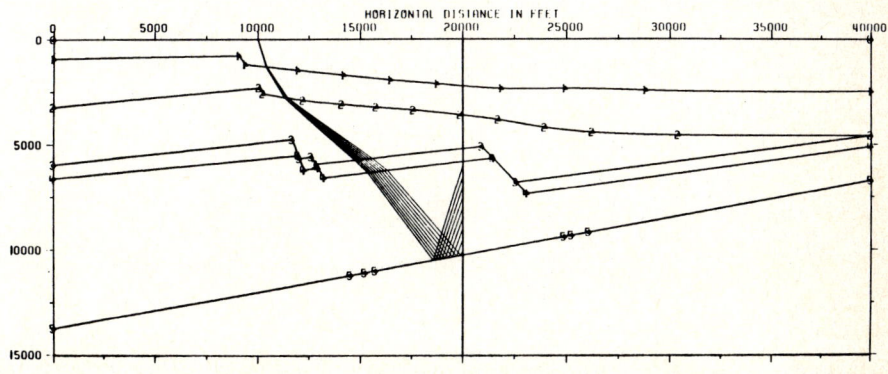

Figure 9(e)   FINAL INVERSION MODEL FOR LAYER 5.

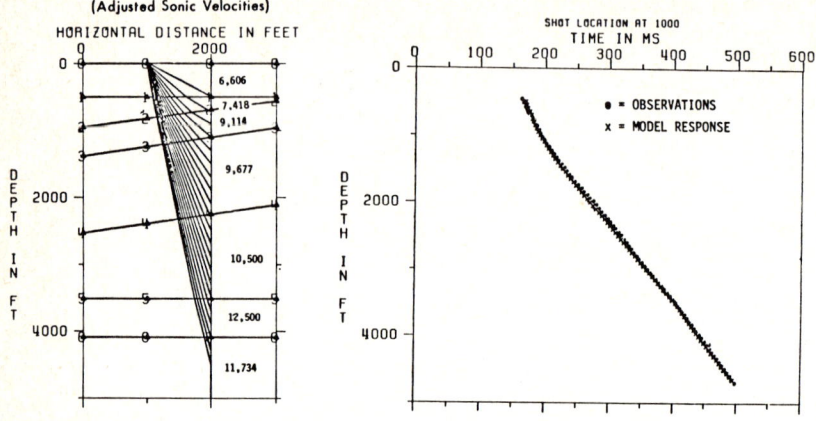

Figure 10 a) SLOPING LAYER MODEL WITH PERTURBED SONIC VELOCITIES.
b) TRAVELTIME RESPONSE FOR MODEL COMPARED WITH OBSERVED TRAVELTIMES.

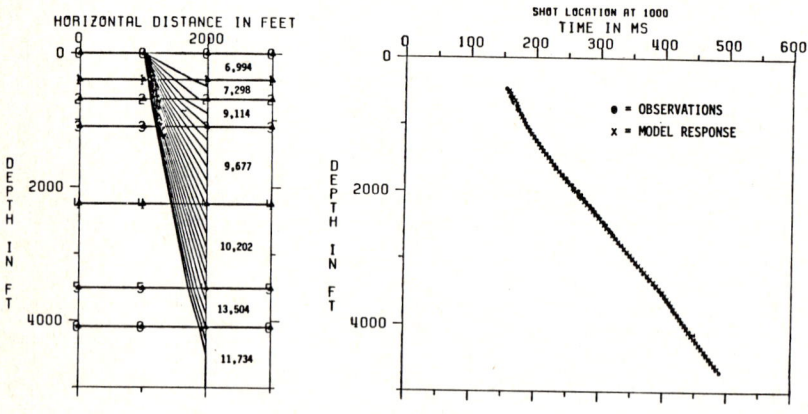

Figure 11 a) FINAL FLAT LAYER MODEL WITH VELOCITIES PERTURBED BY INVERSION.
b) TRAVELTIME RESPONSE FOR MODEL COMPARED WITH OBSERVED TRAVELTIMES.

good fits to the observations. In Figure 10, layer dips were estimated by using velocities deduced from well logs. In Figure 11, layer velocities were estimated by using a flat earth model. Dip meter information from the well tends to favor the second model.

Experience with synthetic and real data testing has indicated that accurate slope estimates can be deduced from wide offset VSP data, providing the traveltime picks are accurate and the traveltime values are sensitive to the estimate of deeper slopes.

5.0 Conclusions

The essence of geophysical data inversion is to provide a suitable earth model from a given set of observations. A possible method of performing this task involves nonlinear regression. The nonlinear regressive approach we chose involves an iterative application of Marquardt's method. It was shown that the SVD solution to the least squares problem has certain numerical advantages over conventional solutions to the normal equations. Moreover, the Marquardt damping factor can be included in the SVD approach by simple adjustment of the singular values. Three geophysical examples, namely seismic deconvolution, gravity inversion, and VSP traveltime inversion, were described to demonstrate the application of the method.

APPENDIX I

An efficient algorithm for performing the Singular Value Decomposition (SVD) has been given by Golub and Reinsch (1970), and involves two orthogonal transformations.

The first includes the left-handed Householder transformation, Q, which transforms a vector w with nonzero elements w_i into a vector having a single nonzero first term, that is,

$$Qw = -\sigma |w| e_1 \tag{42}$$

where $e_1 = \begin{bmatrix} 1 \\ 0 \\ \cdot \\ \cdot \\ \cdot \\ 0 \end{bmatrix}$ and $\sigma = \text{sgn}(w_1)$

Geometrically this implies a coordinate rotation such that the axis e_1 becomes parallel to the rotated vector Qw.

Householder (1958) shows that Q is

$$Q = I - \frac{2yy^T}{y^T y} \tag{43}$$

where $y = w + \sigma |w| e_1$, \tag{44}

where $|w|$ is the magnitude of w. A series of Householder transformations on the columns of Z reduces this matrix to upper triangular form, that is,

$$QZ = R \tag{45}$$

or

$$Z = Q^T R \tag{46}$$

where R is upper triangular, and Q is now the <u>product</u> of successive left-handed Householder transformations. The operation (46) is termed the QR decomposition, and has many important applications in numerical analysis (Lawson and Hanson, 1974). Next, post-multiplication of (45) by a series of right-handed Householder transformations operates on the rows of Z. This yields an upper bidiagonal matrix B

$$QZH = B \tag{47}$$

where H represents a <u>product</u> of successive right-handed Householder transformations.

The second orthogonal transformation employs the orthonormal Givens transformation algorithm, which produces a coordinate system rotation. A series of left-handed and right-handed Givens transformations S and W are applied to the matrix B (the Golub "chasing" algorithm). These steps reduce the off-diagonal terms of B to zero, so that

$$SBW = \begin{bmatrix} \lambda_1 & 0 & \cdots & & 0 \\ 0 & \lambda_2 & & & 0 \\ \vdots & & \ddots & & \vdots \\ 0 & \cdots & & 0 & \lambda_p \end{bmatrix} \tag{48}$$

or

$SBW = \Lambda$.

Combining (47) with (48), and using the orthonormality properties of Q, H, S, and W, we obtain

$Z = Q^T B H^T$

or

$$Z = Q^T S^T \begin{bmatrix} \lambda_1 & 0 & \cdots & & 0 \\ 0 & \lambda_2 & & & 0 \\ \vdots & & \ddots & & \vdots \\ 0 & \cdots & & 0 & \lambda_p \end{bmatrix} W^T H^T$$

that is,

$Z = U \Lambda V^T$

where U and V are the orthonormal matrices $U = (SQ)^T$ and $V = (HW)^T$. This is the desired singular value decomposition.

ACKNOWLEDGMENTS

The authors wish to thank Amoco Production Company for permission to publish this manuscript. We also thank Enders Robinson, Ken West, Pat Riley, Steve Hildebrand, Lajuanta Young, and Freeman Gilbert for suggestions and assistance in the development of various nonlinear regression computer codes. Finally, we thank Karen Bushyhead for typing this manuscript.

REFERENCES

AKI, K. AND RICHARDS, P., 1980, Quantitative Seismology - Theory and Methods, Vol. 2, Freeman Pub. Co., San Francisco, Calif.

BACKUS, G. E. AND GILBERT, J. F., 1967, Numerical Application of a Formalism for Geophysical Inverse Problems: Geophy. J. R. Astr. Soc., V. 13, p. 247-276.

BACKUS, G. E. AND GILBERT, J. F., 1968, The Resolving Power of Gross Earth Data: Geophy. J. R. Astr. Soc., V. 16, p. 169-205.

BILGERI, D. AND CARLINI, A., 1981, Non-Linear Estimation of Reflection Coefficients from Seismic Data: Geophys. Prosp., v. 29, p. 672-686.

COOKE, D. A. AND SCHNEIDER, W. A., 1981, Generalized Inversion of Reflection Seismic Data: paper presented at the 1981 Society of Exploration Geophysicists Meeting in Los Angeles, and published in Geophysics, v. 48, p. 665.

CROSSON, R. W., 1976, Crustal Structure Modeling of Earthquake Data, 1. Simultaneous Least Squares Estimation of Hypocenter and Velocity Parameters, J. Geophys. Res., v. 81, p. 3036-3046.

DENNIS, J., 1977, Nonlinear Least Squares and Equations, in The State of the Art of Numerical Analysis, edited by D. Jacobs, Academic Press, London.

DOBRIN, M., 1976, Introduction to Geophysical Prospecting, Third Edition, McGraw-Hill, New York.

DRAPER, N. R. AND SMITH, H., 1981, Applied Regression Analysis, Second Edition, John Wiley & Sons, New York.

FLETCHER, R., 1980, Practical Methods of Optimization, Vol. 1, Wiley Interscience, New York.

GJOYSTDAL, H. AND URSIN, B., 1981, Inversion of Reflection Times in Three Dimensions, Geophysics, v. 46, p. 972-983.

GOLUB, G. H. AND REINSCH, C., 1970, Singular Value Decomposition and Least Squares Solutions: Handbook for Automatic Computation, II, Linear Algebra, edited by J. Wilkinson and C. Reinsch, Springer-Verlag, New York.

GOUPILLAUD, P. L., 1961, An Approach to Inverse Filtering of Near Surface Layer Effects from Seismic Records: Geophysics, v. 26, p. 754-760.

GRAYBILL, F. A., 1969, Introduction to Matrices with Applications in Statistics: Wadsworth Publishing Co. Inc., Belmont, Calif.

HOUSEHOLDER, A. S., 1958, Unitary Triangularization of a Nonsymmetric Matrix: J. Assoc. Comput. Mach., 5, p. 339-342.

INMAN, J. R., 1975, Resistivity Inversion with Ridge Regression: Geophysics, V. 40, p. 798-817.

JACKSON, D. D., 1972, Interpretation of Inaccurate, Insufficient and Inconsistent Data: Geophys. J. R. Astr. Soc., V. 28, P. 97-109.

JUPP, D. L. B. AND VOZOFF, K., 1975, Stable Iterative Methods for the Inversion of Geophysical Data: Geophys. J. R. Astr. Soc., V. 42, p. 957-976.

KUNETZ, G. AND D'ERCEVILLE, I., 1962, Sur certaines propriétés d'une onde plane de compression dans un milieu stratifié, Am. Geophysique, v. 18, p. 351-359.

LANCZOS, C., 1961, Linear Differential Operators, D. Van Nostrand Co. Ltd., Princeton, New Jersey. p. 665-679.

LAWSON, C. L. AND HANSON, R. J., 1974, Solving Least Squares Problems, Prentice-Hall, Englewood Cliffs, New Jersey.

LEVENBERG, K., 1944, A Method for the Solution of Certain Nonlinear Problems in Least Squares: Quart. Appl. Math, V. 2, p. 164-168.

LINES, L. R. AND ULRYCH, T. J., 1977, The Old and the New in Seismic Deconvolution and Wavelet Estimation, Geophys. Prosp., V. 25, p. 512-540.

LINES, L. R., BOURGEOIS, A., COVEY, J. D., 1983, Traveltime Inversion of Offset VSP's - A Feasibility Study, Geophysics (in press).

MARQUARDT, D. W., 1963, An Algorithm for Least Squares Estimation of Nonlinear Parameters: J. Soc. Indust. Appl. Math., V. 11, p. 431-441.

OLDENBURG, D. W., 1974, The inversion and interpretation of gravity anomalies: Geophysics, V. 39, p. 526-536.

ORISTAGLIO, M. L. AND WORTHINGTON, M. H., 1980, Inversion of Surface and Borehole Electromagnetic Data for Two Dimensional Electrical Conductivity Models: Geophys. Prosp., V. 28, p. 633-657.

PARKER, R. L., 1974, Best Bounds on Density and Depth from Gravity Data: Geophysics, V. 39, p. 644-649.

ROBINSON, E. A., 1967, Multichannel Time Series Analysis with Digital Computer Programs: San Francisco, Holden Day.

ROBINSON, E. A. AND TREITEL, S., 1980, Geophysical Signal Analysis, Prentice-Hall, Englewood Cliffs, New Jersey.

SMITH, F. B. AND SHANNO, D. F., 1971, An Improved Marquardt Procedure for Nonlinear Regressions: Technometrics, V. 13, p. 63-75.

STEWART, G. W., 1973, Introduction to Matrix Computations, Academic Press, New York.

TREITEL, S. AND ROBINSON, E. A., 1966, The Design of High Resolution Digital Filters: IEEE Trans. on Geosci. Electronics, V. GE-4, p. 25-38.

TREITEL, S. AND LINES, L. R., 1982, Linear Inverse Theory and Deconvolution, Geophysics, v. 47, p. 1153-1159.

TREITEL, S., GUTOWSKI, P. R., AND WAGNER, D. E., 1982, Plane wave decomposition of seismograms, Geophysics, v. 47, p. 1375-1401.

TWOMEY, S., 1977, Introduction to the Mathematics of Inversion in Remote Sensing and Indirect Measurements, Elsevier Scientific Publ. Co., Amsterdam.

VIGNERESSE, J. L., 1977, Linear Inverse Problem in Gravity Profile Interpretation: J. Geophys., V. 43, p. 193-213.

WIGGINS, R. A., The Generalized Linear Inverse Problem: Implication of Surface Waves and Free Oscillations for Earth Structure: Reviews of Geophysics and Space Physics, V. 10, p. 251-285.

WIGGINS, R. A., LARNER, K. L., AND WISECUP, R. P., 1976, Residual Static Analysis as a General Linear Inverse Problem: Geophysics, V. 41, p. 922-938.

ARMA MAXIMUM ENTROPY SPECTRAL ANALYSIS USING ITERATIVE PREWHITENING

Charles E. Schmid
Honeywell, 5303 Shilshole Avenue N.W., Seattle, Washington 98107 USA

This parametric approach for estimating a spectrum $S(f)$ is based upon prewhitening with a filter with a squared frequency response denoted by $P(f)^{-1}$. The entropy of the output is given by $\int \ln\{S(f) P(f)^{-1}\} df = \int \ln S(f) df + \int \ln P(f)^{-1} df$. Maximizing the first term produces the standard maximum entropy, autoregressive (AR) spectral estimate of order N, where the autocorrelation values past N are assumed to be completely unknown. Maximizing the second term is accomplished using a concept similar to maximum entropy in conjunction with Cleveland's inverse spectrum and autocorrelation. This reasoning results in a moving average (MA) spectral estimate of order M, where the autocorrelaton values greater than M turn out to be zero, as in the traditional Blackman-Tukey technique. Combining the two spectral estimates gives an ARMA spectrum estimate at the input of the prewhitening filter. The technique is iterated back and forth using the inverse AR prewhitening filter to obtain the MA coefficients, and an inverse MA prewhitening filter to obtain the AR coefficients. The ARMA maximum entropy formulation and iteration have similarities to recent minimum cross-entropy principles. Examples using simulated geophysical processes illustrate the application and advantages of this approach.

I. INTRODUCTION

This paper considers the traditional problem of estimating a spectrum $S(f)$ from only a single channel, finite record length time series, $x(1), x(2), \ldots, x(L)$, which is assumed to be sampled at equal time intervals, Δt. The experienced analyst and the novitiate may read about the long and interesting history of the technical and philosophical aspects of this problem in an engaging IEEE article by E. A. Robinson (1982). Two of the most important and popular techniques reviewed in his article are the so-called "Tukey Empirical Spectral Analysis" and the "Burg Maximum Entropy Spectral Analysis." These two methods may be interpreted through the autocorrelation function, $R(k)$, along with its Fourier transform $S(f)$, both of which are based on the second order statistics of the data $\{x\}$.

$$R(k) = \xi \left\{ X^*(\ell) X(\ell+k) \right\} \tag{1}$$

$$= \int_{-f_s/2}^{f_s/2} S(f) e^{j2\pi f k \Delta t} df \tag{2}$$

where f_s = sampling frequency = $1/\Delta t$.

$$S(f) = \sum_{k=-\infty}^{\infty} R(k) e^{-j2\pi f k \Delta t} \tag{3}$$

Since we only have a limited number of data points $\{x\}$, we must be satisfied with an estimate of the autocorrelation function, such as the biased estimate,

$$\hat{R}(k) = \frac{1}{L} \sum_{\ell=1}^{L} x^*(\ell) \, x(\ell+k) \quad . \tag{4}$$

The "Tukey Empirical Spectral Analysis" technique initiated in the late 1950s makes the assumption that the values of the autocorrelation function, $R(k)$, outside the index range $|k| > M$ are all zero. The analyst can force the unknown autocorrelation estimate to go smoothly to zero by using a lag window on $\hat{R}(k)$ or a data window on $x(\ell)$. The assumption that $R(k) = 0$ for $|k| > M$ is equivalent to modeling the underlying process which generated $x(\ell)$ as a moving average (MA) process of order M [Cadzow (1982)].

Objections to "Tukey's Empirical Spectral Analysis" method, such as a lack of theoretical bases [Jaynes (1982)], poor spectral resolving power for doublet sinusoids, and the convenience of a parametric approach, led to use of "Burg's Maximum Entropy Spectral Analysis." Burg (1967) maximized the entropy rate defined by,

$$h = \int_{-f_s/2}^{f_s/2} \ln S(f) \, df \tag{5}$$

subject to knowing $R(k)$ between the range $-N \leq k \leq N$. All values of $R(k)$ outside of this range ($|k| > N$) are assumed completely unknown. The result of maximizing the entropy rate and an unknown $R(k)$ for $|k| > N$ turns out to be equivalent to modeling the underlying generation of $\{x\}$ as an autoregressive (AR) process of order N. Figure 1 reviews the brief discussion above for "Tukey's Empirical Spectral Analysis" and "Burg's Maximum Entropy Spectral Analysis." Figure 2 shows the underlying generation model, the parameters of which are unknown to the analyst and, except for simulated data, never really existed.

Various proposals have been made to sequentially or simultaneously combine these two popular techniques, often under the topical heading of "ARMA" (Autoregressive Moving Average) spectral estimation [Kay and Marple (1981); Cadzow (1982); Robinson and Treitel (1980)]. This paper also attempts to fuse the two techniques under the principle of maximum entropy combined with the idea of prewhitening. The results of combining maximum entropy and prewhitening are derived in Section II, and indicate that the AR parameters should be estimated after the data $\{x\}$ is initially prewhitened with an inverse MA filter. The problem with the prewhitening result is that the inverse MA filter is not known. Section III proposes that the MA parameters should be estimated after the data $\{x\}$ is initially prewhitened with an inverse AR filter. Section III also discusses an interative approach of alternately carrying out these two prewhitening and estimation steps. Section IV relates that approach to some other methods in the literature. The algorithm in this paper has been programmed on a computer and some simulated geophysical time series are spectrally analyzed in Section V to illustrate the application and advantages of the technique. Section VI finishes the paper with a short discussion on the advantages and disadvantages of the approach.

II. ARMA MAXIMUM ENTROPY VIA PREWHITENING

The term "prewhitening" was introduced by Tukey to describe a general method to pre-emphasize the data in order to improve the subsequent spectral analysis [Blackman and Tukey (1959)]. It has been used in numerous applications for spectral estimation of geophysics [Båth (1974)], speech [Markel and Gray (1976)], wave-guide analysis [Thomson (1977)], and bias reduction [Kay and Marple (1981)]. Figure 3 shows a prewhitening filter with a transfer function having a squared magnitude response of $P(f)^{-1}$. This prewhitening filter pre-emphasizes the time series $x(\ell)$, where $x(\ell)$ has an underlying spectrum $S(f)$, which we eventually would like to estimate. The output time series, $z(\ell)$, will therefore have a spectrum given by $S(f)P(f)^{-1}$.

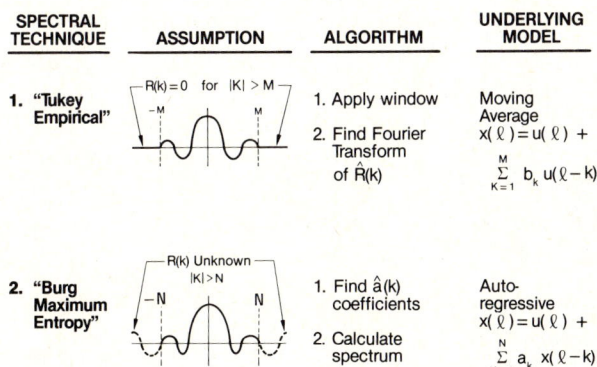

Figure 1. Review of Two Popular Spectral Analysis Techniques

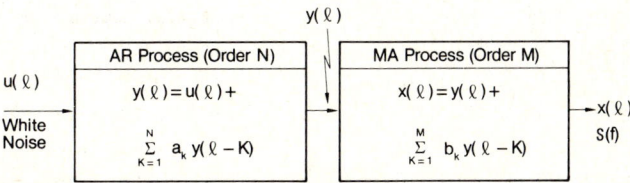

Figure 2. Parametric Model for Generating Process

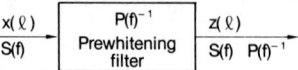

Figure 3. Prewhitening Filter

Following Burg's definition in Equation (5), the entropy rate at the output of the filter is

$$h = \int_{-f_s/2}^{f_s/2} \ln \left| S(f) \, P(f)^{-1} \right| df \quad . \tag{6}$$

Note that this reverts to Burg's definition in Equation (5) if the prewhitening filter is uniform over all frequencies. Expanding (6) gives

$$h = \underbrace{\int_{-f_s/2}^{f_s/2} \ln S(f) \, df}_{h_q} + \underbrace{\int_{-f_s/2}^{f_s/2} \ln P(f)^{-1} \, df}_{h_p} \quad . \tag{7}$$

The first term, h_q, is maximized by the now-standard Burg maximum entropy approach [Burg (1967); Robinson (1982)]. The main constraint, as mentioned above, is that $R(k)$ is completely unknown for $|k| > N$. The result is a spectral estimate of $S(f)$

$$\frac{1}{Q(f)} = \frac{1}{\left| 1 - \sum_{K=1}^{N} \hat{a}_k e^{-j2\pi f K \Delta t} \right|^2} \tag{8}$$

where the $\hat{a}_k$s turn out to be the autoregressive estimates of the underlying generating process. The $\hat{a}_k$ parameters, or the equivalent set of reflection coefficients, are found using Levinson's algorithm on the Yule-Walker equations,

$$\hat{R}(n) = \sum_{k=1}^{N} \hat{a}_k \, \hat{R}(n-k) \qquad n = 1,2,...N \tag{9}$$

or by Burg's algorithm [Robinson and Treitel (1980); Kay and Marple (1981)].

The second entropy term in Equation (7), h_p, will be maximized in a similar fashion to Burg's maximization approach for h_q; the variation from Burg's maximization is based on the inverse autocorrelation function, initially alluded to by Durbin (1959), formalized by Cleveland (1972), and discussed by Makhoul (1975) and Kopec, Oppenheim, and Tribolet (1977). Cleveland's notation of $RI(k)$ will be used to define the inverse of the autocorrelation function as the Fourier transform of the inverse spectrum.

$$P(f)^{-1} = \sum_{K=-\infty}^{\infty} RI(k) \, e^{-j2\pi f K \Delta t} \tag{10}$$

$$RI(k) = \int_{-f_s/2}^{f_s/2} P(f)^{-1} e^{j2\pi f K \Delta t} \, df \tag{11}$$

The second entropy term can now be stated by placing Equation (10) in the definition for h_p in (7)

$$h_p = \int_{-f_s/2}^{f_s/2} \ln \left\{ \sum_{K=-\infty}^{\infty} RI(k) \, e^{-j2\pi f K \Delta t} \right\} df \quad . \tag{12}$$

Burg's development for maximizing the entropy [Burg (1967); Robinson (1982)] can be applied to Equation (12). First, we differentiate with respect to the inverse correlation function.

$$\frac{\delta h_p}{\delta RI(m)} = \int_{-f_s/2}^{f_s/2} \left[\Sigma \, RI(k) \, e^{-j2\pi f k \Delta t} \right]^{-1} e^{-j2\pi f m \Delta t} \, df \tag{13}$$

The result is set to zero to maximize it for inverse autocorrelation values for $|m| > M$ which are assumed unavailable. This assumes we know the values of $RI(k)$ for $|m| \leq M$.

$$\int_{-f_s/2}^{f_s/2} \left[P(f)^{-1} \right]^{-1} e^{-j2\pi f m \Delta t} \, df = 0 \qquad |m| > M \tag{14}$$

If we expand $[P(f)^{-1}]^{-1}$ in a Fourier series,

$$\left[P(f)^{-1} \right]^{-1} = \sum_{m=-\infty}^{\infty} \lambda(m) \, e^{-j2\pi f m \Delta t} \qquad -\frac{f_s}{2} \leq f \leq \frac{f_s}{2} \tag{15}$$

then, using one of various techniques discussed such as analytic continuation or z transform theory, Burg shows the limits are finite.

$$[P(f)^{-1}]^{-1} = \sum_{m=-M}^{M} \lambda(m) e^{-j2\pi f m \Delta t} \tag{16}$$

Rewriting this in terms of MA parameters,

$$P(f) = \left| 1 + \sum_{k=1}^{M} \hat{b}_k e^{-j2\pi f k \Delta t} \right|^2 \quad . \tag{17}$$

The moving average parameters, $\hat{b}_k$, and the equivalent reflection coefficients are once again solutions to the Yule-Walker equations,

$$\widehat{RI}(m) = \sum_{k=1}^{M} \hat{b}_k \, \widehat{RI}(m-k) \qquad m = 1, 2, \ldots M \tag{18}$$

which may be obtained by Levinson's algorithm, or can be found directly via Burg's algorithm.

The parametric model and spectrum at the input to the prewhitening filter has now been achieved by obtaining the estimates to the two sets of parameters $\{\hat{a}_k; k = 1, 2, \ldots, N\}$ and $\{\hat{b}_k; k = 1, 2, \ldots, M\}$.

$$S(f) = \frac{P(f)}{Q(f)} = \frac{\left| 1 + \sum_{k=1}^{M} \hat{b}_k e^{-j2\pi f k \Delta t} \right|^2}{\left| 1 - \sum_{k=1}^{N} \hat{a}_k e^{-j2\pi f k \Delta t} \right|^2} \tag{19}$$

If the two processes are stable, then the poles corresponding to $\{a_k\}$ and the zeros corresponding to $\{b_k\}$ will be inside the unit circle and thus the expression is minimum phase. Furthermore, it is instructive to rewrite Equation (14).

$$R(k) = \int_{-f_s/2}^{f_s/2} P(f) e^{-j2\pi fk\Delta t} df = 0 \quad |k| > M \tag{20}$$

Note that this is exactly the assumption for Tukey's empirical method. Hence, maximum entropy produces a prewhitening filter $P(f)^{-1}$ which initially de-emphasizes the MA components in $S(f)$, and then extracts the AR or pole components. This de-emphasis for MA components concentrates on compensating for the zeros which primarily affect the "valleys" of the spectrum and is not just "prewhitening" per se. This allows the AR spectral estimate to more effectively match the poles or "peaks" of the spectrum.

III. ITERATIVE PREWHITENING

The discussion above is incomplete in that the prewhitening filter $P(f)^{-1}$ is unknown and needs to be estimated. This is part of the general dilemma of spectral analysis, in that we want to know the general spectral model before actually carrying out the spectral analysis. A logical sequence for estimating $P(f)^{-1}$ is to calculate the MA spectra of the time series, $x(\ell)$, after it has been prewhitened to de-emphasize the AR components.

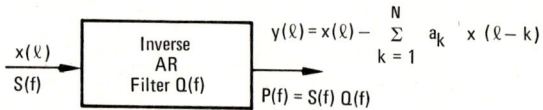

Figure 4. Prewhitening with an Inverse AR Filter

The structure, $Q(f)$, has been called an "inverse" filter, and the output, $y(\ell)$, is referred to as the residual [Markel and Gray (1976); Kopec, Oppenheim, and Tribolet (1977); Thomson (1977)]. The MA coefficients will be estimated in this paper from the residual data, $y(\ell)$, using the inverse autocorrelation or LPC method used in the entropy derivation in Section II [Cleveland (1972); Makhoul (1975); Kopec, Oppenheim, and Tribolet (1977); Rice and Siegel (1983); Mills, Mullis, and Roberts (1980)]. Other methods can also be used [Cadzow (1982); Kay and Marple (1981)].

Once the MA coefficients are found, the prewhitening filter shown in Figure 3 can be utilized. The iteration is continued by alternately applying: (1) the process in Figure 3 to obtain the AR parameters after first using an inverse MA prewhitening filter, and (2) the process in Figure 4 to obtain the MA parameters after first using an inverse AR filter. Figure 5 summarizes this iterative process, and Figure 6 shows a flow chart outlining the various steps in the iterative operation, with references to equations in the text.

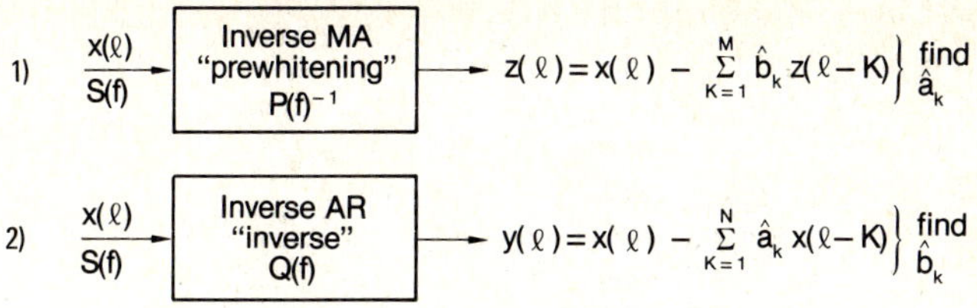

ARMA Maximum Entropy Spectral Analysis

1) $\dfrac{x(\ell)}{S(f)}$ → [Inverse MA "prewhitening" $P(f)^{-1}$] → $z(\ell) = x(\ell) - \sum_{K=1}^{M} \hat{b}_k z(\ell-K)$ } find $\hat{a}_k$

2) $\dfrac{x(\ell)}{S(f)}$ → [Inverse AR "inverse" $Q(f)$] → $y(\ell) = x(\ell) - \sum_{K=1}^{N} \hat{a}_k x(\ell-K)$ } find $\hat{b}_k$

Figure 5. Interative Prewhitening

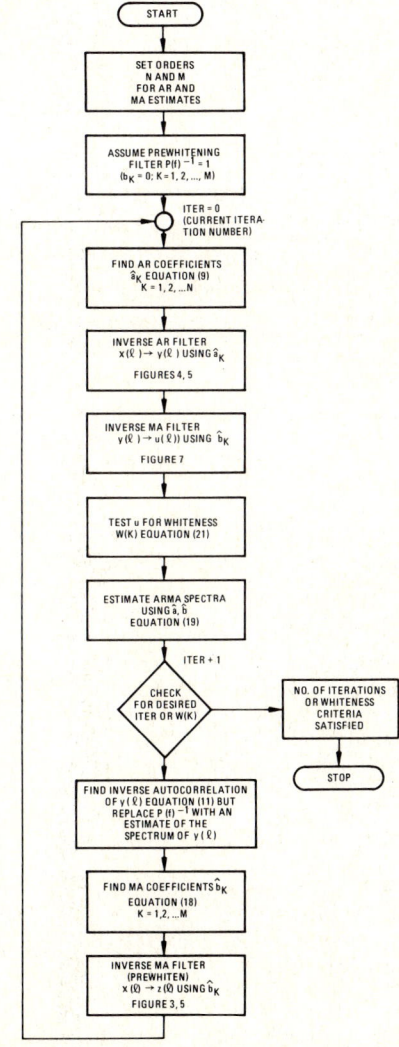

Figure 6. Flow Chart Outlining Iterative Prewhitening Sequence

When the AR inverse filter and MA prewhitening filter are placed in cascade, they should whiten the input time series, $x(\ell)$. In fact, the whiteness of the output, $u(\ell)$, is a measure of how well the input spectra is whitened by the ARMA inverse filter.

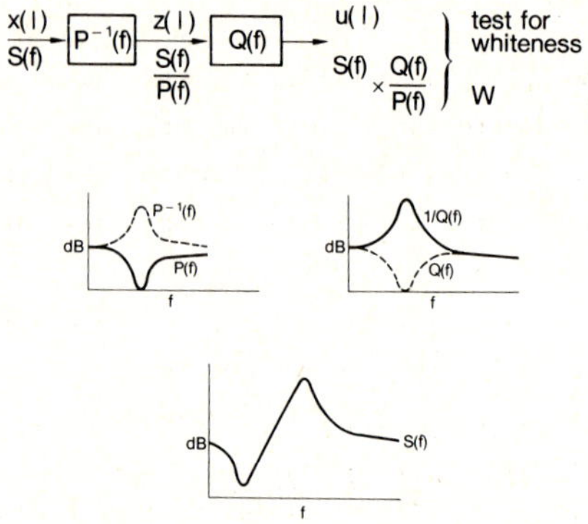

Figure 7. Cascaded Prewhitening and Inverse Filters

The whiteness of $u(\ell)$ can be estimated from its autocorrelation function, $R(k)$, and a test statistic proposed by Stoica (1977).

$$W(K) = \frac{L}{R^2(0)} \sum_{k=1}^{K} R^2(k) \qquad (21)$$

This value provides a check on convergence and rates of convergence.

It is interesting to note that the entropy rate of the process, $u(\ell)$, is maximum if it is white and Gaussian [Papoulis (1981)]. If infinite orders of M or N were permitted, then either the inverse AR or inverse MA filters would independently whiten the output. But we, unfortunately, have noisy and finite length data to provide estimates for these filters, which leads to the natural question of orders for M and N. I will, like most other authors, avoid this question and only point out that the technique advocated above should, in general, allow a more parsimonious estimate, since the analyst can use both M and N to decrease the total (M + N) order. Furthermore, the test statistic, W(K), provides a mechanism to evaluate and trade off the effects of altering the orders of M and N.

The concept of iteratively using AR and MA estimates is a natural approach and has been used by a number of investigators [Konvalinka and Mataušek, (1979); Mills, Mullis, and Roberts (1980); Robinson and Treitel, (1980); Thomson, (1977)]. It has also been used for ARMA filter design by Schmid (1983) which, in fact, provided the impetus for the present investigation.

IV. RELATED SPECTRAL ESTIMATION TECHNIQUES

Recent work on cross-entropy spectral analysis [Shore (1981)] contains some connections with this paper's approach and, so, cross-entropy is briefly covered here. The basic definition of cross-entropy between two amplitude probability densities p(x) and q(x) is given by

$$H(p,q) = \int q(X) \ln \left\{ \frac{q(X)}{p(X)} \right\} dx \quad . \tag{22}$$

The density p(x) is the prior probability density and the density q(x) is the posterior density. We desire an estimate to the true density $q^t(x)$ by minimizing cross-entropy H. In addition to having p(x), we are given new information, I, to arrive at the estimate of the posterior density, q(x). According to Shore, cross-entropy provides a logically consistent approach for incorporating this information, I, with the prior density, p(x), to arrive at the posterior estimate q(x). Shore converts the cross-entropy amplitude density form above into a spectral density approach using a Fourier series representation. He assumes a prior spectrum estimate which resembles a prewhitening filter, and new information in the form of an autocorrelation function. Musicus (1983) has extended the cross-entropy approach to include iterations back and forth between p(x) and q(x), using each individual estimate to independently and iteratively estimate the other. Although Shore's and Musicus' specific results are not in the same form as this paper's, their general approaches to iteration and priors are similar to this paper's concept of iterative prewhitening.

Another recent paper by Steinhardt and Roberts (1983) shows that ARMA spectral estimates maximize entropy when cepstral matching constraints are included. Thus, ARMA maximim entropy spectral estimation can be derived by an approach different from this paper's prewhitening approach.

V. SPECTRAL ESTIMATION USING SIMULATED TIME SERIES

Two examples using simulated time series from geophysical processes are presented to illustrate the application of the approach presented in Sections II and III. The time series were generated on a digital computer by passing white Gaussian, pseudorandom noise samples, $u(\ell)$, through the ARMA filters shown in Figure 2. The coefficients for the simulations for both examples were calculated using modeled and measured spectra given in Båth (1974), in conjunction with an ARMA filter design program [Schmid (1983)]. Each simulation used six poles and six zeros, giving an ARMA process of order (6, 6). The simulated time samples, $x(\ell)$, consisted of 1,024 values (L = 1024). The spectral analysis followed the iteration algorithm shown in Figure 6.

The solid lines on Figure 8 show the true underlying spectra for the simulated time series that represents ocean wave spectra with a 12-second wave period. Brettscheider's theoretical spectra were used to determine the simulation filter coefficients. These spectra have the same shape as Pierson-Moskowitz spectra [Båth 1974)].

The dotted lines on Figure 8 represent three spectral density estimates for different ARMA orders. The top estimate used an ARMA order of (6, 6) which is exactly what was used to generate the spectra. The middle estimate used an AR order-12 maximum entropy estimate. The estimate understandably has trouble matching the zeros that were used in generating the original time series. The bottom estimate used an AR order-6 maximum entropy estimate, and demonstrates even more difficulty matching the spectra. The obvious point of these three graphs is that the ARMA (6, 6) maximum entropy spectra produce a better match for spectra that come from an ARMA time series, than the AR order alone (6) or sums of the two orders (12).

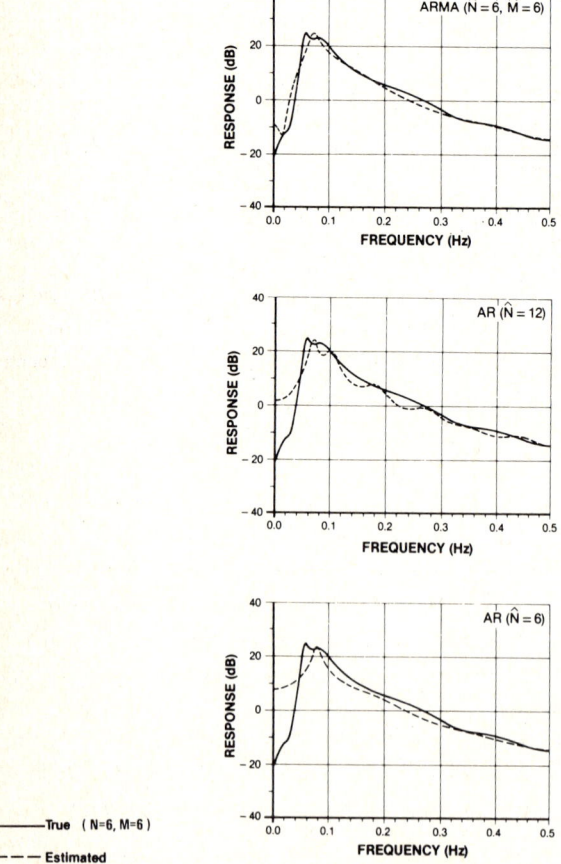

Figure 8. ARMA Maximum Entropy Spectrum Estimation Using a Simulated Ocean Wave Time Series

The results of the second simulation are shown in Figure 9. Again, the solid lines indicate the true underlying spectra of the generated time series, which in this case were microseism measurements [Båth (1974), Figure 94]. The lower peak at .075 Hz is referred to as a primary microseism, and the higher peak at .135 Hz is referred to as a secondary microseism. The primary microseism is thought to result from ocean waves, since it peaks at the same point that Figure 8 does. The top graph, which employed the ARMA approach discussed in this paper, follows the true spectra closely. The middle graph, which uses a 12th order AR Maximum Entropy estimate, does not accurately estimate the notch between the primary and secondary microseisms. Without our prior knowledge of the true spectra, we would probably have overlooked the primary microseism using this estimate. Finally, the bottom graph, with only a sixth order AR estimate, completely overlooks the low-power primary microseism. Once again, this sequence points out the effectiveness of using poles and zeros for estimating spectra with underlying ARMA spectra.

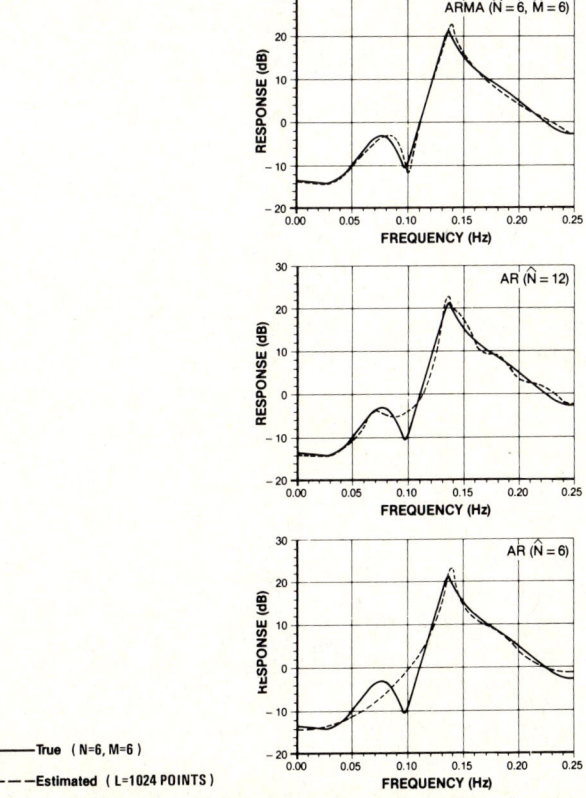

Figure 9. ARMA Maximum Estimation Using a Simulated Microsiesm Time Series

VI. SUMMARY AND DISCUSSION

Sections II and III develop the theoretical and intuitive aspects of ARMA maximum entropy spectral estimation via iterative prewhitening. Section IV relates these ideas to recent results using cross-entropy.

Finally, Section V gives results using simulated geophysical time series. The results indicate that the ARMA maximum entropy approach is superior to AR analysis. However, the reader with experience in time series analysis has probably noted that the simulation and analysis both utilize ARMA models with identical orders, and naturally would be expected to perform better. A generating process that was strictly AR would probably be best matched by an AR analysis. Hence, we have once more returned to the basic dilemma of spectral analysis, which is the need to know something about the spectra before we perform analysis. Not knowing this, the ARMA approach at least permits us to vary M and N and measure the goodness of our fit with the whiteness test. However, the use of iterations is costly in computation time and does not always converge when the underlying spectra have steep slopes or wide spectral dynamic range. But if the analyst is willing to carefully carry out these iterations, the ARMA approach will give parsimonious spectral estimates and also produce AR and MA coefficients that could be useful for future simulations.

This approach will have to be tried on various types of underlying models to evaluate its potential. A variation on this approach has been compared separately to the "Tukey Empirical Approach" and the "Burg Maximum Entropy Method," using shaped broadband noise and sinusiods [Mordojovich and Roberts (1981)]. They found varying results, with none of the approaches working best all of the time. However, they found the iterative ARMA approach detailed by Mills, Mullis, and Roberts [1980] to be the best in most cases.

REFERENCES

BÅTH, M. (1974). <u>Spectral Analysis in Geophysics</u>, Elsevier Scientific Publishing Company, Amsterdam.

BLACKMAN, R.B. and TUKEY, J.W. (1959). <u>The Measurement of Power Spectra from the Point of View of Communication Engineering</u>, Dover, New York.

BURG, J.P. (1967). Maximum Entropy Spectral Analysis in the <u>Proceedings of the 37th Meeting of the Society of Exploration Geophysicists</u>. Oklahoma City, 1967. (This paper is reprinted in an IEEE press collection, <u>Modern Spectrum Analysis</u>, edited by D.G. Childers, 1978. Burg's 1975 dissertation at Stanford, <u>Maximum Entropy Spectral Analysis</u>, elaborates on his original 1967 paper.)

CADZOW, J.A. (1982). Spectral Estimation: An Overdetermined Rational Model Equation Approach in the <u>Proceedings of the IEEE</u>, 70, 907-939.

CLEVELAND, W.S. (1972). The Inverse Autocorrelations of Time Series and Their Applications. Technometrics, 14, 277-298.

DURBIN, J. (1959). Efficient Estimation of Parameters in Moving Average Models. Biometrica, 46, 306-316.

JAYNES, E.T. (1982). On the Rationale of Maximum-Entropy Methods in the Proceedings of the IEEE, 70, 939-952.

KAY, S.M. and MARPLE, Jr., S.L. (1981). Spectrum Analysis--A Modern Perspective in the Proceedings of the IEEE, 69, 1380-1419, (corrections to this paper may be found in October 1982 issue).

KONVALINKA, I.S. and MATAUŠEK, M. (1979). Simultaneous Estimation of Poles and Zeros in Speech Analysis and ITIF--Iterative Inverse Filtering Algorithm, IEEE Transactions of ASSP, 27, 485-495.

KOPEC, G., OPPPENHEIM, A., and TRIBOLET, J. (1977). Speech Analysis by Homomorphic Prediction, IEEE Transactions of ASSP, 25, 40-49.

MAKHOUL, J. (1975). Linear Prediction: A Tutorial Review, Proceedings of the IEEE, 63, 561-580

MARKEL, J. and GRAY, Jr., A.H. (1976) Linear Prediction of Speech, Springer-Verlag, Chapter 6.3.

MILLS, W.L., MULLIS, C.T., and ROBERTS, R. (1980). An Iterative Estimation Technique for Power Spectra by an ARMA Model, (Proceedings of the IEEE for ICASSP, 622-625, held in Denver, April 1980).

MORDOJOVICH, A, and ROBERTS, R. (1981). A Comparison of Spectral Estimators for Real Data, (Proceedings of the IEEE for ICASSP, 492-495, held at Atlanta, May 1981).

MUSICUS, B.R. (1983). Iterative Algorithms for Optimal Signal Reconstruction and Parameter Identification Given Noisy and Incomplete Data, (Proceedings of the IEEE for ICASSP, 235-238, held at Boston, 1983).

PAPOULIS, A. (1981). Maximum Entropy and Spectral Estimation: A Review, IEEE Transactions of ASSP, 29, 1176-1186.

RICE, T. and SIEGEL, L. (1983). Parallel Processing for Computationally Intensive Speech Analysis Operations, (Proceedings of the IEEE for ICASSP, 471-474, held at Boston, April).

ROBINSON, E.A. and TREITEL, S. (1980). Geophysical Signal Analysis, Prentice-Hall, New Jersey, 1980, Chapter 16. (Reviews contents of 1978 paper, Spectral Estimation: Fact or Fiction, IEEE Transactions on Geoscience Electronics, 16).

ROBINSON, E.A. (1982). A Historical Perspective of Spectrum Estimation, Proceedings of the IEEE, 70, 885-907.

SCHMID, C.E. (1983). Design of IIR/FIR Filters Using a Frequency Domain Bootstrapping Technique and LPC Methods, IEEE Transactions for ASSP, 31, 999-1006.

SHORE, J., (1981). Minimum Cross-Entropy Spectral Analysis, IEEE Transactions for ASSP, 29, 230-237.

STEINHARDT, A. and ROBERTS, R.A. (1983). An Optimization Theoretic Framework for Spectral Estimation, (Proceedings of the IEEE for ICASSP, 1430-1433, held at Boston, April).

STOICA, P. (1977). A Test for Whiteness, IEEE Transactions on Automatic Control, 22, 992-993.

THOMSON, D.J. (1977). Spectrum Estimation Techniques for Characterization and Development of WT4 Waveguide-1, Bell System Technical Journal, 56, 1769-1815.

RECENT DEVELOPMENTS IN SEISMIC MIGRATION

Irshad R. Mufti
Superior Oil Co, Geoscience Lab, 12401 Westheimer, Houston, Texas 77077, USA

A seismic section can be regarded as a distorted image of the subsurface. As a consequence of propagation, the energy arriving at the surface is spread out over a broader domain than the physical extension of the corresponding reflectors. A numerical procedure designed to correct for such propagation effects is known as migration. Earlier methods of migration are based on simple concepts of ray theory. More advanced algorithms utilize the wave equation and involve continuation of seismic data down to the depth of the corresponding reflector. Downward continuation can be accomplished either by transforming the data into the frequency domain or by using the method of finite-differences. Although the latter algorithm is less accurate, it yields more satisfactory results on real data. Thus the attempt to answer old questions raises some new ones.

1. INTRODUCTION

Shortly after the introduction of the seismic reflection method, it was realized that the various events identified in a seismic section do not, in general, represent the true locations of the subsurface structures. The reflected energy reaching the surface propagates in the form of ever-expanding wavefronts. In a nonuniform medium, the variations in velocity and density tend to influence and modify the direction of propagation and are accompanied by mutual interference of wavefronts. In the zones of sudden changes along the geologic interfaces, such as faults, a portion of this energy undergoes diffraction. Consequently, the record of events constituting a seismic section represents a distorted image of the subsurface reflectors which has undergone a complicated process of focusing, defocusing, interference and diffraction. A numerical procedure aimed at correcting for these propagation effects is known as migration.

The first comprehensive and practically meaningful algorithm for doing migration was put forward by Hagedoorn (1954). He developed a method of repositioning the various events identified in a time section such that the transformed section would correctly represent the geometry of the subsurface reflectors. Hagedoorn's method utilizes the concepts of ray theory and yields reasonably good results in the absence of strongly dipping events or lateral changes in the subsurface velocities. Moreover, it is quite efficient and can be easily extended to three dimensions.

In 1971, Claerbout introduced a new method of migration which makes use of an approximate version of the acoustic wave equation (Claerbout, 1971; Claerbout and Johnson, 1971; Claerbout and Doherty, 1972). Such an approach is directly based on the distribution of energy of the seismic record, and it opens the possibility of taking into account all types of phenomena associated with the propagation of waves such as diffraction and interference. The importance of Claerbout's work was readily recognized; it followed a period of intense effort of research and development in the area of wave equation migration.

It is not our intention to review and evaluate every article written on this subject. Rather, we shall restrict ourselves to the most common migration algorithms in current use. We shall start with some of the basic ideas and assumptions underlying the subject of seismic migration.

2. SOME SIMPLIFYING ASSUMPTIONS

The propagation of seismic energy into the subsurface is a very complicated phenomenon which depends on a large number of factors. Any attempt to migrate the seismic data observed at the surface that would take into account all these factors becomes hopelessly involved. In order to make things tractable, migration is accomplished by introducing a number of simplifying assumptions mentioned below:

1. <u>The earth behaves like an acoustic medium</u>. This assumption is pretty close to reality in the case of offshore measurements where the topmost layer is water and the quantity measured is pressure. In the case of land measurements when the depth of the reflectors is larger than the geophone spread, and we measure only the vertical component of the particle velocity, the assumption of the acoustic medium is justified. If we go one step further and ignore variations in density, we can use a much simpler form of the wave equation for investigating the migration problem; it is given by

$$u_{xx} + u_{yy} + u_{zz} = c^{-2} u_{tt} \qquad (1)$$

where $u(x,y,z,t)$ = pressure (newton m^{-2}), $c(x,y,z)$ = velocity of the medium (ms^{-1}), and the subscripts denote partial derivatives.

2. <u>The real earth can be treated as a two-dimensional medium</u>. It is admittedly true that in some situations, this assumption can lead to serious errors and that includes migration (Hagedoorn, 1954). On the other hand, the acquisition of field data is usually done along linear profiles. Moreover, the assumption of a two-dimensional earth is almost invariably inherent at all stages of seismic data processing. Therefore, (1) can be replaced by

$$u_{xx} + u_{zz} = c^{-2} u_{tt} \qquad (2)$$

with x = distance along the measurement profile (m), and z = depth (m).

3. <u>Every subsurface reflector is made up of a set of point diffractors</u>. This means that the superposition of the wavefields generated by the individual point diffractors is approximately equivalent to the energy reflected by the corresponding subsurface horizon (Trorey, 1970).

4. <u>A CDP section is the product of upward propagating waves only</u>. In the process of stacking the observed seismic data, the multiples tend to cancel. We shall assume that a CDP section is completely free from multiples. Under these conditions, the seismic sources which are actually located at or near the surface of the ground can be replaced by the subsurface reflectors which explode simultaneously at an initial time t = 0 (Loewenthal et al, 1974). The energy emanating from the reflectors travels only in the upward direction <u>without</u> generating any multiples on its way to the surface. This simplification reduces the travel time by one-half. On a seismic section, it amounts to reducing the value of the time sampling interval by one-half. An alternative approach is to reduce the velocity of the medium everywhere by one-half and leave the two-way travel time unchanged. But since it necessitates a modification of the wave equation and all other relations dependent upon it, we shall prefer to leave the velocity unchanged.

The concept of exploding reflectors is a rather crude way of explaining things and it seems to have no counterpart in other disciplines; however, it works reasonably well for migrating the CDP sections.

3. SEISMIC IMAGES OF SIMPLE STRUCTURES

Before we go into the details of migration, it would be useful to know how some simple structures appear on a seismic section. The apparent shape of a given structure as seen on a seismic section will be referred to as its image.

The Point Diffractor

The simplest example of a seismic image is due to a point diffractor. But since a point diffractor can be used as a building block to construct a reflector of any arbitrary geometry, this example is also the most important.

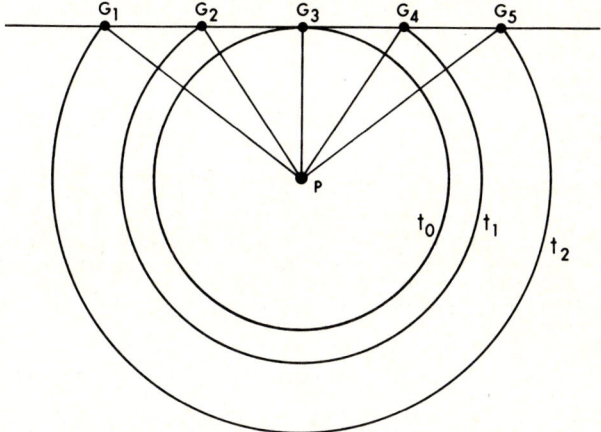

Figure 1. Various Stages of Evolution of an Expanding Wavefront Originating at the Point P.

Figure 1 shows a point diffractor P and a set of geophones $G_1,\ldots,G_5$ located along the surface. The energy emanating from P propagates upwards in the form of an expanding wavefront. Three different positions of the front corresponding to times t_0, t_1 and t_2 are indicated as partially drawn concentric circles. In accordance with the well known Fermat's principle, the transfer of energy from P to the various geophones will take place along a set of linear paths or rays as indicated in Figure 1. Let the location of the middle geophone G_3 correspond to the origin of coordinates ($x = 0$, $z = 0$) with positive z-axis pointed downward. If z denotes the depth of the point diffractor and x represents the horizontal position of any of the geophones, the time needed for the energy to travel from the point P to the various geophones is given by

$$t = c(x^2 + z^2)^{\frac{1}{2}} \qquad (3)$$

where c denotes the velocity of the medium. Relation (3) represents a hyperbola. Let us use this relation to compute the arrival times of the front to the various

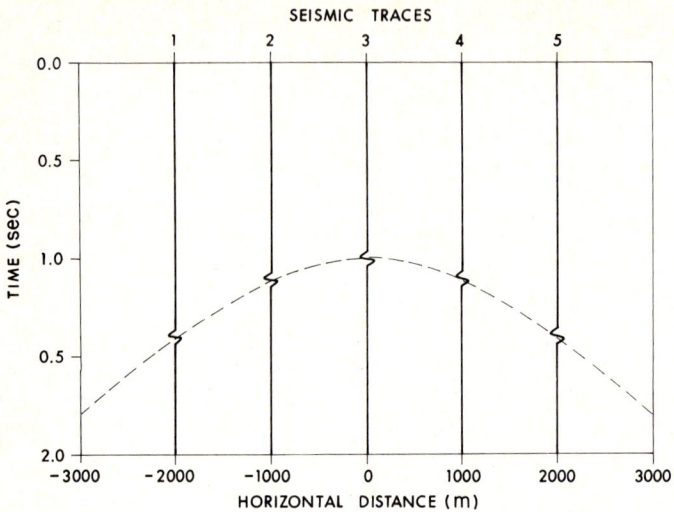

Figure 2. Seismic Image of a Point Diffractor Located at a Depth of 2000 Meters in a Medium of Velocity 4000 m/sec.

geophones by setting z = 2000 m and c = 4000 m/s. The results are shown in Figure 2; it represents the seismic image of the point diffractor.

It is obvious that as we increase the depth of the point diffractor, it will take the front longer to reach the surface, losing more and more of its curvature during its upward path. A flatter front implies that the arrival times to the various geophones will differ more slowly and the corresponding seismic image will appear as a broader hyperbola. By the same argument, the shallower the depth of the diffractor, the narrower its image. In the limiting case which corresponds to z = 0, the hyperbola will collapse to a point and there will be no difference left between the image and the shape of its object.

A Horizontal Linear Reflector

Figure 3 shows a horizontal reflector approximated by a set of equidistant point diffractors $P_1,\ldots,P_5$. The corresponding seismic image consists of five identical but laterally displaced hyperbolas. It is obvious that a proper representation of the reflector requires a much larger number of point diffractors. In that case, the various hyperbolas lose their identity due to destructive interference of the corresponding wavefronts and the resulting image is reduced to a continuous line or the envelope running tangentially along the apices of the hyperbolas. But note that as we approach either end of the reflector, the effect of interference gradually reduces. Consequently, the seismic image of the reflector is identical to the shape of the object except for a gradual tapering of energy along the hyperbolic arcs.

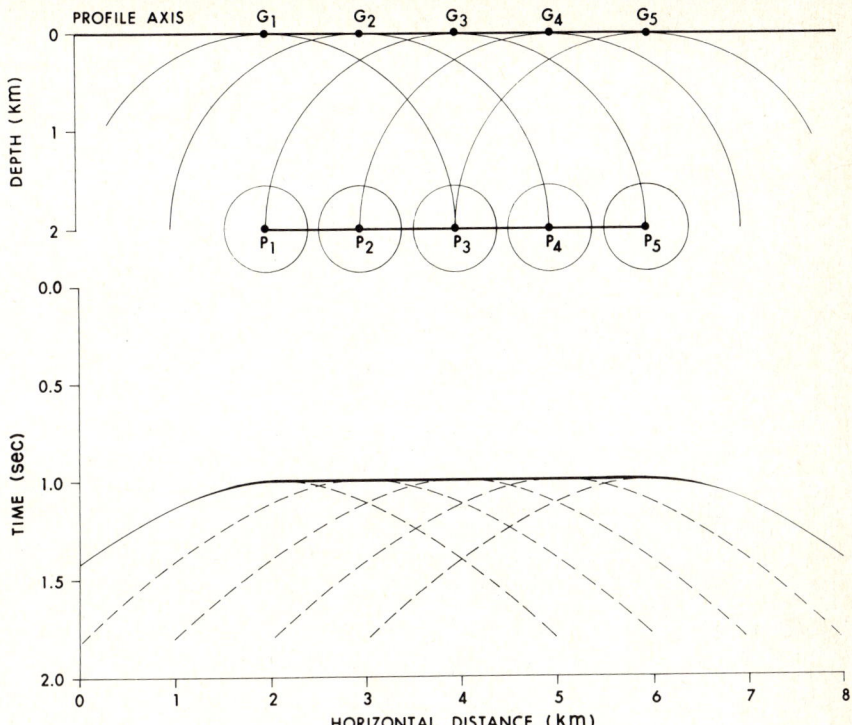

Figure 3. Approximate Representation of Horizontal Linear Reflector and the Corresponding Seismic Image.

A Linear Dipping Reflector

Figure 4 shows a linear reflector P_1P_2 dipping at an angle $\alpha = 30°$. We shall employ the above mentioned procedure for constructing its image. Since it is a linear reflector, it would suffice to compute only two hyperbolas associated with the end points P_1 and P_2. Note that the hyperbola associated with the point P_1 is broader than that for P_2 due to differences in depth. The line $P_1'P_2'$ which meets the two hyperbolas tangentially at P_1' and P_2' is the required envelope and represents the image of the reflector.

Since most of the energy leaving the reflector P_1P_2 will propagate to the surface along the direction normal to the line P_1P_2, we can also construct the image by simple considerations based on ray theory. Thus, the energy arriving at the geophone G_1' will be mostly due to the point source P_1 and will reach the surface at the time which corresponds to the two-way travel along the ray path between P_1 and G_1'. Now since on a time section, the traces are plotted at right angles to the direction of the profile, the event associated with P_1 will appear on the time trace of the geophone G_1' such that $P_1G_1' = P_1'G_1'$. A simple way to determine

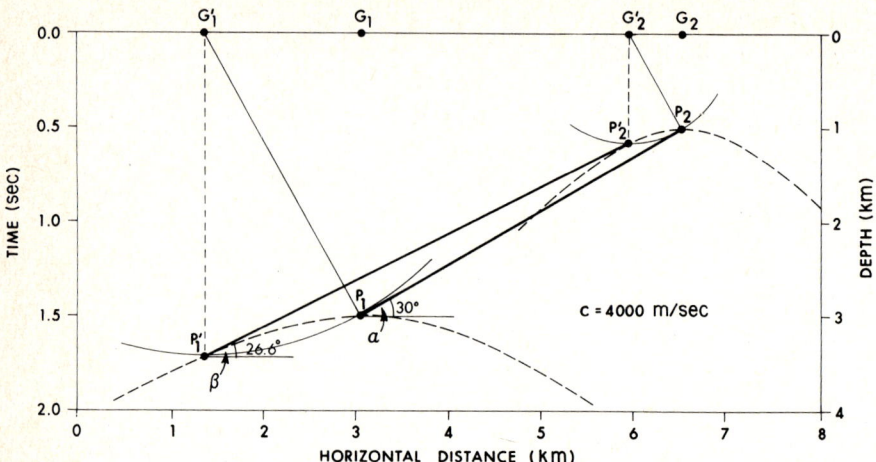

Figure 4. A Dipping Linear Reflector P_1P_2 and its Seismic Image $P_1'P_2'$ Obtained by Ray-Theory Considerations.

P_1' is to draw a circular arc through P_1 with G_1 as its center; P_1' corresponds to the intersection of this arc and the vertical line passing through G_1'. By a similar procedure, we can determine the image P_2' for the point P_2.

By using simple geometric considerations, it can be easily established that the dip angle β of the image is related to that of the reflector by

$$\sin \alpha = \tan \beta. \qquad (4)$$

Since every other structure can be visualized as made up of linear segments dipping at different angles, the importance of the case under consideration cannot be overemphasized.

Some Useful Conclusions

1. Every point diffractor on the object space (the geological depth model) gives rise to a hyperbola on the corresponding image space (the seismic section). Consequently, as long as we ignore multiples, a one-to-one correspondence exists between a given set of point diffractors and the corresponding set of hyperbolas. Except in the zones of sudden changes in reflectivity such as faults and termination of reflectors, the individual hyperbolas lose their identity due to destructive interference.

2. Every point diffractor belonging to the object space represents the apex of its hyperbola in the image space. The seismic image of such a diffractor obtained by simple considerations involving rays is one of the points belonging to this hyperbola. Thus in Figure 4, the image point P_1' belongs to the hyperbola associated with the point P_1.

3. For any given combination of velocity and depth, the corresponding hyperbola represents a curve of maximum convexity (Hagedoorn, 1954). This implies that the two hyperbolas shown in Figure 4 cannot be interchanged. Either of them is as convex as it could be for the respective depth. In the case of real data obtained in an area where reliable velocity control exists, one can identify the diffraction patterns by measuring the curve of the hyperbolic type features. Conversely, one can look for and use such features for estimating subsurface velocities (Taner and Koehler, 1969).

4. Formation of the seismic image involves a rolling down of the object points along their hyperbolas. Thus in Figure 4, the point P_1 rolls down to the point P'_1. The distance along which this rolling down takes place increases with depth.

5. The seismic image can be regarded as a more or less distorted version of the object, and usually appears longer than the object. This apparent increase in length amounts to a loss of lateral resolution. Moreover, its displacement is approximately along the downdip direction and takes place in such a manner that it tends to overlie the object. Consequently, the image of a salt dome is broader than the dome and fully encloses it from above. Similarly, by visualizing a synclinal structure as made up of linear reflectors dipping at different angles, it can be easily explained why its image gives rise to the familiar pattern of a buried focus.

We have familiarized ourselves with some of the basic features of the seismic image and how they differ from the corresponding features in the object space. We shall now discuss the various methods of transforming the seismic images into the true geometry of the corresponding geologic structures.

4. MIGRATION BY SUMMATION METHODS

Let us use our newly acquired skill to compute a synthetic seismic section showing a dipping reflector underlying a point diffractor. For convenience, we shall use the same reflector which is shown in Figure 4. The results are shown in Figure 5. The time length of the section is 2.5 seconds and it consists of 17 traces. The image of the linear reflector appears as a chain of events along the dashed line connecting the points P'_1 and P'_2. The point diffractor can be identified as a hyperbola with its apex on trace 5. For subsequent discussion, we have added a number of hyperbolas passing through the various points as well as some random noise on trace 12 at time 1.95 sec. and on trace 14 at time 1.75 sec.

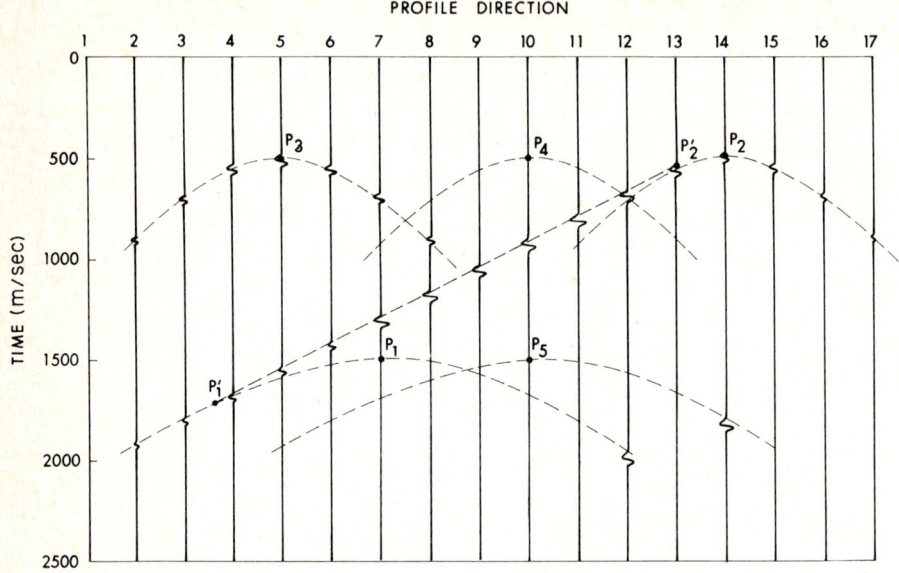

Figure 5. A Simple Example of a Synthetic Seismic Section.

The Simple Summation Method

A set of dashed hyperbolic curves shown in Figure 5 were obtained by setting c = 5000 m/s in (3). Let us start with the hyperbola passing through the point P_3. The chain of events which falls on this curve constitute the image of the point diffractor whose true location corresponds to the point P_3. Consequently, the process of summing the energy associated with these events and assigning the sum value to the point which corresponds to the apex of the hyperbola will undo the effect of diffraction and the image will collapse to the point diffractor. But this is exactly what migration is expected to accomplish. This procedure can also be used for migrating the linear reflector. We observe that the curve passing through P_1 touches the image of the linear reflector tangentially and passes through the events on traces 2, 3 and 4. Thus if we sum up the energy along this curve and assign it to the point P_1, we are able to migrate the left end of the reflector from its apparent position P_1' to the correct position P_1. A similar procedure enables us to migrate P_2' to P_2. In this way we are able to migrate the linear reflector from position $P_1'P_2'$ to the correct position P_1P_2. The ingenious method described above was introduced by Hagedoorn (1954). In the absence of the digital computers, he used a large number of precomputed curves for different velocities and depths. When it comes to dealing with the real size seismic sections, such a procedure is extremely laborious and is not worth the effort even at today's prices of oil and gas. Subsequent workers were able to automate Hagedoorn's procedure by making the assumption that every point on the seismic section can be treated as the apex of a hyperbola (Larner and Hornsby, 1972). In the absence of a reflector, the sum signal results in zero energy except for any residual energy associated with noise.

The summation process described above usually yields good results. However, since it is based on the ray theory approximation, it fails to take into account small changes in phase and can give rise to a variety of computational noise. As an example, the sum signal assigned to the point P_1 will also include the noise at the lower portion of trace 12, whereas point P_5 may indicate a weak diffracting region due to the noise present on trace 14. Similarly, point P_4 will derive its energy from the linear reflector below it. These are some simple examples of the so-called migration noise.

The Kirchhoff Summation Method

The idea of summing up the diffraction energy recorded at the surface and assigning it to the corresponding source of diffraction is equivalent to deriving the strength of the wavefield at the location of the source from the value of the field along a surface influenced by the source. This invokes the possibility of doing migration by using some of the well-known results from the theory of optics.

Consider a wavefield $u(x, y, z, t)$ defined as a function of space and time over a closed surface. Then the value of the field at any point inside the surface and at time $t = 0$ can be computed by means of an integral expression associated with the name of Kirchhoff. We shall not go into mathematical details which can be obtained elsewhere (see, e.g., Baker and Copson, 1950). It may be mentioned, however, that the results obtained by the Kirchhoff's integral remain unaffected by the shape of the closed surface which may be chosen to suit the problem under investigation. For our purposes, a good choice would be as shown in Figure 6; it consists of a hemispherical bowl whose flat surface coincides with the surface of the ground. A portion of the flat surface represents the zone of field measurements. Let us make the radius of the bowl infinitely large. In that case, the curved surface would be too far away to be influenced by any subsurface reflectors. Since the field measurements are confined within a finite area, we must introduce an additional assumption that the value of the field outside the surveyed area becomes negligibly small. It is not a good assumption, but we cannot avoid it. In any case, the integration over the closed surface is now confined to a finite area where the field measurements are available. Keeping in view the concept of the exploding reflector, if a subsurface point (x, y, z) belongs to a reflector, the field $u(x, y, z, t = 0)$ will be anomalously strong as

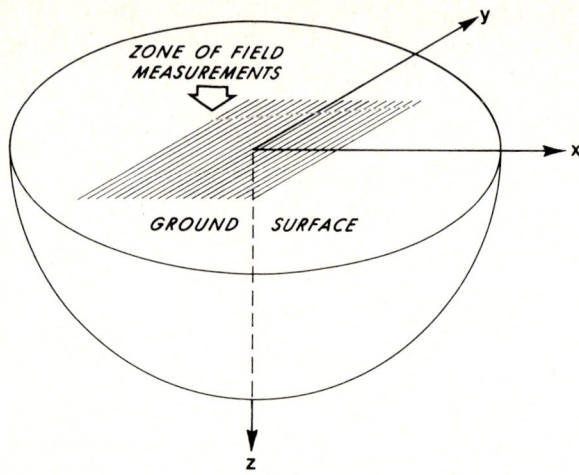

Figure 6. A Hemispherical Surface Over Which the Kirchhoff Summation is Carried Out.

compared to the points where no reflectors are present. This is an elegant approach to migration (French, 1975; Schneider, 1978) and in principle, it should lead to superior results as compared to the migration methods based on straight summation. In practice, however, the quality of migrated data obtained by this method is often inferior to that obtainable by other schemes in current use (Johnson and French, 1982). At least some of the problems associated with this method may be the direct consequence of too much compromise between the theory and the practical limitations. The theory implies that the data to be migrated are based on a 3-D seismic survey. In actual practice, the field measurements are usually obtained along a profile rather than over an area. This necessitates further compromise with the theory. Sometime in the future when 3-D acquisition of field data has become a common practice, this approximation could be eliminated.

5. DOWNWARD CONTINUATION OF WAVEFIELDS

A Conceptual Experiment

Figure 7(a) shows three point diffractors located in a uniform medium. The depth of the diffractors is indicated as z_1, z_2 and z_3. It would be convenient to choose these depths as multiples of z_1. Then $z_2 = 2z_1$ and $z_3 = 3z_1$. Figure 7(b) is a schematic representation of the corresponding seismic section as one would obtain from field measurements along the surface of the ground ($z = 0$). The section consists of three hyperbolic shaped images of the diffractors. The time coordinates of the apices of these hyperbolas are given by $t_1 = z_1/c$, $t_2 = z_2/c$

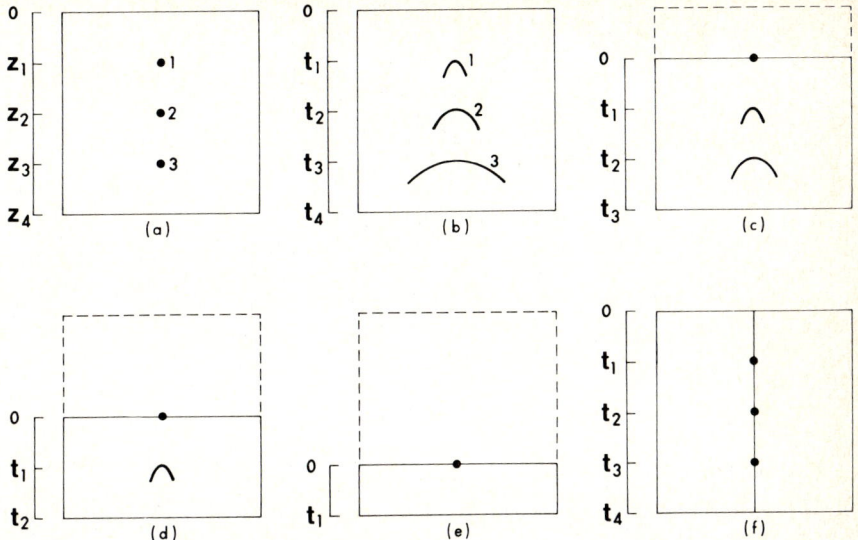

Figure 7. The Results of a Conceptual Experiment Involving the Recording of Seismic Data at Different Planes.

and $t_3 = z_3/c$, where c is the velocity of the medium. As expected, the image due to the shallowest source is the narrowest of the three hyperbolas. We also recall from a previous discussion that in the limiting case when the depth of the point diffractor becomes zero, its image collapses to a point. Bearing that in mind, let us carry out a conceptual experiment in which we record the upward propagating waves at different subsurface levels. We start with the level $z = z_1$. The results are shown in Figure 7(c). Since the plane of measurements now corresponds to the depth of diffractor 1, its image is collapsed to a point. Moreover, due to change in the level of the recording plane, the hyperbolas 1 and 2 of Figure 7(b) are now associated with diffractors 2 and 3, respectively. For the same reason, there is also a reduction in the arrival times as indicated by a shift in the time scale. Figure 7(d) shows the results of a similar experiment for the plane of measurement $z = z_2$. Since we record only the upward propagating waves, the influence of diffractor 1 is absent in this case. Moreover, due to change of depth, the image of diffractor 2 is now collapsed to a point and that of diffractor 3 is identical to that of diffractor 1 in Figure 7(b). In Figure 7(e) which corresponds to the plane of measurement $z = z_3$, the only event recorded is a point image due to diffractor 3. Figure 7(f), shows the field of the three diffractors measured at their respective depth planes and plotted at times which correspond to the apices of the hyperbolas of Figure 7(b). These results are exactly the same as one would have obtained by migrating the seismic section shown in Figure 7(b). Therefore, the conceptual experiment is a means of migrating a given section.

In real life, the measurements of the wavefield must be restricted along the surface of the ground. Therefore, the conceptual experiment described by us is not a physically realizable process. It is possible, however, to develop mathematical procedures which enable one to transform a wavefield at a given level to its configuration corresponding to another level. Such a transformation is usually referred to as the continuation of a wavefield.

We are particularly interested in the continuation of the wavefield in the downward direction. When the data obtained along the surface of the ground is continued down to the various subsurface reflectors, the image field is transformed into the corresponding object field which shows the true geometry of the reflectors.

Migration Methods Based on Downward Continuation

Let us assume that reflectors exist everywhere in the subsurface. Those locations which do not give rise to reflections represent reflectors of zero strength (reflectivity). This amounts to saying that every time instant along each of the traces of the unmigrated section corresponds to the apex of a hyperbola. Therefore, we can perform downward continuation of the section in uniform successive steps. Ideally, the size of each step should be equal to the time sampling interval. In practice, however, one can obtain virtually identical results by using larger steps. Further operational details can be better understood with the help of a concrete example.

The box shown in Figure 8 represents a finite wavefield domain. The vertical scale $j = 0, 1, ..., 10$ denotes different depths. In particular, $j = 0$ represents the surface of the ground. The top face of the box shows the unmigrated section. The position of the various traces is determined by the index i ($i = 0, 1, ..., I$). Let us choose the length of the section to be 2 seconds and divide this length into 10 equal intervals. Each of these intervals will be referred to as the migration step. The various steps are numbered as $\tau = 0, 1, ..., 10$. Suppose that $\tau = 2$ corresponds to the two-way travel time of 400 msec on the section. Let the velocity of the medium be 5000 meters/sec. Remembering that in migration we consider only one-way travel time, the continuation of the wavefield by a migration step of 200 msec corresponds to a change in depth of 500 meters. Thus the index j represents downward continuation intervals of 500 meters.

When the section is continued from the plane $j = 0$ to the plane $j = 1$, the output section corresponds to another horizontal surface possessing the same size as the

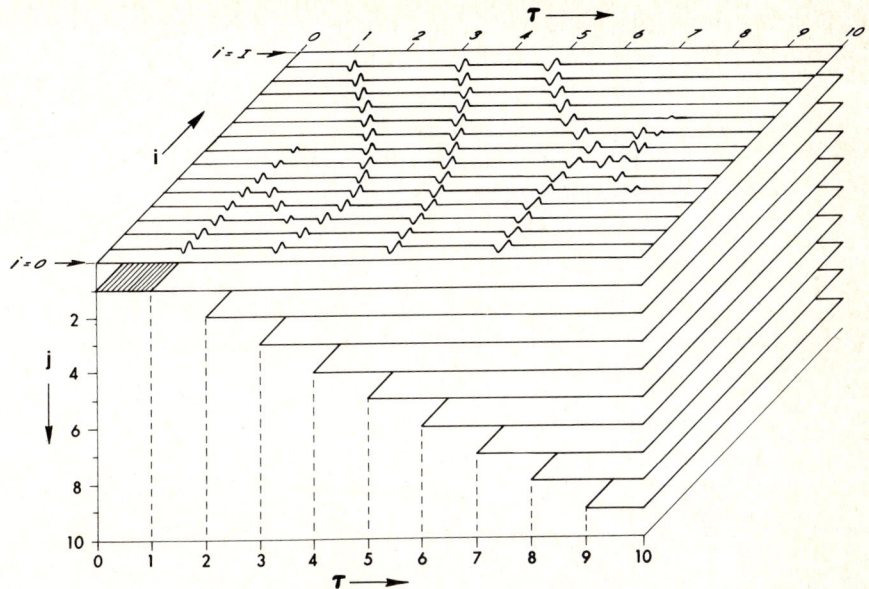

Figure 8. Successive Steps Involved in the Migration of a Seismic Section by Downward Continuation.

input section, and lying immediately below it. In Figure 8, it is shown as partially visible. The earlier portion of the output section ($\tau = 0$ to $\tau = 1$) is fully migrated and is indicated as a shaded strip of data. The rest of the section is said to be partially migrated. We use this partial section ($\tau = 1$ to $\tau = 10$) as the new input and repeat the continuation process. The output section corresponds to the plane $j = 2$ and ranges from $\tau = 1$ to $\tau = 10$, including the fully migrated portion from $\tau = 1$ to $\tau = 2$. Continuing in this fashion, it can be easily seen that at the completion of the tenth migration step, the last strip of the section ($\tau = 9$ to $\tau = 10$) would be fully migrated. If we collect all the fully migrated strips and place them together in their original sequence, we get the migrated section. The numerical method employed for this kind of migration is known as the finite-difference method.

It may be pointed out that at the end of each step, the input section is never needed again and may be replaced by the output section. This is exactly what is done in practice. Thus at the completion of step 1, the original section is replaced entirely by the output section although it corresponds to the lower level $j = 1$. At the end of the second step, the partial section ($\tau = 2$ to $\tau = 10$) is continued and placed back into its original location and so on.

Figure 8 is an oversimplified picture of the migration process. In actual cases, the migration steps are much smaller. In the limiting case when the step size becomes very small, the strips of data which become fully migrated during the various migration steps fall along the diagonal plane ABGH shown in Figure 9. As an example, the set of data along the line P'Q' of the unmigrated section is projected along the line PQ which belongs to the plane ABGH. After that, the migrated output is stored back along the line P'Q'. Thus the migration process can be regarded as a means of projecting a set of data from the horizontal plane ABCD onto the diagonal plane ABGH and storing the output back into the original plane.

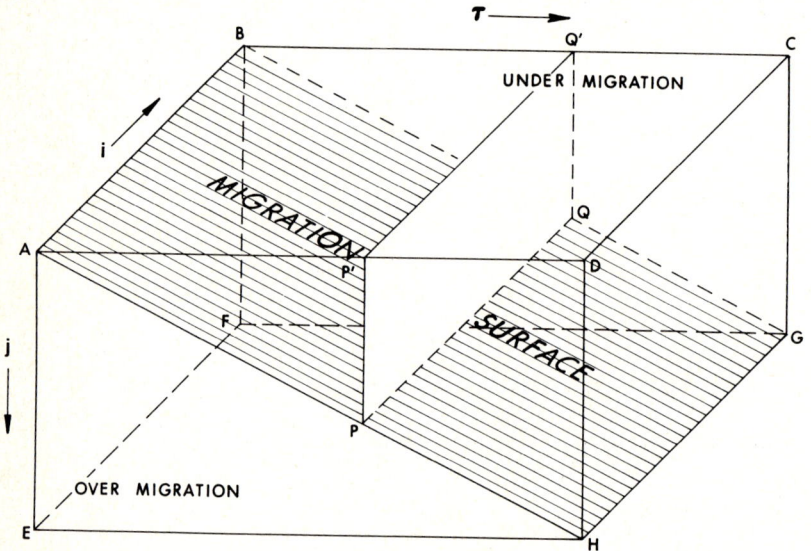

Figure 9. Orientation of the Migration Surface in a Constant Velocity Medium.

Figure 9 suggests that if we continue the wavefield by using higher (or lower) than the correct velocity, the migrated output will lie below (or above) the diagonal plane ABGH. Thus the diagonal plane divides the rectangular box into two different zones. The migrated data which belongs to the wedgeshaped zone ABCDGH is said to be undermigrated; similarly, the data belonging to the zone ABEFGH will be overmigrated. Unfortunately, the velocity is an unknown parameter which must be estimated in most cases. The better this estimate, the closer will be the migrated data to the diagonal plane.

Hitherto, we have assumed that the velocity of the medium remains uniform. In the case of variable velocity, the migration surface is no longer a plane; instead, it is a complex surface characterized by arbitrary topography depending upon the variations in velocity. Let us consider a somewhat simpler situation in which the velocity varies only with depth. With respect to the unmigrated section, it means that the velocity may vary along the time axis but it must remain the same for all traces. Under these conditions, the migration surface consists of a number of strips of uniform width joined together as shown in Figure 10. The changes in velocity occur at times τ_1, τ_2, τ_3, τ_5 and τ_6. There is no change in velocity at time τ_4, but this value of time corresponds to the intersection of the migration surface and the plane ABGH along the line LM. It implies that one would have obtained the same migration output at time τ_4 either by proceeding along the plane ABGH or by taking into account all the variations in velocity over the time interval $t = 0$ to $t = \tau_4$. This interesting observation suggests the possibility of using an effectively equivalent single value of velocity to obtain virtually identical results as one would obtain by taking into account the actual velocities (Dix, 1955). This powerful method works only when there are no variations of velocity in the horizontal direction.

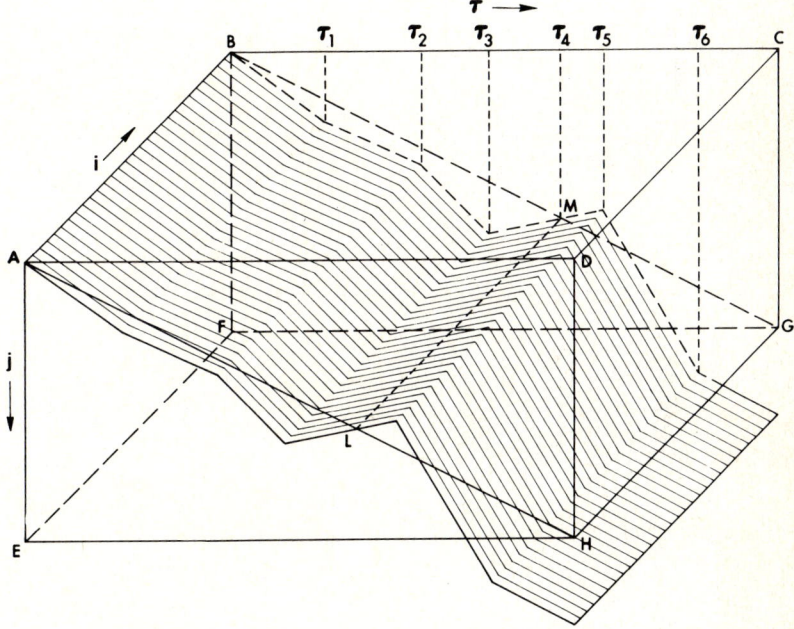

Figure 10. Orientation of the Migration Surface when the Velocity of the Medium Varies with Depth Only.

A Computational Shortcut

Referring back to Figure 8, let us consider the strip of data (from $\tau = 2$ to $\tau = 3$) which becomes fully migrated during the third migration step. The original data belonging to this time interval undergoes downward continuation in 3 steps described below.

Step No.	Continuation Levels	Output Status
1	j = 0 to j = 1	Partially migrated
2	j = 1 to j = 2	Partially migrated
3	j = 2 to j = 3	Fully migrated

In general one needs n migration steps for fully migrating the data belonging to the interval $(n-1)\tau$ to τn. When n is very large as is usually the case, this amounts to a huge amount of numerical effort. A far more efficient procedure would be to accomplish full migration in a single step by using a variable step size. Thus, if the step size for migrating the first strip of data ($\tau = 0$ to $\tau = 1$) is $\Delta\tau$, one could use a single step of size $n\Delta\tau$ for the strip $(n-1)\tau$ to $n\tau$. This can be accomplished provided the data to be migrated have been transformed to the frequency domain. This explains why the migration methods which work in the frequency domain (Stolt, 1978; Gazdag, 1978) are extremely efficient; unfortunately they do not permit variations in velocity.

This completes our introductory discussion on the migration methods involving the continuation of wavefields. We shall now describe in some detail two of the most popular methods of doing migration.

6. MIGRATION IN THE FREQUENCY DOMAIN

Some Basic Concepts

One of the basic theorems about Fourier transforms is known as the shift theorem (see, e.g., Bracewell, 1965 p.104); it is stated below:

If $F\{g(t)\} = G(\omega)$

then $F\{g(t-a)\} = e^{-ia} G(\omega)$. (5)

In (5), $g(t)$ is an arbitrary function of the independent variable t and $G(\omega)$ denotes its Fourier transform. The constant a may assume any value positive or negative.

Another useful relation of interest to us is the derivative theorem:

If $F\{g(t)\} = G(\omega)$

then $F\{\partial g/\partial t\} = i\omega G(\omega)$ (6)

Since (2) involves the derivatives of the wavefield, a direct application of (6) enables one to derive the frequency domain version of the wave equation; it is given by

$$k_x^2 + k_z^2 = \omega^2/c^2 \qquad (7)$$

where k_x = wave number in the x-direction, k_z = wave number in the z-direction, and ω = temporal frequency. Equation (7) is commonly known as the dispersion relation. An amazing similarity between (3) and (7) suggests that the two relations must be very closely related--and they are.

Downward Continuation in the Frequency Domain

We have already seen that the continuation of a wavefield recorded at the surface (z = 0) to a depth z is accompanied by a reduction of the one-way travel time t by an amount t = z/c, where c is the velocity of the medium. We shall try to accomplish this transformation in the frequency domain.
Let us take the Fourier transform of (2) with respect to x and t. We get

$$(ik_x)^2 U + \frac{d^2U}{dz^2} = (\frac{i\omega}{c})^2 U. \tag{8}$$

In (8), $U(k_x, z, \omega)$ is the Fourier transform of the wavefield $u(x, z, t)$ taken with respect to x and t. In view of (7), this relation can be expressed as

$$\frac{d^2U}{dz^2} = (ik_z)^2 U. \tag{9}$$

Equation (9) is an ordinary differential equation. It can be readily verified by direct substitution that (9) is satisfied by

$$U(k_x, z, \omega) = U_0 \exp(ik_z z) \tag{10}$$

where U_0 denotes the value of U at z = 0. Let us choose U_0 as the spectrum of the unmigrated section. Then the inverse transform of U_0 will be given by

$$u(x, z=0, t) = \sum_{k_x} \sum_{\omega} U(k_x, z = 0, \omega) \exp[i(k_x x + \omega t)]. \tag{11}$$

Consequently, the inverse transform of (10) can be written as

$$u(x, z, t) = \sum_{k_x} \sum_{\omega} U(k_x, z = 0, \omega) \exp[i(k_x x + k_z z + \omega t)]. \tag{12}$$

A direct application of the shift theorem (Relation (5)) to (12) leads to

$$u(x, z, t - \Delta t) = \sum_{k_x} \sum_{\omega} U(k_x, z = 0, \omega) \exp[i(k_x x + k_z z) + i\omega (t - \Delta t)].$$

If the depth z corresponds to the depth of the reflector, the one-way travel time will vanish. This corresponds to a time shift of $\Delta t = t$ and we get

$$u(x, z, t = 0) = \sum_{k_x} \sum_{\omega} U(k_x, z = 0, \omega) \exp[i(k_x x + k_z z)]. \tag{13}$$

The exponential term appearing in (13) depends on the discrete values of k_z. This is not compatible with the summation which is carried over a set of discrete values of ω and k_x. In order to eliminate this discrepancy, let us rewrite (7) in the form

$$\omega = c(k_x^2 + k_z^2)^{\frac{1}{2}}.$$

Therefore

$$\Delta\omega = c k_z (k_x^2 + k_z^2)^{-\frac{1}{2}} \Delta k_z.$$

Consequently, (13) can be replaced by

$$u(x, z, t = 0) = c \sum_{k_x} \sum_{k_z} U(k_x, z = 0, k_z) k_z (k_x^2 + k_z^2)^{-\frac{1}{2}} \exp [i(k_x^2 + k_z^2)] \quad (14)$$

Relation (14) is identical to that derived by Stolt (1978).

The evaluation of $U(k_x, z = 0, k_z)$ appearing in (14) involves interpolation of the unmigrated data in the frequency domain. This problem of data resampling can be avoided by eliminating the k_z-term appearing in (13). In view of (7), k_z is given by

$$k_z = \frac{\omega}{c} [1 - (k_x c/\omega)^2]^{\frac{1}{2}}. \quad (15)$$

Substituting (15) into (13) leads to

$$u(x, z, 0) = \sum_{k_x} \sum_{\omega} U(k_x, 0, \omega) \exp [i \{k_x x + \frac{\omega}{c} [1 - (k_x c/\omega)^2]^{\frac{1}{2}} z \}] \quad (16)$$

Relation (16) is identical to that derived by Gazdag (1980).

The Scaling Factor

As long as the medium remains uniform, we can set $z/c = t$. Consequently, the exponential term in (16) can be expressed as

$$\exp [i \{k_x x + \omega [1 - (k_x c/\omega)^2]^{\frac{1}{2}} t \}]. \quad (17)$$

Let us compare (17) with the exponential term appearing in the expression for the unmigrated section (Eq. 11). We note that the coefficient of k_x remains unchanged, whereas the temporal frequency ω is replaced by a lower frequency

$$\omega' = \omega [1 - (k_x c/\omega)^2]^{\frac{1}{2}}.$$

The accompanying change of scale can be expressed as

$$\omega'/\omega = \frac{c}{\omega} (\frac{\omega^2}{c^2} - k_x^2)^{\frac{1}{2}} = c k_z / \omega.$$

Therefore

$$\omega'/\omega = k_z (k_x^2 + k_z^2)^{-\frac{1}{2}}.$$

Simple considerations related to the mapping of seismic data in the frequency domain (Chun and Jacewitz, 1979) indicate that the scaling factor can also be expressed as

$$\omega'/\omega = \cos \theta = k_z (k_x^2 + k_z^2)^{-\frac{1}{2}} \quad (18)$$

where θ denotes the dip angle of the reflecting surface. Relation (18) may be written as

$$\omega'/\omega = [1 + a(\tan \theta)^2]^{-\frac{1}{2}}$$

with $a = 1$. Even when we set $a \neq 1$, we note that $\lim_{\theta \to 0} \frac{\omega'}{\omega} = 1$ for all finite values of a.

Suppose we deliberately try to distort the amplitudes of the migrated section by setting $a \neq 1$. Thus if we set $a = 10$, the amplitudes of the more or less flat events will remain virtually unaffected but the steeply dipping events will be dimmed out. On the other hand, if we subtract the migrated data obtained in this manner from the corresponding migrated data obtained by using a = 1, the resulting section will show only dipping events and all the flat events will be dimmed out. Such cosmetic measures are sometimes used as an aid to interpretation.

7. FINITE-DIFFERENCE APPROACH TO MIGRATION

In 1971, Claerbout and Johnson introduced a finite-difference method for doing migration. In order to bypass some numerical difficulties as well as to increase computational efficiency, they replaced (2) by a simplified version of the one-way wave equation; it can be expressed as

$$u_{xx} = -\frac{2}{c} u_{tz} \tag{19}$$

For further details about the derivation of this equation, the reader is referred to Claerbout and Johnson (1971) and Claerbout (1976).

In order to derive a difference relation based on (19), let us discretize the wavefield $u(x, z, t)$ by introducing

$$\begin{aligned} x &= i \Delta x & i &= 0, 1, \ldots, I \\ z &= j \Delta z & j &= 0, 1, \ldots, J \\ t &= n \Delta t & n &= 0, 1, \ldots, N \end{aligned} \tag{20}$$

where Δx = trace interval (m), Δz = continuation step (m), Δt = time sampling interval (s), $I + 1$ = number of traces in a given section, and $N + 1$ = number of time samples in each trace.

We shall derive a difference relation for a fictitious grid point $(i, j+\frac{1}{2}, n+\frac{1}{2})$. The value of u_{xx} for the point $(i, j + \frac{1}{2}, n + \frac{1}{2})$ can be expressed as

$$(u_{xx})_{i,j+\frac{1}{2},n+\frac{1}{2}} = \tfrac{1}{4}[(u_{xx})_{i,j,n} + (u_{xx})_{i,j+1,n} + (u_{xx})_{i,j,n+1} + (u_{xx})_{i,j+1,n+1}] \tag{21}$$

where

$$(u_{xx})_{i,j,n} = \frac{1}{(\Delta x)^2} [u_{i-1,j,n} - 2u_{i,j,n} + u_{i+1,j,n}], \text{ etc.}$$

Similarly,

$$(u_{tz})_{i,j+\frac{1}{2},n+\frac{1}{2}} = \frac{1}{\Delta z} [(u_t)_{i,j+1,n+\frac{1}{2}} - (u_t)_{i,j,n+\frac{1}{2}}]$$

where

$$(u_t)_{i,j,n+\frac{1}{2}} = \frac{1}{\Delta z} (u_{i,j,n+1} - u_{i,j,n}), \text{ etc.}$$

Therefore

$$(u_{tz})_{i,j+\frac{1}{2},n+\frac{1}{2}} = \frac{1}{\Delta t\, \Delta z} [u_{i,j+1,n+1} - u_{i,j+1,n} - u_{i,j,n+1} + u_{i,j,n}]. \tag{22}$$

Substitution of (21) and (22) into (19) and rearrangement of terms leads to

$$u_{i-1,j+1,n+1} - (2+\alpha) u_{i,j+1,n+1} + u_{i+1,j+1,n+1} = b_i \qquad i = 2, \ldots, (I-1) \quad (23)$$

where

$$\alpha = -8(\Delta x)^2/(c\Delta t \, \Delta z) \tag{24}$$

and

$$b_i = -u_{i-1,j,n} + (2+\alpha)u_{i,j,n} - u_{i+1,j,n} - u_{i-1,j+1,n} + (2-\alpha)u_{i,j+1,n} - u_{i+1,j+1,n}$$
$$- u_{i-1,j,n+1} + (2+\alpha)u_{i,j,n+1} - u_{i+1,j,n+1} \tag{25}$$

Let us introduce the boundary conditions:

$$\left. \begin{array}{l} u_{0,j,n} = 0 \\ u_{I,j,n} = 0 \end{array} \right\} \quad \text{for all values of } j \text{ and } n. \tag{26}$$

The system of equations (23) and (26) constitutes a tridiagonal implicit system; it enables one to compute the wavefield at level (j+1) from the field at level j. In the migration process, one uses the unmigrated data $u_{i,o,n}$ (n=0, 1, ..., N) as the starting point and computes $u_{i,1,n}$ for various values of n, the computations being carried out in the increasing order of n from n = 1 to n = N. In general, the set of data $u_{i,j,n}$ for the level j are computed over the range n = j to n = N from the knowledge of the data $u_{i,j-1,n}$ from n = j-1 to n = N. Further details can be understood with the help of Figure 8.

The method presented by us usually breaks down when the dip of the subsurface reflectors exceeds about 15 degrees. Attempts have been made to extend the usefulness of this method to steeper dips (Berkhout, 1981), but they all lead to enormous increase in computational effort and the outcome is not worth the effort. More recent investigations carried out by Robinson (1983) indicate that there may be some fundamental theoretical limitations underlying this method. We shall attempt to determine the usefulness of both methods of migration by some numerical tests.

8. NUMERICAL EXPERIMENTS ON MIGRATION ALGORITHMS

A series of migration experiments were carried out on synthetic seismic sections by using the frequency-domain and the finite-difference methods. We deliberately chose the synthetic data for this purpose because the migrated output as one would expect from the ideal migration program is completely known. Moreover, the migration process can be tested under controlled environments by keeping some of the parameters fixed while allowing others to vary.

A two-dimensional earth model (Figure 11) consisting of seven different layers was used for setting up two bench mark seismic models. In the first model, the velocity of the various layers was held constant but the density was allowed to vary. For the second model, the roles of density and velocity were reversed.

Further details about the two models which will be referred to as models 1 and 2 are given in Tables 1 and 2.

The synthetic sections were computed by using a modeling program which is based on a simplified theory of diffraction (Trorey, 1970). The output generated by this program is in the form of spike data which may be scaled and convolved with a wavelet of arbitrary form.

TABLE 1. IMPEDANCE PARAMETERS OF MODEL 1

Layer No.	Density (gm/cm^3)	Velocity (m/s)
1	2.00	
2	2.17	
3	2.35	2500;constant
4	2.54	for all layers
5	2.75	
6	3.00	
7	3.23	

TABLE 2. IMPEDANCE PARAMETERS OF MODEL 2

Layer No.	Velocity (m/s)	Density (gm/cm^3)
1	1600	Arbitrary; constant
2	1950	for all layers
3	2390	
4	2920	
5	3570	
6	4360	
7	5330	

Tests on Model 1 (Constant Velocity, Variable Density)

A synthetic section based on the parameters given in Table 1 is shown in Figure 12. It corresponds to the trace interval $\Delta x = 25$ m and the time sampling interval $\Delta t = 2$ msec. It consists of 285 traces and includes reflections from all the interfaces shown in Figure 11. The migrated version of this section obtained by using the frequency-domain method is shown in Figure 13. A comparison of Figures 11 and 13 indicates that as long as the velocity of the medium remains constant, this method yields excellent results even in the presence of very steep dips.

It is commonly reported that the finite-difference algorithm breaks down when the dips of the reflectors exceed 15°. Therefore, for the purpose of testing this method, we used a curtailed version of the input section by retaining reflections from all the interfaces except the lowest horizontal interface and the steeply dipping portion of the interface immediately above it. Figure 14 shows the migrated section which was obtained by using a migration step of 80 msec. Since $\tau = n$ implies a time interval of $n\Delta t$, this migration step corresponds to $\tau = 40$. In an attempt to improve the results, we repeated this test by reducing the step size to 40 msec. To our surprise, we obtained inferior results (Figure 15). The same thing happened when we increased the step size to 100 msec. These results led us to suspect that the effectiveness of a method widely known to work only for dips not exceeding 15°, might be strongly dependent on the frequency content of the input section. The input section used by us was obtained by convolving the original spike section with a Ricker wavelet of 40 msec duration. In a real seismic section, the events originating at greater depths are accompanied by a successive loss of higher frequencies. In view of this fact, we decided to prepare a new input section by convolving the spike section with a time varying filter. The details of this filter are given in Table 3.

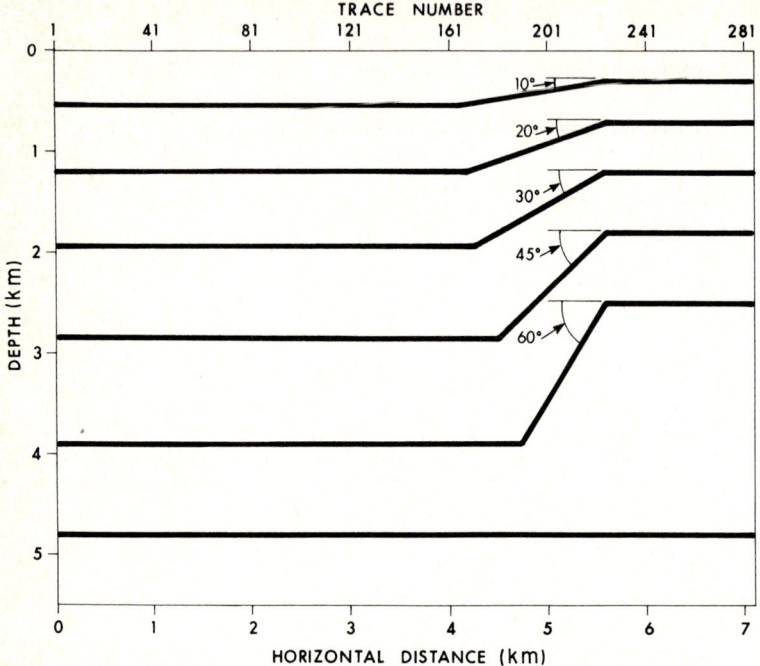

Figure 11. Two-Dimensional Depth Model Used for Computing Synthetic Sesimic Sections.

TABLE 3. SPECIFICATIONS OF THE TIME VARYING FILTER

Ricker Wavelet Duration (msec)	Range of Application (msec)
25	0-500
50	800-1100
75	1400-2300
100	2600-3700

Over the unspecified portion of the section such as 500-800 msec, the duration of the wavelet was obtained by interpolation. In essence, we convolved the spike section with an expanding Ricker wavelet. The section obtained in this manner was migrated by using the step size of 80 msec. The results obtained are excellent (Figure 16) for all dips up to 30°, whereas the 45°- interface is undermigrated. Moreover, we are able to preserve the higher frequency content of the shallower events.

Tests on Model 2 (Variable Velocity, Constant Density)

A CDP spike section based on the earth model shown in Figure 11 and the impedance parameters given in Table 2 was convolved with a 40 msec Ricker wavelet. The resulting section is shown in Figure 17. It consists of 285 traces and is based on Δx = 25 m and Δt = 2 msec.

It was mentioned earlier that the frequency-domain method is only valid when the velocity of the medium is constant. In real-life situations, variations in velocity always exist. In order to account for that, the usual procedure is to

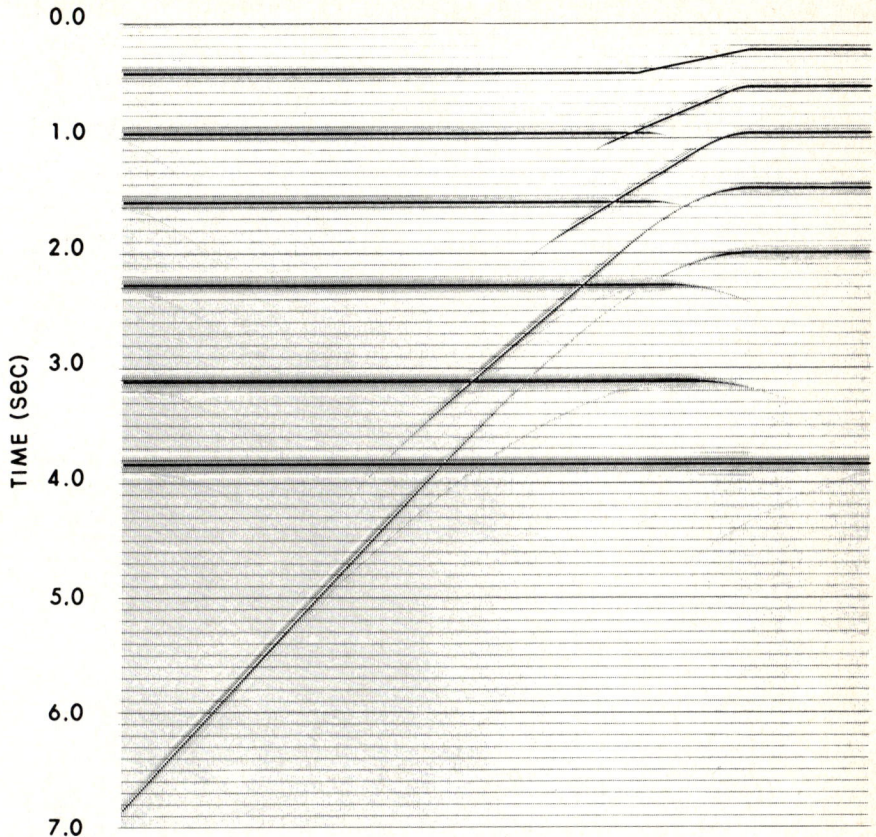

Figure 12. Synthetic CDP Section for Model 1.

"stretch out" the unmigrated section in accordance with the variations in velocity used for migrating the data. The migration velocity functions used by us were derived by taking the average values of the interval velocities over different parts of the section including those where the interfaces are dipping. The migrated section obtained in this manner is shown in Figure 18. The results are reasonably good as long as the dips do not exceed $30°$. More steeply dipping reflectors which were originally linear appear bent and misplaced. We repeated this experiment by using the finite-difference method and the same velocity functions. The migrated output is shown in Figures 19. The results are reasonably good up to a dip of $20°$. The reflector dipping at $30°$ is also correctly migrated, but the corresponding events have undergone a significant loss of energy. The events associated with steeper dips are almost obliterated.

9. CONCLUDING REMARKS

We have presented a concise introduction to the subject of migration with special emphasis on those methods which are now considered routine tools of seismic data processing. In order to keep this material as simple as possible, it became necessary to bypass a number of excellent contributions made by various authors.

Figure 13. Frequency Domain Migration of the Data Shown in Figure 12.

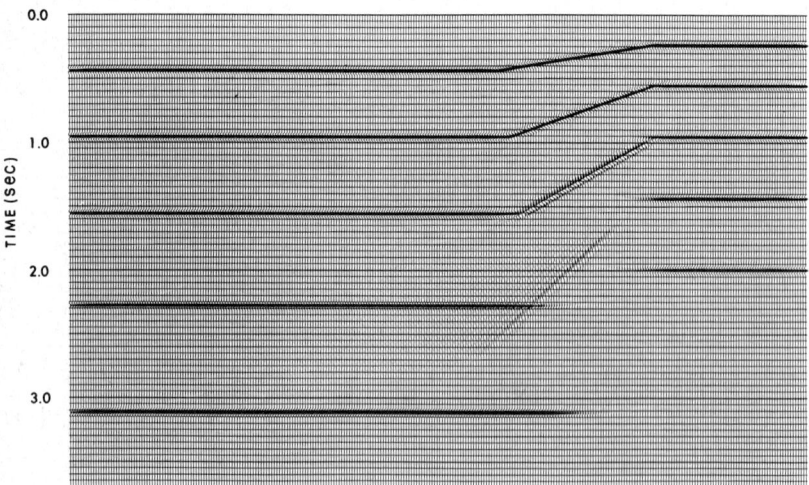

Figure 14. Finite-Difference Migration of the Data Shown in Figure 12, Obtained by Using $\tau = 40$.

The unmigrated section is a function of space and time. In the process of migration, one tries to determine the value of this function which is shifted both in space and time. During our discussion on the frequency domain method, we exploited this fact by using the shift theorem of Fourier transforms. Not only did it render the theoretical treatment much simpler but led to an interesting conclusion that the expressions derived by Stolt (1978) and Gazdag (1980) are almost equivalent. Stolt's method requires resampling of the input data in the

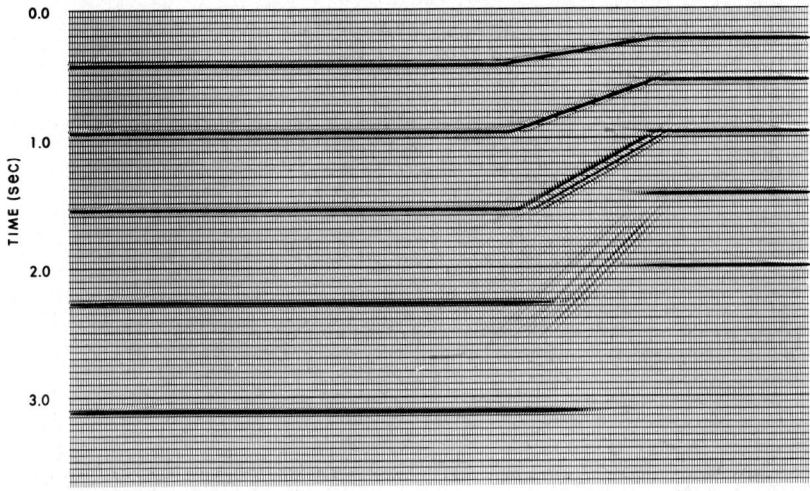

Figure 15. Finite-Difference Migration of the Data Shown in Figure 12, Obtained by Using $\tau = 20$.

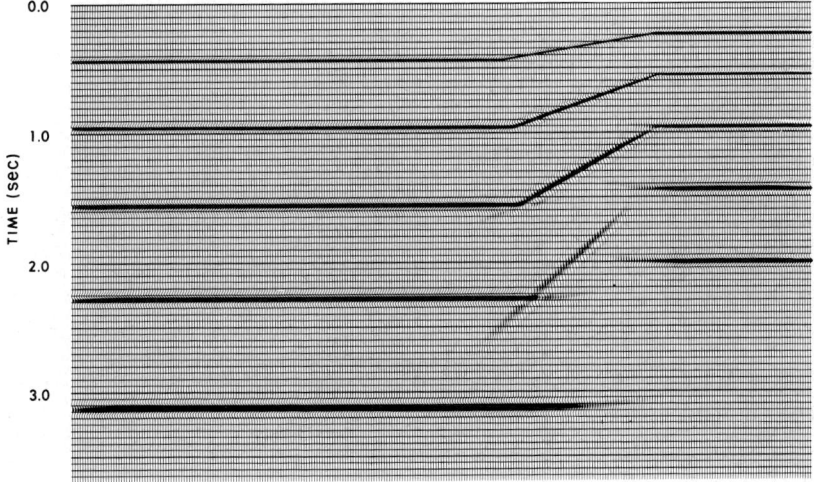

Figure 16. Finite-Difference Migration of the Data of Model 1, Convolved with a Time-Varying Filter.

frequency domain, whereas for the expression derived by Gazdag, it is not necessary. For reasons mentioned above, our treatment on finite-difference migration is restricted to the 15 degree method introduced by Claerbout (1976).

We carried out a series of tests to migrate two synthetic sections. One of these sections was based on a medium characterized by constant velocity. For computing the second section, the velocity of the medium was allowed to vary both vertically and horizontally.

218   I.R. Mufti

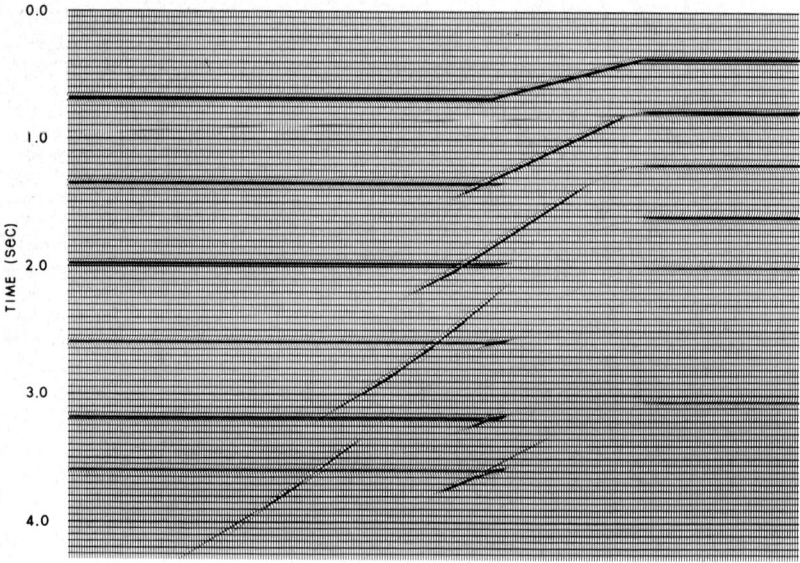

Figure 17. Synthetic CDP Section for Model 2, Obtained by Convolving the Spike Data with a 40-msec Ricker Wavelet.

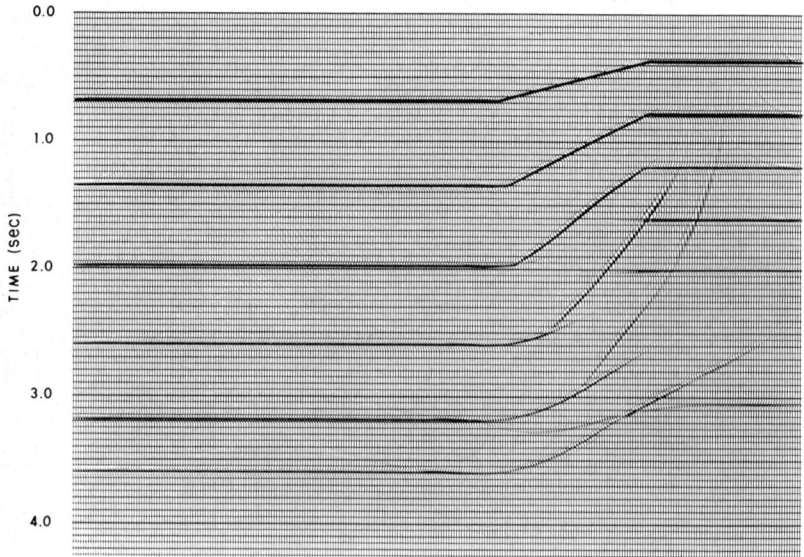

Figure 18. Frequency Domain Migration of the Data Shown in Figure 17. The Migration Velocity Function Corresponds to Average Velocities Along Horizontal and Dipping Events.

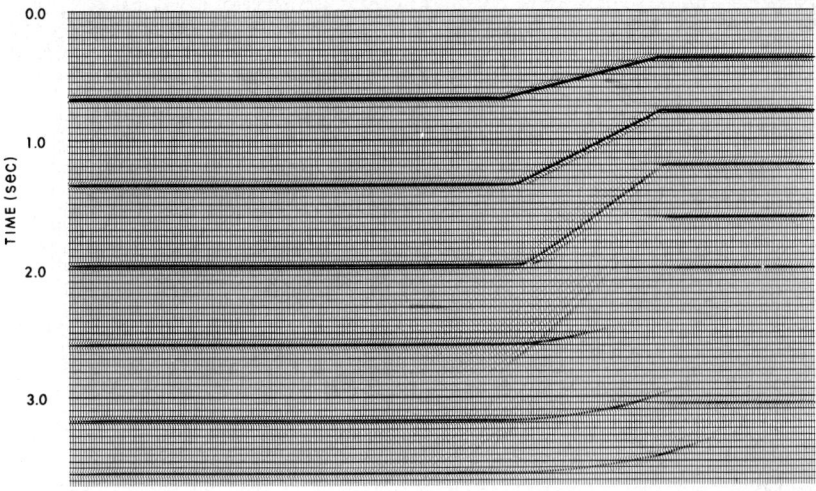

Figure 19. Finite-Difference Migration of the Data Shown in Figure 17, Obtained by Using $\tau = 60$. The Migration Velocity Function Corresponds to the Average Velocities Along Horizontal and Dipping Events.

For the constant velocity medium, the frequency domain method yielded amazing results. All events including those originating from reflectors dipping at $60°$ were correctly migrated. On the other hand, the finite-difference method which is usually referred to as the 15 degree method yielded reasonably good results for dips ranging up to $30°$. This statement may not be valid for other reflector geometrics. In order to further improve the quality of results, we reduced the migration step size. To our disappointment, we obtained inferior results. This implies that the output from the finite-difference method is strongly dependent on the frequency content of the input section.

For the variable velocity medium, the dispersion relation which forms the basis of the frequency domain method is not valid. The usual procedure to overcome this problem involves a stretching of the input section by appropriate amounts that would compensate for variations in velocity. An immediate consequence of stretching is connected to the fact that the various points along a dipping reflector are no longer normal incidence points after stretching. Thus the concept of exploding reflector for which the condition of normal incidence serves as a linch pin falls apart. One possible way to preserve the condition of normal incidence would be to introduce some kind of bending and curving of the reflectors before migrating the data. Any attempt to ignore that would give rise to bent and curved reflectors in the output section. Moreover, in the case of 3D migration, this bending could also be accompanied by twisting of the reflectors.

The frequency domain migration results obtained by us for the variable velocity medium were reasonably good up to a dip of about $30°$. More steeply dipping reflectors which were originally linear appeared curved. This kind of artifacts which are the direct consequence of stretching have rejuvenated interest of the explorationists in favor of the finite-difference method. Among other things, the latter algorithm tends to have a smoothing effect on the noise which is invariably present in the real data. In our tests, related to the finite-difference algorithm, the events originating from the reflector dipping at $30°$ were correctly migrated but they underwent a significant loss of energy. The events associated with steeper dips were almost completely obliterated.

The progress made in recent years in the area of migration has been impressive. However, the present state of the art dictates serious need for further work to accomodate lateral variations in velocity. In view of that, the studies undertaken by Berkhout (Berkhout and Van Wulfften Palthe, 1979; Berkhout, 1980) cannot be overemphasized. For the same reason, any new developments in the area of ray-theory migration (Sattlegger, 1982; Hubral, 1975, 1976, 1977; Hubral and Krey, 1980) should be fully exploited for solving practical problems.

ACKNOWLEDGMENTS

The author wishes to thank Dr Louis R. Castro and Dr Enders A. Robinson for making many valuable suggestions and comments. Permission from Superior Oil Company to publish the results of this study is appreciated.

REFERENCES

BAKER, B. B. and COPSON, E. T. (1950). The Mathematical Theory of Huygen's Principle. Oxford University Press, London.

BERKHOUT, A. J. (1980). Seismic Migration. Elsevier, Amsterdam.

BERKHOUT, A. J. (1981). Wavefield extrapolation techniques in seismic migration, a tutorial. Geophysics 46, 1638-1656.

BERKHOUT, A. J. and Van Wulfften Palthe, D. W. (1979). Migration in terms of spatial deconvolution. Geophys. Prosp. 27, 261-291.

CLAERBOUT, J. F. (1976). Fundamentals of Geophysical Data Processing. McGraw-Hill Book Co., New York.

CLAERBOUT, J. F. and DOHERTY, S. (1972). Downward continuation of moveout-corrected seismograms. Geophysics 37, 741-768.

CLAERBOUT, J. F. and JOHNSON, A. G. (1971). Extrapolation of time dependent waveforms along their path of propagation. Geophys. J. Roy. Astr. Soc. 26, 285-293.

CHUN, J. H. and JACEWITZ, C. H. (1981). Fundamentals of frequency domain migration. Geophysics 46, 717-733.

DIX, C. H. (1955). Seismic velocities from surface measurements. Geophysics 20 68-86.

FRENCH, W. S. (1975). Computer migration of oblique seismic reflection profiles. Geophysics 40, 961-980.

GAZDAG, J. (1978). Wave equation migration with the phase shift method. Geophysics 43, 1342-1351.

GAZDAG, J. (1980). Wave equation migration with the accurate space derivative method. Geophys. Prosp. 28, 60-70.

HAGEDOORN, J. G. (1954). A process of seismic reflection interpretation. Geophys. Prosp. 2, 85-127.

HUBRAL, P. (1975). Locating a diffractor below plane layers of constant interval velocity and varying dip. Geophys. Prosp. 23, 313-322.

HUBRAL, P. (1977). Time migration--Some ray theoretical aspects. Geophys. Prosp. 25, 728-745.

HUBRAL, P. and KREY, TH. (1980). Interval Velocities from Seismic Reflection Time Measurements. Society of Exploration Geophysicists, Tulsa.

JOHNSON, J. D. and FRENCH, W. S. (1982). Migration--the inverse method. In Concepts and Techniques in Oil and Gas Exploration. Ed: K. C. Jain, Society of Exploration Geophysicists, Tulsa, 115-157.

LARNER, K. L. and HORNSBY, J. M. (1972). Full-waveform migration of seismic data, Western Geophysical Co., Houston.

LOEWENTHAL, D., LU, L., ROBERSON, R. and SHERWOOD, J. (1976). The wave equation applied to migration. Geophys. Prosp. 24, 380-399.

ROBINSON, E. A. (1983). Migration of Geophysical Data. International Human Resources Development Corporation, Boston.

SATTLEGER, J. (1982). Migration of seismic interfaces. Geophys. Prosp. 30, 71-85.

SCHNEIDER, W. A. (1978). Integral formulation for migration in two and three dimensions. Geophysics 43, 49-76.

STOLT, R. H. (1976). Migration by Fourier transform. Geophysics 43, 23-48.

TANER, M. T. and KOEHLER, F. (1969). Velocity spectra--digital computer derivation and applications of velocity functions. Geophysics 34, 859-881.

TROREY, A. W. (1970). A simple theory of seismic diffractions, Geophysics 35, 762-784.

AN APPROACH TO SEISMIC INFORMATION EXTRACTION

Ferial El-Hawary
Technical University of Nova Scotia, P.O. Box 1000, Halifax, Nova Scotia
Canada B3J 2X4

The paper treats approaches to modeling the response of the ocean subsurface and underlying media to acoustic or seismic excitation assuming that the source wave form is known. An important issue is the extent of information that can be extracted about the subsurface from the received time series record on the basis of the chosen model. Signal processing on the return seismic data is performed to obtain estimates of the amplitude and delay parameters. A minimum variance approach is detailed. Practical problems of relating the physical properties of the layered media to the estimated parameters are also discussed.

1. INTRODUCTION

The time series record of the echo signal from the ocean subsurface and underlying media contains valuable information about the nature of the media. A successful signal processing algorithm relies on appropriate models of the earth response to acoustic or seismic excitation inputs. The paper reviews a number of important approaches to the modeling task. Models of this nature are known as synthetic seismograms and there is a wealth of literature in the area. The mathematical form of the input excitation function is known. The modeling task can be generally stated as that of tracing the input wave as it travels through the media to the receiving equipment.

Several modeling assumptions are common to all models here. A flat completely elastic layered medium structure with plane acoustic wave propagation is assumed. Normal wave incidence is assumed. This is appropriate for the available field results where the source and the down looking receivers are in close proximity. The models considered differ in the degree of detail and mathematical refinement.

Many factors have contributed to the advances in synthetic seismograms development. Among these factors is the introduction of the digital computers. The field has benefitted from this tool as have virtually all areas of scientific and technological activity.

Models of acoustic wave propagation in lossless layered media systems reviewed in this paper can be classified into two broad categories, according to the degree of detail involved. Models in the first category utilize a parametric representation involving two signals for each layer. The basis for this type of model is the decomposition of the solution to the wave particle displacement, (particle velocity, and pressure) into forward and backward travelling components in each layer. In the second category are models which are based on the assumption that the system is described by the lossless wave equation and boundary conditions and the nature of the solution of the wave equation requiring that the output be the sum of time shifted and scaled replicas of the source signals. The model in this case is a convolution summation, which is sometimes referred to as a non-parametric representation, Mendel et al. (1980). The models reviewed differ mainly in their approach to modeling the reflection coefficients at the interfaces.

The following two sections review the two modeling approaches. A minimum variance approach to estimating the amplitude and delay parameters is given in Section 4. The practical problem of relating these parameters to the physical properties of the media is discussed in Section 5.

2. PARAMETRIC MODELS OF EARTH RESPONSE

This category of models gives rise to cascade type matrix model structure, which is a feature common to all parametric type models reviewed. The models differ in the choice of the two signals associated with each interface. Another difference is in the choice of representation either in discrete-time (Z-transform), continuous time (Laplace transform) or the frequency domain.

Wuenschel's model described in section (2.1) employs the pressure and particle velocity in arriving at a transmission matrix type model in terms of the Laplace transform. Section (2.2) describes the model due to Goupillaud which uses the upward and downward propagating pressure wave amplitudes as the model variables. A scattering matrix type representation in the frequency domain is the feature of this model. Robinson's model discussed in Section (2.3) is among the most widely accepted models. It is a scattering-matrix type model developed in the Z-domain for equal time delay. Robinson's model is the basis for the predictive convolution method. The quasi-state space model of Mendel is discussed in Section (2.4). This is a time-domain representation in terms of a dynamical equation with multiple time delays referred to as a casual functional equation.

2.1 Wuenschel's Model

Wuenschel (1960) approaches the modeling problem starting with the wave equation in terms of displacement ζ given in terms of the Laplace operators by

$$\frac{\partial^2 \zeta(S,X)}{\partial X^2} = \frac{S^2}{C^2} \zeta(S,X). \qquad (2.1.1)$$

The solution of the above equation is given by

$$\zeta(S,X) = \zeta_i e^{-SX/C} + \zeta_r e^{SX/C}.$$

The indicent and reflected displacement amplitudes are denoted by ζ_i and ζ_r respectively. As a result the particle velocity $U(S,X)$ and pressure $P(S,X)$ are given by

$$U(S,X) = (\zeta_i e^{-SX/C} + \zeta_r e^{SX/C}) S \qquad (2.1.2)$$

$$P(S,X) = (-\zeta_i e^{-SX/C} + \zeta_r e^{SX/C}) ZS. \qquad (2.1.3)$$

Wuenschel considered two layers (K-1) and K with interface at X=0. The depth of layer K is denoted by d_K. The particle velocity and pressure at X=0, are thus obtained as:

$$U_{K-1} = S(\zeta_i + \zeta_r) \qquad (2.1.4)$$

$$P_{K-1} = Z_K S(\zeta_r - \zeta_i). \qquad (2.1.5)$$

At $X = d_K$, we obtain

$$U_K = S[\zeta_i e^{-Sd_K/C_K} + \zeta_r e^{Sd_K/C_K}] \qquad (2.1.6)$$

$$P_K = Z_K S[-\zeta_i e^{-Sd_K/C_K} + \zeta_r e^{Sd_K/C_K}]. \qquad (2.1.7)$$

Eliminating ζ_i and ζ_r, one obtains

$$\begin{bmatrix} U_k \\ P_k \end{bmatrix} = \alpha_{-k} \begin{bmatrix} U_{k-1} \\ P_{k-1} \end{bmatrix}. \qquad (2.1.8)$$

The matrix $\underline{\alpha}_K$ is the Wuenschel's layer matrix given by

$$\underline{\alpha}_K = \begin{bmatrix} \left(\dfrac{e^{(Sd_K/C_K)} + e^{-(Sd_K/C_K)}}{2}\right) & \left(\dfrac{e^{(Sd_K/C_K)} - e^{-(Sd_K/C_K)}}{2Z_K}\right) \\ Z_K\left(\dfrac{e^{(Sd_K/C_K)} - e^{-(Sd_K/C_K)}}{2}\right) & \left(\dfrac{e^{(Sd_K/C_K)} + e^{-(Sd_K/C_K)}}{2}\right) \end{bmatrix} \quad (2.1.9)$$

For an N-layer medium, assuming continuity of velocity and pressure along the interfaces one can write

$$\begin{bmatrix} U_N \\ P_N \end{bmatrix} = \underline{\alpha}_N \underline{\alpha}_{N-1} \cdots \underline{\alpha}_1 \begin{bmatrix} U_0 \\ P_0 \end{bmatrix}. \quad (2.1.10)$$

Wuenschel's formulation can be used to express the particle velocity in terms of the forcing function of the source as follows. Assume that the (N+1)th layer has an acoustic impedance of Z_{N+1}. The variables emerging from the Nth layer are thus related by
$P_N = -Z_{N+1} U_N$.
As a result the fundamental expression of the equation (2.1.10) reduces to

$$\begin{bmatrix} U_N \\ -Z_{N+1} U_N \end{bmatrix} = \underline{A} \begin{bmatrix} U_0 \\ P_0 \end{bmatrix} \quad (2.1.11)$$

where

$$\underline{A} = \begin{bmatrix} A_{11} & A_{12} \\ A_{21} & A_{22} \end{bmatrix} = \underline{\alpha}_N \underline{\alpha}_{N-1} \cdots \underline{\alpha}_1. \quad (2.1.12)$$

It is easy to verify that

$$U_0 = \dfrac{(A_{22}/Z_{N+1}) + A_{12}}{(A_{21}/Z_{N+1}) + A_{11}} (-P_0) \quad (2.1.13)$$

Thus the particle velocity at the source is expressed in terms of the forcing function of the source with a single factor of proportionality which is a function of the characteristic impedances of the individual layers and the transit time across the layers.

2.2 Goupillaud's Model

The variables used in Goupillaud's (1961) model are down and up-going wave amplitudes D and U as shown in Figures (2.2.1). It can be shown that be defining the matrix

$$M_K = \dfrac{1}{T_K} \begin{bmatrix} e^{j\omega\zeta_K} & R_K e^{j\omega\zeta_K} \\ R_K e^{-j\omega\zeta_K} & e^{-j\omega\zeta_K} \end{bmatrix} \quad (2.2.1)$$

the layer equation is given by

$$\begin{bmatrix} D_K \\ U_K \end{bmatrix} = M_K \begin{bmatrix} D_{K+1} \\ U_{K+1} \end{bmatrix}. \quad (2.2.2)$$

The transmission and reflection coefficients in the Kth layer are T_K and R_K, with
$$T_K = 1 + R_K \qquad (2.2.3)$$
The travel time in layer K is denoted by τ_K.

For a system of N interfaces, one obtains
$$\begin{bmatrix} D_0 \\ U_0 \end{bmatrix} = M_0^{(N-1)} \begin{bmatrix} D_N \\ U_N \end{bmatrix} \qquad (2.2.4)$$
where the matrix product is defined by
$$M_0^{(N-1)} = M_0 M_1 \cdots M_{N-1} . \qquad (2.2.5)$$

The reflection response U_0 for a unit input and for a completely absorptive Nth interface $U_N = 0$ is given by

$$U_0 = [\sum_{i=0}^{N} R_i \exp\{-2j\omega \sum_0^i \tau_i\} + \sum_{i=0}^{N-2} \sum_{\substack{j=1 \\ j \neq i}}^{N-1} \sum_{\substack{k=2 \\ k \neq i}}^{N} R_i R_j R_k) \exp\{-2j\omega (\sum_0^K \tau_K$$
$$-\sum_0^j \tau_j + \sum_0^i \tau_i\} + \ldots]/(1 + \Delta_1 + \Delta_2 + \ldots) . \qquad (2.2.6)$$

It should be noted that U_0 is a frequency response expression and in a manner similar to that mentioned in Wuenschel's treatment, the total response is obtained through the Fourier integral.

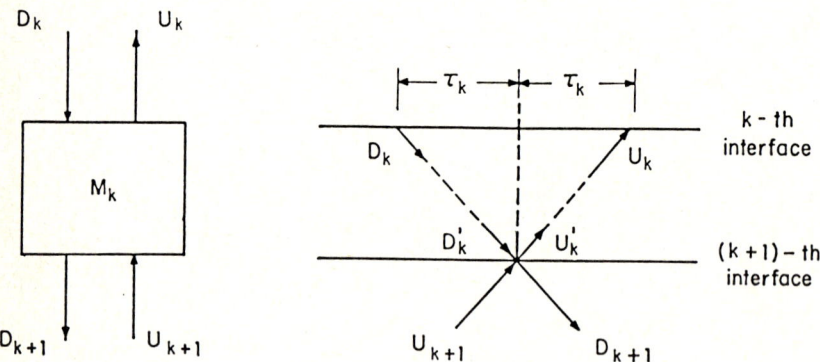

Figure (2.2.1) Variable definitions for Goupillaud's model.

2.3 Robinson's Model

Robinson's earth model of 1967 is in terms of the Z or shift operator and is obtained in terms of the basic relation

$$\begin{bmatrix} D_{K+1}(Z) \\ U_{K+1}(Z) \end{bmatrix} = \frac{Z^{-\frac{1}{2}}}{T_K} N_K \begin{bmatrix} D_K(Z) \\ U_K(Z) \end{bmatrix} \qquad (2.3.1)$$

where

$$\underline{N}_K = \begin{bmatrix} Z & -R_K \\ -R_K Z & 1 \end{bmatrix}. \qquad (2.3.2)$$

As a result, the overall model is obtained as

$$\begin{bmatrix} D_{K+1}(Z) \\ U_{K+1}(Z) \end{bmatrix} = \frac{Z^{-K/2}}{T'_K \cdots T'_2 T'_1} \underline{N}^{(K)} \begin{bmatrix} D_1(Z) \\ U_1(Z) \end{bmatrix} \qquad (2.3.3)$$

where the matrix $\underline{N}^{(K)}$ denotes the product

$$\underline{N}^{(K)} = \underline{N}_K \cdots \underline{N}_2 \underline{N}_1. \qquad (2.3.4)$$

An important feature of Robinson's formulation is that the prevailing matrix properties are exploited to arrive at an efficient implementation. It can be shown that

$$\underline{N}^{(K)} = \underline{N}_K \underline{N}_{K-1} \cdots \underline{N}_1 = \begin{bmatrix} Z^K n_{22}^{(K)}(Z^{-1}) & n_{12}^{(K)}(Z) \\ Z^K n_{12}^{(K)}(Z^{-1}) & n_{22}^{(K)}(Z) \end{bmatrix}. \qquad (2.3.5)$$

Thus only n_{22} and n_{12} are needed, and can be generated recursively using

$$n_{12}^{(K)}(Z) = Z n_{12}^{(K-1)}(Z) - R_K n_{22}^{(K-1)}(Z) \qquad (2.3.6)$$

$$n_{22}^{(K)}(Z) = n_{22}^{(K-1)}(Z) - R_K Z n_{12}^{(K-1)}(Z). \qquad (2.3.7)$$

If the (N+1) layer is apsorptive, then the impulse response transfer function of the earth model can be obtained as

$$G^{(N)}(Z) = \frac{E^{(N)}(Z)}{A^{(N)}(Z)} \qquad (2.3.8)$$

where

$$A^{(N)}(Z) = n_{22}^{(N)}(Z) - R_0 Z^N n_{12}^{(N)}(Z^{-1}) \qquad (2.3.9)$$

$$E^{(N)}(Z) = -Z^N n_{12}^{(N)}(Z^{-1}) \qquad (2.3.10)$$

with

$$n_{12}^{(1)}(Z) = -R_1$$

$$n_{12}^{(1)}(Z) = 1$$

$$G^{(0)}(Z) = 0$$

The numerator and denominator $E^N(Z)$ and $A^N(Z)$ can be expressed in polynomial forms as

$$E^{(N)}(Z) = \sum_{j=1}^{N} \varepsilon_j^{(N)} Z^j \qquad (2.3.11)$$

$$A^{(N)}(Z) = \sum_{j=0}^{N} a_j^{(N)} Z^j \qquad (2.3.12)$$

with
$a_0^{(N)} = 1$

If one denotes the output by Y(K) and the input by M(K), then the signal time domain representation is

$$Y(K) + \sum_{j=1}^{N} a_j^{(N)} Y(K-j) = \sum_{j=1}^{N} \varepsilon_j^{(N)} M(K-j) \qquad (2.3.13)$$

Using the defining relations, the coefficients of the characteristic polynomial $a_j^{(N)}$ can be obtained. Silvia (1977), gives the values in terms of the serial correlation defined by

$$\phi_j^{(N)} = \sum_{i=0}^{N} R_i R_{i+j} \qquad N \geq j \geq 1 \qquad (2.3.14)$$

$$\phi_0^{(N)} = 1 \qquad (2.3.15)$$

Table (1) gives the expressions corresponding to N=1, ..., 4. Note that residual terms such as $R_0 R_1 R_2 R_3$ are present. If one assumes a small reflection coefficient approximation, then these residuals may be neglected giving

$$a_j^{(N)} \approx \phi_j^{(N)} \qquad (2.3.16)$$

$$a_j^{(N)} \approx \sum_{i=0}^{N} R_i R_{i+j} \qquad (2.3.17)$$

In a similar fashion, the coefficients of the reflection polynomial E(Z) are found, as given in Table (2). If one uses the same small reflection approximation one has

$$\varepsilon_j^{(N)} \approx R_j. \qquad (2.3.18)$$

It can thus be seen that polynomial $E^{(N)}(Z)$ contains the reflection information required for seismic data processing.

TABLE (1)

N	$a_1^{(N)}$	$a_2^{(N)}$	$a_3^{(N)}$	$a_4^{(N)}$
1	$\phi_1^{(1)}$	—	—	—
2	$\phi_1^{(2)}$	$\phi_2^{(2)}$	—	—
3	$\phi_1^{(3)}$	$\phi_2^{(3)} + R_0 R_1 R_2 R_3$	$\phi_2^{(3)}$	—
4	$\phi_1^{(4)}$	$\phi_2^{(4)} + R_4 R_3 R_2 R_1 + R_4 R_3 R_1 R_0 + R_3 R_2 R_1 R_0$	$\phi_3^{(4)} + R R R R_{4320} + R R R R_{4210}$	$\phi_4^{(4)}$

TABLE (2)

N	$\Sigma_1^{(N)}$	$\Sigma_2^{(N)}$	$\Sigma_3^{(N)}$	$\Sigma_4^{(N)}$
1	R_1	—	—	—
2	R_1	R_2	—	—
3	R_1	$R_2 + R_3 R_2 R_1$	R_3	—
4	R_1	$R_2 + R_4 R_3 R_1 + R_3 R_2 R_1$	$R_3 + R_4 R_2 R_1 + R_4 R_3 R_2$	R_4

2.4 Mendel's Model

Mendel (1960) developed a quasi-state space model for layered media which uses the ideas presented in the models discussed above. The model form introduced by Mendel is given by the layer-ordered relations

$\tilde{d}_1(t+\tau_1) = -R_0 u_1(t) + (1+R_0) m(t)$
$u_1(t+\tau_1) = R_1 d_1(t) + (1-R_1) u_2(t)$
.
.
.
$\tilde{d}_K(t+\tau_K) = (1+R_{K-1}) \tilde{d}_{K-1}(t) - R_{K-1} u_K(t)$
$u_K(t+\tau_K) = R_K \tilde{d}_K(t) + (1-R_K) u_{K+1}(t)$ } K=2, 3, ..., N-1
$\tilde{d}_N(t+\tau_N) = (1+R_{N-1}) \tilde{d}_{N-1}(t) - R_{N-1} u_N(t)$
$u_N(t+\tau_N) = R_N \tilde{d}_N(t)$.

Note that
$\tilde{d}_K(t) \stackrel{\Delta}{=} d_K(t-\tau_K)$.

The above set of equations are dynamic with multiple time delays (generally incrementally non-equal). The equations are referred to as casual functional equations.

A quasi-state space representation is obtained by defining
$\underline{X}(t) = \text{col. } [u_1(t), \tilde{d}_1(t), ..., u_N(t), \tilde{d}_N(t)]$. (2.4.1)

The 2Nx2N matrix operator $\tilde{\underline{Z}}$ is defined by
$\tilde{\underline{Z}} \stackrel{\Delta}{=} \text{diag. } [Z_1, Z_1, Z_2, Z_2, ..., Z_K, Z_K]$ (2.4.2)

where $\tilde{Z}_i$ is a scalar operator denoting a τ_i sec. time delay.

$\tilde{Z}_i f(t) = f(t-\tau_i)$. As a result one has
$\tilde{\underline{Z}}^{-1} \underline{X}(t) = \underline{A} \underline{X}(t) + \underline{b} m(t)$ (2.4.3)

$$y(t) = \underline{C}^T \underline{X}(t) + R_0 \underline{m}(t). \tag{2.4.4}$$

This equation provides a conceptual solution given by

$$\underline{X}(t) = [\tilde{Z}^{-1} - \underline{A}]^{-1} \underline{b} \, m(t) \tag{2.4.5}$$

and

$$y(t) = [\underline{C}^T [\tilde{Z}^{-1} - \underline{A}]^{-1} \underline{b} + R_0] \, m(t). \tag{2.4.6}$$

The transfer function of the system will be a ratio of two polynomials in $z_1, z_2, \ldots, z_N$ which should reduce to Robinson's formula when $z_1 = z_2 = \ldots = z_N$. As an example, the two layer case Nahi, et al. (1978) gives

$$\frac{Y(\tilde{Z})}{M(\tilde{Z})} = \frac{R_0 + R_0 R_1 R_2 z_2^2 + R_1 z_1^2 + R_2 z_1^2 z_2^2}{1 + R_1 R_2 z_2^2 + R_0 R_1 z_1^2 + R_0 R_2 z_1^2 z_2^2} \tag{2.4.7}$$

for the case of equal z_i's, i.e. $\tau_1 = \tau_2 = \ldots \tau_N \overset{\Delta}{=} \tau$, a state space solution is possible, since the model reduces to

$$\underline{X}[(K+1)\tau] = \underline{A}\,\underline{X}(K\tau) + \underline{b}\,m(K\tau) \tag{2.4.8}$$

whose solution is

$$\underline{X}[(K+1)\tau] = \underline{A}^K \underline{X}(0) + \sum_{i=0}^{K} \underline{A}^{K-1} \underline{b}\, m(i\tau). \tag{2.4.9}$$

It appears that no solution to the general case has been documented in the literature.

3. NON-PARAMETRIC MODELS

Synthetic seismograms of the non-parametric type are developed with emphasis on the nature of the reflection coefficients. Three modeling approaches are reviewed here.

3.1 PFC Approach

A non-parametric earth model based on an approximation of the reflection coefficients is proposed in the work of Peterson, Fillippone and Croker (PFC) in (1955). For normal incidence, plane wave propagation between two media, the reflection coefficient is given in terms of acoustic impedances by

$$R = z_2 - z_1 / z_2 + z_1. \tag{3.1.1}$$

Assuming incremental changes, one sets

$$\Delta z = z_2 - z_1 \tag{3.1.2}$$

$$z \approx z_1 \approx z_2 \tag{3.1.3}$$

thus

$$R = \Delta z / 2z. \tag{3.1.4}$$

As a result an approximate expression for the reflection coefficient between the two media is given by

$$R = \tfrac{1}{2}\Delta[\ln(z)]. \tag{3.1.5}$$

In multi-layer media, the ith reflection coefficient is given by

$$R_i = \tfrac{1}{2}[\ln(z_{i+1}) - \ln(z_i)]. \tag{3.1.6}$$

The Seismograph representing the signal $y(t)$ received by the hydrophones is given by the summation

$$y(t) = m(t) + \sum_{i=1}^{N} R_i\, m(t - \tau_i) \tag{3.1.7}$$

where m(t) is the input signal and the delays τ_i are given by the recursive relation

$$\tau_i = \tau_{i-1} + 2C_i^{-1}[d_i - d_{i-1}]. \quad (3.1.8)$$

The velocity of sound propagation in medium i is denoted by C_i. The vertical distance of the ith interface from reference is denoted by d_i.

3.2 SLM Approach

In addition to the PFC approximation, Sengbush, Lawrence, and McDonald (SLM, 1961) assume that the layer's density is velocity dependent according to this relation.

$$\rho = KC^m. \quad (3.2.1)$$

As a result the reflection coefficient in terms of velocity of propagation is obtained as

$$R = \frac{m+1}{2} \Delta[\ln C]. \quad (3.2.2)$$

SLM obtains an expression for a reflectivity function $\tilde{r}(t)$ by extending the concept to the continuous case, which is given by

$$\tilde{r}(t) = \frac{d}{dt} \ln C(t). \quad (3.2.3)$$

A synthetic seismogram can be obtained from a velocity log C(t) as the output of a filter due to an input $\tilde{r}(t)$. The filter's impulse response is simply that of the shot pulse m(t).

3.3 BGW Approach

A method for computing the synthetic seismogram assuming variable sound propagation velocity profiles is due to Berryman, Goupillaud, and Waters (BGW, 1958). In their paper, transformation of variables is introduced to the particle displacement equation.

$$\frac{\partial^2 \zeta(x,t)}{\partial t^2} = c^2 \frac{\partial^2 \zeta(x,t)}{\partial x^2} + 2c \frac{\partial c}{\partial x} \frac{\partial \zeta(x,t)}{\partial x}. \quad (3.3.1)$$

The main BGW equation is given by

$$c^2 \frac{d^2\pi}{dx^2} + \omega^2 \pi = 0. \quad (3.3.2)$$

The new variable $\pi(x)$ is given by

$$\pi(x) = c^2 \, d\zeta(x,t)/dx. \quad (3.3.3)$$

The displacement is assumed to be in sinusoidal form. The pressure p(x,t) can be obtained in terms of the new variable as

$$p(x,t) = \rho\pi(x) \, e^{j\omega t}. \quad (3.3.4)$$

A closed-form solution to equation (3.3.2) can be obtained if the velocity profile in a layer K is assumed to vary linearly with x according to

$$C(x) = C_K + b_K(x - x_K). \quad (3.3.5)$$

The constants C_K and b_K are characteristics of the medium. The solution is of the form

$$\pi_K(x) = \pi_K^+ \exp\{\tfrac{1}{2}(1-\beta_K)\gamma_K\} + \pi_K^- \exp\{\tfrac{1}{2}(1+\beta_K)\gamma_K\}. \quad (3.3.6)$$

The constant β_K is given by the expression

$$\beta_K^2 = 1 - (4\omega^2/b_K). \quad (3.3.7)$$

The new variable γ_K is defined as

$$\gamma_K = \ln[1 + \{b_K(x - x_K)/C_K\}]. \quad (3.3.8)$$

A layer reflection coefficient denoted by R_K is the ratio
$$R_K = \pi_K^- / \pi_K^+ . \tag{3.3.9}$$
BGW shows that a recursive form to obtain this coefficient is given by:
$$R_{K-1} = \frac{b_{K-1}\beta_{K-1} + (b_K - b_{K-1}) - b_K\beta_K(1 - R_K/1 + R_K)}{b_{K-1}\beta_{K-1} - (b_K - b_{K-1}) + b_K\beta_K(1 - R_K/1 + R_K)} e^{-\beta_{K-1} \ln(C_K/C_{K-1})} . \tag{3.3.10}$$

To produce a synthetic seismogram, the bottom layer is assumed to have zero reflection and thus
$R_N = 0$.
The computations are then carried out backward to obtain $R_{N-1}, R_{N-2}, \ldots, R_0$ for a specific frequency ω. The overall synthetic seismogram is obtained through the use of the Fourier integral.

4. PARAMETER ESTIMATION

This section presents two procedures for processing acoustic return signals. The objective is to provide reliable estimates of the amplitude and delay parameters associated with source signal replicas present in the return signal. The procedures rely on some fundamental, yet powerful tools from communications and estimation theory.

The input-output expression relating the signal $y(t)$ received by the hydrophone to the source signal $m(t)$ is made more realistic by including the measurement noise $v(t)$. This accounts for physical effects not accommodated by the noise free, simple reflection model; as well as for sensor and instrumentation inaccuracies and acoustical disturbances, primarily scattering and reverberation. Thus on the basis of Section 2 the model is given by

$$y(t) = \sum_{i=1}^{N} a_i m(t - \tau_i) + v(t) \tag{4.1}$$

where each term in the summation is a delayed and scaled replica of the source wavelet. These arise from primary, secondary and higher order reflections. Thus a_i are functionally related to the reflection coefficients at the layer interface and τ_i are related to the layer travel times.

4.1 Formulation

Identification of sub-bottom structure and geometry reduces to a problem in parameter estimation. This is due to the functional relations between the parameters of Equation (4.1) and the structure and geometry of the sub-bottom. The estimation task is to determine, in an optimal sense, the amplitude parameters a_i and the delay times τ_i.

The observation equation can be expressed in a number of ways to facilitate the estimation procedure. The first formulation proceeds assuming the availability of an observation sequence $\{y(t_i), i=0,\ldots,N-1\}$. In this case the following vector-matrix equation is obtained.

$$\begin{bmatrix} y(t_0) \\ y(t_1) \\ \vdots \\ y(t_{N-1}) \end{bmatrix} = \begin{bmatrix} m(t_0-\tau_1) & m(t_0-\tau_2) & \cdots & m(t_0-\tau_n) \\ m(t_1-\tau_1) & m(t_1-\tau_2) & & m(t_1-\tau_n) \\ \vdots & \vdots & & \vdots \\ m(t_{N-1}-\tau) & m(t_{N-1}-\tau_2) & & m(t_{N-1}-\tau_n) \end{bmatrix} \begin{bmatrix} a_1 \\ a_2 \\ \vdots \\ a_n \end{bmatrix} + \begin{bmatrix} v(t_0) \\ v(t_1) \\ \vdots \\ v(t_{N-1}) \end{bmatrix}. \quad (4.1.1)$$

More compactly this is of the form
$$\underline{y} = \underline{H}\,\underline{p} + \underline{v}. \qquad (4.1.2)$$

A second formulation assumes that for a given time interval $[t_{q_i}, t_{q_{i+w-1}}]$, only one signal replica appears corresponding to the ith component.
$$y(t) = a_i\, m(t-\tau_i) + v(t) \quad t_{q_i} < t < t_{q_{i+w-1}}. \qquad (4.1.3)$$

As a result, the vector matrix equation takes on the form
$$\begin{bmatrix} y(t_{q_i}) \\ y(t_{q_{i+1}}) \\ \vdots \\ y(t_{q_{i+w-1}}) \end{bmatrix} = \begin{bmatrix} m(t_{q_i}-\tau_i) \\ m(t_{q_{i+1}}-\tau_i) \\ \vdots \\ m(t_{q_{i+w-1}}-\tau_i) \end{bmatrix} a_i + \begin{bmatrix} v(t_{q_i}) \\ v(t_{q_{i+1}}) \\ \vdots \\ v(t_{q_{i+w-1}}) \end{bmatrix}. \qquad (4.1.4)$$

This can be represented by the compact form
$$\underline{y}_i = \underline{m}_i\, \underline{a}_i + \underline{v}_i. \qquad (4.1.5)$$

The above discussion indicates that the estimation task using any of the proposed formulations corresponds to finding an estimate $\hat{\underline{p}}$ of the random parameter vector $\underline{p}$ in some optimal sense.

The additive noise vector v is assumed to have the following statistical properties
$$E\{\underline{v}\} = \underline{0} \qquad (4.1.6)$$
$$E\{\underline{v}\,\underline{v}^T\} = \underline{V}. \qquad (4.1.7)$$
The parameter vector's statistical properties are assumed to be
$$E\{\underline{p}\} = \bar{\underline{p}} \qquad (4.1.8)$$
$$E\{(\underline{p}-\bar{\underline{p}})(\underline{p}-\bar{\underline{p}}^T)\} = \underline{P}. \qquad (4.1.9)$$
The noise and parameter vectors are assumed to be uncorrelated
$$E\{\underline{p}\,\underline{v}^T\} = \underline{0}. \qquad (4.1.10)$$
The following section summarizes some useful results from the theory of linear parameter estimation.

4.2 Relationships for Parameter Estimation

A basic problem in the theory of linear parameter estimation is that of finding the optimal estimate $\hat{\underline{p}}$ of a random parameter vector $\underline{p}$ from observation $\underline{y}$ related to $\underline{p}$ according to
$$\underline{y} = \underline{H}\,\underline{p} + \underline{v} \qquad (4.2.1)$$
where $\underline{H}$ is a given matrix and $\underline{v}$ is a vector of additive random noise with statistical properties as given in (4.1.6) through (4.1.10).

The estimate vector $\hat{\underline{p}}$ is assumed to be of the unbiased linear form
$$\hat{\underline{p}} = \bar{\underline{p}} + \underline{K}[\underline{y} - \underline{H}\,\bar{\underline{p}}] \qquad (4.2.2)$$

$\underline{K}$ is a gain matrix to be evaluated according to the estimation criterion. As a result the covariance of the estimation error can be shown to reduce to

$$\hat{\underline{P}} = [\underline{I} - \underline{KH}] \ \underline{P} \ [\underline{I} - \underline{KH}]^T + \underline{K} \ \underline{V} \ \underline{K}^T. \qquad (4.2.3)$$

Consider an estimation criterion which is a compound least squares error minimization, with weighting matrices $\underline{A}$ and $\underline{B}$, given by

$$J_{WLS} = (\underline{p} - \bar{\underline{p}})^T \ \underline{A} (\underline{p} - \bar{\underline{p}}) + (\underline{y} - \underline{Hp})^T \ \underline{B} (\underline{y} - \underline{Hp}). \qquad (4.2.4)$$

This criterion penalizes both the deviations from the a-priori estimate $\bar{\underline{p}}$ and the observation error. $\underline{A}$ and $\underline{B}$ are positive definite matrices such that $(\underline{A} + \underline{H}^T \ \underline{B} \ \underline{H})$ is non-singular. In this case the gain matrix $\underline{K}$ is given by

$$\underline{K}_{WLS} = [\underline{A} + \underline{H}^T \ \underline{B} \ \underline{H}]^{-1} \ \underline{H}^T \ \underline{B}. \qquad (4.2.5)$$

Another estimation criterion is that of minimum variance

$$J_{MV} = \text{trace} \ E\{(\underline{p} - \bar{\underline{p}}) \ (\underline{p} - \bar{\underline{p}})^T\}. \qquad (4.2.6)$$

The gain matrix $\underline{K}$ in this instance is given by

$$\underline{K}_{MV} = \underline{P} \ \underline{H}^T \ [\underline{H} \ \underline{P} \ \underline{H}^T + \underline{V}]^{-1}. \qquad (4.2.7)$$

Comparison of (4.2.5) and (4.2.7) shows that the particular least square weighting

$$\underline{A} = k \ \underline{P}^{-1}$$
$$\underline{B} = k \ \underline{V}^{-1}$$

for an arbitrary k leads to the minimum variance condition, Vetter (1971). Using the matrix inversion lemma, the gain $\hat{\underline{K}}$ is found to be

$$\hat{\underline{K}} \stackrel{\Delta}{=} \underline{K}_{MV} = [\underline{P}^{-1} + \underline{H}^T \ \underline{V}^{-1} \ \underline{H}]^{-1} \ \underline{H}^T \ \underline{V}^{-1}. \qquad (4.2.8)$$

The error covariance matrix is given by

$$\hat{\underline{P}} = [\underline{P}^{-1} + \underline{H}^T \ \underline{V}^{-1} \ \underline{H}^{-1}]. \qquad (4.2.9)$$

Combining (4.2.1) and (4.2.8), the minimum variance estimate of $\underline{p}$ is given by

$$\hat{\underline{p}} = \bar{\underline{p}} + (\underline{P}^{-1} + \underline{H}^T \ \underline{V}^{-1} \ \underline{H})^{-1} \ \underline{H}^T \underline{V}^{-1} \ (\underline{y} - \underline{H}\bar{\underline{p}}) \qquad (4.2.10)$$

In the absence of a-priori information one would set $\bar{\underline{p}} \to \underline{0}$ and $\underline{P}^{-1} \to \underline{0}$ (or $\underline{P} \to \infty$), which would reflect the complete prior ignorance for estimates of the parameters. Entries for the covariance matrix of the disturbance must be chosen to reflect the disturbance activity.

4.3 Sequential Estimation of Delay and Amplitude Parameters

A procedure for estimating the amplitude and delay parameters is now reviewed. The procedure is sequential in the sense that the delay parameters τ_i are estimated first using the cross-correlation of the source signal $m(t)$ and the received signal $y(t)$. The second step is to obtain the amplitude parameter estimates, a_i, using the minimum variance estimator results. The cross-correlation calculated in the first step is used in the second step to simplify the computational procedure.

The use of the cross-correlation function provides a simple, yet powerful means for obtaining time delays of signals see Stremler (1977). Assume that a signal $f(t)$ is cross-correlated with a function $g(t)$ which is a delayed replica of $f(t)$, plus a random wave form

$$g(t) = f(t - t_0) + n(t) \qquad (4.3.1)$$

The cross-correlation function $\Phi_{fg}(\tau)$ will exhibit a large peak at the value of the time delay t_0.

The application of the above principle to estimate the delay parameter in Equation (4.1) is straight forward. The cross-correlation $\Phi_{my}(\tau)$ is generated, then the extrema of $\Phi_{my}(\tau)$ are detected by a direct search. The extrema times are the required estimates of the delay parameters, $\hat{\tau}_i$.

In the practical application there is a need to establish a threshold on the magnitude of the extrema to be considered. A threshold might be chosen as a small margin on the magnitude of the extremum of the cross-correlation of signal with a return-pulse-free portion of $y(t)$ but excluding the portion near $\tau = 0$.

The second step involves the application of the minimum variance estimator results to the problem of estimating the amplitude parameter a_i using the model of Equation (4.1.4). The minimum variance estimate $\hat{a}_i$ is, by Equation (4.2.10),

$$\hat{a}_i = \bar{a}_i + [1/(1/\sigma_{a_i}^2) + (m_i^T m_i/\sigma_v^2)] \; [(m_i^T/\sigma_v^2)(y_i - m_i \bar{a}_i)]. \tag{4.3.2}$$

This reduces to

$$\hat{a}_i = [(\sigma_v^2/\sigma_{a_i}^2)\bar{a}_i + (m_i^T y_i)]/[(\sigma_v^2/\sigma_{a_i}^2) + m_i^T m_i] \tag{4.3.3}$$

$\sigma_{a_i}^2$ and σ_v^2 are the variances of the parameter a_i and the noise v respectively.

The interpretation of the above result in terms of the correlation functions can be obtained by considering the following terms:

$$m_i^T y_i = \sum_{K=0}^{W-1} m(t_K - \hat{\tau}_i) \; y(t_K) \tag{4.3.4}$$

$$m_i^T m_i = \sum_{K=0}^{W-1} m(t_K - \hat{\tau}_i) \; m(t_K - \hat{\tau}_i). \tag{4.3.5}$$

The first expression can be interpreted as the cross-correlation of the signal $m(t)$ and $y(t)$ at correlation time $\tau = \hat{\tau}_i$. Thus using correlation notation

$$\Phi_{my}(\hat{\tau}_i) = m_i^T y_i. \tag{4.3.6}$$

The second can be interpreted as the autocorrelation of $m(t)$ at $\tau = 0$. Thus

$$\Phi_{mm}(0) = m_i^T m_i. \tag{4.3.7}$$

As a result the estimate $\hat{a}_i$ is given by

$$\hat{a}_i = [(\sigma_v^2/\sigma_{a_i}^2) \; \bar{a}_i + \Phi_{my}(\hat{\tau}_i)]/[(\sigma_v^2/\sigma_{a_i}^2) + \Phi_{mm}(0)]. \tag{4.3.8}$$

In the absence of a-priori information $\bar{a}_i$ and $\sigma_{a_i}^2$, $\sigma_{a_i}^2 \to \infty$, the above simplifies to

$$\hat{a}_i = \Phi_{my}(\hat{\tau}_i)/\Phi_{mm}(0). \tag{4.3.9}$$

While the error is $\tilde{a}_i \overset{\Delta}{=} a_i - \hat{a}_i$ and the associated (minimum variance) estimation error variance is

$$\hat{\sigma}_{\tilde{a}_i}^2 = [1/\{(1/\sigma_{a_i}^2) + (m_i^T m_i/\sigma_v^2)\}] + [\sigma_v^2/\{(\sigma_v^2/\sigma_{a_i}^2) + \Phi_{mm}(0)\}]. \tag{4.3.10}$$

The simplification without a-priori information leaves us with

$$\hat{\sigma}_{\tilde{a}}^2 = \sigma_v^2/\Phi_{mm}(0). \tag{4.3.11}$$

The minimum variance estimate can be obtained from a matched-filter implementation. The matched-filter maximizes the signal-to-noise ratio at the filter output. The filter's impulse response is a reverse time replica of the known signal $m(t)$,

given by
$$h(t) = k \, m \, (\tau_d - t). \tag{4.3.12}$$

The scaling parameter k is often chosen to provide a unity gain for the filter. The delay τ_d is inserted to make the filter realizable. The filter output is the convolution of h(t) with y(t) and can be shown to reduce to the cross-correlation of the signal m(t) with the received signal y(t).

The practical implementation of the above procedure and computations to detect a sequence of well separated response pulses on a given observation record consists of the following steps:

(1) Generate $\phi_{mm}(\tau)$, the autocorrelation of the source signal [or determine $\phi_{mm}(0) = \sigma_m^2$]

(2) Determine an estimate for σ_v^2 from a response-pulse-free portion of return signal, e.g. as $\sigma_v^2 = \phi_{vv}(0)$.

(3) Generate the cross-correlation $\phi_{my}(\tau)$. Detect by search the extrema of $\phi_{my}(\tau)$ which occur at $\tau = \tau_i$ and use these correlation time values as estimates $\hat{\tau}_i$. At the respective $\hat{\tau}_i$ determine peak values $\phi_{my}(\hat{\tau}_i)$, and use these together with σ_v^2, σ_m^2, $\sigma_{a_i}^2$ and $\bar{a}_i$ to evaluate estimate return parameter values $\hat{a}_i$ and associated error variance $\hat{\sigma}_{a_i}^2$.

A refinement on the above procedure can be made by considering the observation sequence y(t) in the form given in Equation (4.1.1). Suppose that we have detected an amplitude parameter $\hat{a}_i$ at a delay $\hat{\tau}_1$ by the procedure given above based on Equation (4.1.4). We may then rearrange (4.1.1) to the form

$$\begin{bmatrix} y(t_0) - \hat{a}_1 m(t_0 - \hat{\tau}_1) \\ y(t_1) - \hat{a}_1 m(t_1 - \hat{\tau}_1) \\ \vdots \\ y(t_N) - \hat{a}_1 m(t_N - \hat{\tau}_1) \end{bmatrix} = \begin{bmatrix} m(t_0 - \tau_2) & m(t_0 - \tau_3) \cdots \\ m(t_1 - \tau_2) & m(t_1 - \tau_3) \cdots \\ \vdots & \vdots \\ m(t_N - \tau_2) & m(t_N - \tau_3) \cdots \end{bmatrix} \begin{bmatrix} a_2 \\ a_3 \\ \vdots \\ a_N \end{bmatrix} + \begin{bmatrix} v_1(t_0) \\ v_1(t_1) \\ \vdots \\ v_1(t_N) \end{bmatrix}. \tag{4.3.13}$$

Denoting this by
$$y^{(1)} = H^{(1)} p^{(1)} + v^{(1)}. \tag{4.3.14}$$

We may use (4.3.14) to repeat exactly the earlier procedure. Equations (4.3.8) to (4.3.10) are used to find estimates of $\hat{a}_i$ and $\hat{\sigma}_{a_i}^2$ with Equation (4.3.14) as the observation model. The cross-correlation $\phi_{my}(\hat{\tau}_i)$ in this case is replaced by $\phi_{my}^{(1)}(\hat{\tau}_i)$. Using the defining relationships (4.3.13) and (4.3.14) this cross-correlation is given by

$$\phi_{my}^{(1)}(\tau) = \sum_{K=0}^{N-1} m(t_K + \tau) [y(t_K) - \hat{a}_1 m(t_K - \hat{\tau}_1)] \tag{4.3.15}$$

or

$$\phi_{my}^{(1)}(\tau) = \phi_{my}(\tau) - \hat{a}_1 \phi_{mm}(\tau - \hat{\tau}_1). \tag{4.3.16}$$

Equation (4.3.16) specifies that the new estimates $\hat{\tau}_i$, a_i, with $\hat{\sigma}_{a_i}^2$ are to be obtained from the ith pulse of the cross-correlation residue $\phi_{my}^{(1)}(\tau)$, i.e. from the original cross-correlation record $\phi_{my}(\tau)$ after subtracting $\hat{a}_i$ times the shifted autocorrelation $\phi_{mm}(\tau - \hat{\tau}_i)$.

An obvious practical implementation is to find in sequence the dominating peaks in the respective residuals obtained by successive subtraction of $\hat{a}_i \phi_{mm}(\tau-\tau_i)$ from $\phi_{my}(\tau)$, and $\hat{\tau}_k$, $\hat{a}_k$, and $\hat{\sigma}^2_{a_k}$ associated therewith. In portions of the y(t) record where return pulse overlap occurs, one should expect to obtain estimates of the delay and return parameters τ_i and a_i which are less accurate than those at well separated signals. The variance associated with $\hat{a}_i$, according to (4.3.8), is not affected by the presence of the return pulse overlap, El-Hawary and Vetter (1977). If greater accuracy than that attendant with the above procedure were required, the estimates $\hat{\tau}_i$, $\hat{a}_i$ with $\hat{\sigma}^2_{a_i}$ for a composite of return pulses might be used in a linearized form of the equation set for a window containing the composite signal, El-Hawary and Vetter (1978).

4.4 Estimating Physical Parameters

The estimates of the amplitude and delay parameters can be combined with the relations of the non-parametric models to obtain the media's physical parameters. As an example, using the PFC assumption we have $R = \frac{1}{2}\Delta \ln[Z]$. As a result, the coefficients a_1, and a_2, ... are related to the acoustic impedance by

$$a_1 = R_1 = \tfrac{1}{2}\ln \frac{Z_2}{Z_1}$$

$$a_2 = R_2 = \tfrac{1}{2}\ln \frac{Z_3}{Z_2}$$

.
.
.

or:

$$Z_2 = Z_1 e^{2a_1}$$
$$Z_3 = Z_2 e^{2a_2}$$

.
.
.

Thus, if Z_1 is known, knowledge of a_1, a_2, ... is sufficient to provide an estimate of the acoustic impedance profile. In underwater applications Z_1 corresponds to propagation in water with known ρ and C. The problem is more complex when multiple reflections are accounted for, El-Hawary (1982).

5. SUMMARY

The paper reviews a number of approaches to problems of modelling and parameter estimation for the response of the ocean subsurface and underlying media to acoustic input signals. Parameteric and non-parametric types of models are reviewed. This is followed by a discussion of the parameter estimation task. The practical problem of finding the actual physical properties of the media is also discussed in this paper.

6. REFERENCES

BERRYMAN, L.H., GOUPILLAUD, P.L. and WATERS, K.H. (1958). Reflections from Multiple Transition Layers. Geophysics 23, 223-243.

EL-HAWARY, F. and VETTER, W.J. (1977). Estimation of Subsurface Layer Parameters by Use of a Multiple-Reflection Model for Layered Media. In POAC (Proceedings of the Fourth International Conference on Port and Ocean Engineering Under Arctic Conditions held in St John's, Newfoundland, 1977). 1087-1099.

EL-HAWARY, F. and VETTER, W.J. (1978). Subsurface Layered Media Parameter Estimation Using a Linearized Multiple Reflection Model. In OCEANS '78 (Proceedings of IEEE Oceans Conference held in Washington, D.C., 1978). 713-717.

EL-HAWARY, F. and VETTER, W.J. (1980). Spatial Parameter Estimation for Ocean Subsurface Layered Media. Canadian Electrical Engineering Journal 5, 28-31.

EL-HAWARY, F. (1982). Seismic Signal Processing for Geotechnical Properties Applied to Marine Sediments. In Associate Committee for Research on Shoreline Erosion and Sedimentation, National Research Council of Canada (Proceedings on Workshop on Atlantic Coastal Erosion and Sedimentation, 1982).

GOUPILLAUD, P.L. (1961). An Approach to Inverse Filtering of Near-Surface Layer Effects from Seismic Records. Geophysics 26, 754-760.

MENDEL, J.M. and ASHRAFI, F.H. (1980). A Survey of Approaches to Solving Inverse Problems for Lossless Layered Media Systems. IEEE Transactions on Geoscience and Remote Sensing GE-18, 320-330.

PETERSON, R.A., FILLIPPONE, W.R. and CROKER, F.B. (1955). The Synthesis of Seismograms from Well Log Data. Geophysics XX, 516-538.

ROBINSON, E.A. (1967). Multichannel Time Series Analysis with Digital Computer Programs. Holden-Day, San Francisco.

SENGBUSH, R.L., LAWRENCE, P.L. and MCDONAL, F.J. (1961). Interpretation of Synthetic Seismograms. Geophysics XXIV, 138-157.

SILVIA, M.T. (1977). Deconvolution of Geophysical Time Series. Doctoral Dissertation, Northeastern University.

STREMLER, F.G. (1977, 1982). Introduction to Communication Systems. Second Edition, Addison-Wesley.

WUENSCHEL, P.E. (1960). Seismogram Synthesis Including Multiples and Transmission Coefficients. Geophysics XXV, 106-129.

Part 3
Spatial Time Series

TIME SERIES IN M DIMENSIONS: PAST, PRESENT, AND FUTURE

L.A. Aroian
Institute of Administration & Management, Union College, Schenectady, NY 12308 USA

This paper presents a history of time series in m dimensions; past and present main results; applications; and possible future worthwhile researches. The usual time series results are obtained for m = 0.

1. INTRODUCTION

Consider the characteristics of an event in time and space $z_{x,t}^p$, single valued where p indicates the pth characteristic of the event $x = (x_1, x_2, \ldots, x_m)$ at time t in any coordinate system. The event, identified as a time series in m dimensions, may be scientific, social, economic, business, industrial, geographic, political, physical, geologic, or meteorologic. Examples are storms, floods, air pollution, incidence of disease, income distribution, population problems, and social welfare. Modeling of time series in m dimensions provides insight into the nature of the event and may be used for estimation and forecasting. The statistical parameters may be assumed to be either constant or not through time and space dependent on the type of modeling. Spatial models are special cases where the time is not a variable, but some changes may be made depending on whether the spatial models are one-sided or two-sided. If m = 0 the usual time series at a point are obtained. Further general discussion of the main ideas are given in Aroian (1980), and for purely spatial processes, Aroian and Gebizlioglu (1981), and Gebizlioglu (1981), (1983).

Others who have engaged in spatial time series are Larimore (1977), Bennett (1979), and Pfeifer and Deutsch (1980a), (1980b), (1980c). Haugh (1984) analyzes the relationship between the models of this paper and those of Pfeifer and Deutsch.

2. EXAMPLES

Some simple examples of MA, AR, and ARMA models are given before formal mathematical definitions.

$$z_{x,t} = -\theta_1 a_{x,t-1} - \theta_2 a_{x-1,t-1} + a_{x,t}, \quad -\infty < t < t_0, \qquad (2.1)$$

$z_{x,t}$ has mean zero and variance $\sigma^2 > 0$, $a_{x,t}$ is a random error with mean zero and variance $\sigma_a^2 > 0$, θ's are constants, m = 1, a moving average model, MA of order one, represented by ——$\overset{x-1}{\cdot}$———$\overset{x}{\cdot}$———. An MA example, m = 2, is

$$z_{x,y,t} = -\theta_1 a_{x,y,t-1} -\theta_2 a_{x-1,y,t-1} -\theta_3 a_{x+1,y,t-1} -\theta_4 a_{x,y-1,t-1} -\theta_5 a_{x,y+1,t-1} +a_{x,y,t} \qquad (2.2)$$

represented by

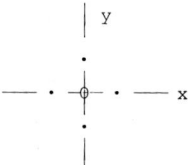

Two autoregressive models, AR, for m = 1 and 2 respectively are:

$$z_{x,t} = \phi_1 z_{x,t-1} +\phi_2 z_{x-1,t-1} +a_{x,t} \qquad (2.3)$$

$$z_{x,y,t} = \phi_1 z_{x,y,t-1} +\phi_2 z_{x-1,y,t-1} +\phi_3 z_{x,y-1,t-1} +\phi_4 z_{x-1,y-1,t-1} +a_{x,y,t} , \qquad (2.4)$$

with corresponding diagram,

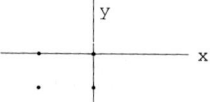

The autoregressive moving average model, ARMA, is the combination of the two models for any m, but for m = 1, a complete analysis will be made of this case

$$z_{x,t} = \phi_1 z_{x,t-1} +\phi_2 z_{x-1,t-1} -\theta_1 a_{x,t-1} -\theta_2 a_{x-1,t-1} +a_{x,t} . \qquad (2.5)$$

3. ASSUMPTIONS AND DEFINITIONS

The necessary formal definitions and assumptions are essential. The assumptions are: The characteristic of an event is $z_{x,t}$, $x = (x_1, x_2, \ldots, x_m)$, t, $-\infty < x < \infty$, $x-\ell = (x_1-\ell_1, x_2-\ell_2, \ldots, x_m-\ell_m)$. Weak stationarity is assumed in time and space as a minimum assumption:

$$\mu_z = E(z_{x,t}) = 0, \quad \sigma_z^2 = E(z_{x,t}-\mu_z)^2$$

$$Ea_{x,t} = 0, \quad \sigma_a^2 > 0, \quad \rho_{\ell,j} = \{E(z_{x,t} z_{x-\ell,t-k})\}/\sigma_z^2 , \qquad (3.1)$$

$\ell = (\ell_1, \ell_2, \ldots, \ell_m)$.

All second order moments exist. Note x may be any coordinate system; if x is dropped, the time series is the usual one at a point; if t is dropped, then the series is purely spatial. Note ℓ may be positive or negative but k is positive.

The MA model is defined:

$$z_{x,t} = \sum_{n=-\infty}^{\infty} \sum_{k=1}^{\infty} \psi_{n,k} a_{x-n,t-k} +a_{x,t} \qquad (3.2)$$

$n = (n_1, n_2, \ldots, n_m)$, $\sum_{n=-\infty}^{\infty} \equiv \sum_{n_m=-\infty}^{\infty} \cdots \sum_{n_1=-\infty}^{\infty}$

$a_{x,t}$ is an i.i.d. variable with mean zero, $\sigma_a^2 > 0$, $Ea_{x,t} z_{x-\ell, t-k} = 0$, unless $\ell = k = 0$. Usually $-p \leq n \leq q$, $1 \leq k \leq r$, so that

$$z_{x,t} = \sum_{n=-p}^{q} \sum_{k=1}^{r} \psi_{n,k} a_{x+n, t-k} + a_{x,t} \quad (3.3)$$

an MA model of temporal order r, spatial order $p_i + q_i$ in each spatial variable x_i, $1 \leq i \leq m$. As an example $m = r = 1$, replace the ψ's by θ's, to obtain (2.1).

The AR model is defined as:

$$z_{x,t} = \sum_{n=-\infty}^{\infty} \sum_{k=1}^{\infty} \phi_{n,k} z_{x+n, t-k} + a_{x,t} \quad (3.4)$$

temporal order r if $1 \leq k \leq r$, spatial order $p_i + q_i$, $-q \leq n \leq p$, in each variable, x_i, $1 \leq i \leq m$. An example is (2.3).

The ARMA model is defined as:

$$z_{x,t} = \sum_{n=-p}^{q} \sum_{k=1}^{r} \phi_{n,k} z_{x+n, t-k} - \sum_{n=-u}^{v} \sum_{k=1}^{s} \theta_{n,k} a_{x+n, t-k} + a_{x,t} \quad (3.5)$$

of order $r+s$ in the temporal domain, $q+p$, and $u+v$ in each spatial variable.

The general case would be $-\infty < n < \infty$, $-\infty < t < \infty$. This model is denoted by ARMA $(r,s;p,q;u,v)$. An example is (2.5).

All the preceding are univariate cases. Multivariate cases are considered later.

4. PROPERTIES OF MA, AR, AND ARMA MODELS

An MA model is stationary, and invertible dependent on $\psi_{n,k}$, and its representation as an infinite AR model. Define

$$B_t z_{x,t} = z_{x,t-1}, \quad F_t = B_t^{-1}, \quad B_{x_i} z_{x,t} = z_{x-\delta_i, t}, \quad (4.1)$$

$$F_{x_i} = B_{x_i}^{-1}, \quad \delta_i = (\delta_{i1}, \delta_{i2}, \ldots, \delta_{im}), \quad \delta_{ij} = \begin{cases} 1 & i=j \\ 0 & i \neq j \end{cases}.$$

Rewrite (3.3) in terms of (4.1):

$$z_{x,t} = (1 + \sum_{n=-p}^{q} \sum_{k=1}^{r} \psi_{n,k} F_x^n B_t^k) a_{x,t} \quad (4.2)$$

where $F_x^n = (F_{x_1}^{n_1}, F_{x_2}^{n_2}, \ldots, F_{x_m}^{n_m})$.

The characteristic function is

$$\Psi(B_x, B_t) = 1 + \sum_{n=-p}^{q} \sum_{k=1}^{r} \psi_{n,k} F_x^n B_t^k \quad (4.3)$$

and $a_{x,t} = \Psi^{-1}(B_x, B_t) a_{x,t}$.

For invertibility (4.3) must converge over all values of S_i, $i = \pm 1, \ldots, \pm m$, where $S_0 = \{B_t : |B_t| < 1\}$, $S_{-1} = \{B_{x_i} : |B_{x_i}| < 1\}$,

$$S_i = \{F_{x_i} : |F_{x_i}| > 1\}, \quad 1 \leq i \leq m. \tag{4.5}$$

Define $\gamma_{\ell,k} = E(z_{x,t} z_{x-\ell, t-k})$, the autocovariance between $z_{x,t}$ and $z_{x-\ell, t-k}$.

Theorem 4.1. The autocovariance function of an MA process may be found by multiplying (3.3) by $z_{x-\ell, t-k}$, where $\ell = (\ell_1, \ldots, \ell_m)$, and taking expectations; or better, the autocovariance function

$$\Gamma(B_x, B_t) = \sigma_a^2 \Psi(B_x, B_t) \Psi(F_x, F_t) \tag{4.6}$$

and $\gamma_{\ell,k}$ is the coefficient of both $B_x^\ell B_t^k$ and $B_x^{-\ell} B_t^{-k}$, with $\gamma_{00} = \sigma_a^2$ being the coefficient of $B_x^0 B_t^0$. Theorem 4.1 may be used to find $\rho_{\ell,k}$ for MA, AR, and ARMA models. The autocorrelation function is not symmetric in m dimensions; it is symmetric to the origin $x = 0$, $t = 0$. Thus $\rho_{\ell 0} = \rho_{-\ell 0}$, $\rho_{0k} = \rho_{0-k}$, $\rho_{\ell k} = \rho_{-\ell -k}$, $\rho_{\ell m} \neq \rho_{-\ell m}$, and $\rho_{0k} \neq \rho_{k 0}$. For $m = 1$, $\rho_{\ell k}$ will have the same value in the first and third quadrants, and second and fourth quadrants. For $m = 2$, four sets of equal ρ's occur. For m variables similar results hold, as reported in Perry's Ph.D. thesis. An important cut-off property for the MA models is given by:

Theorem 4.2. The autocorrelation function for a finite MA model is finite. Use theorem 4.1 for the proof. This cut-off property is an important way of determining where a process is MA, AR, or ARMA.

Theorem 4.3. If the conditions of invertibility are satisfied, every finite MA process in m dimensions may be expressed as an infinite AR model. From (4.4)

$$a_{x,t} = \sum_{d=0}^{\infty} \left(- \sum_{n=-q}^{p} \sum_{k=1}^{r} \Psi_{n,k} F_x^n B_t^k \right)^d z_{x,t} \tag{4.7}$$

and condition (4.5) must be satisfied.

Theorem 4.4. (The power spectrum). Let $B_t = \exp{-2\pi i f}$, $B_{x_j} = \exp{-2\pi i g_j}$, in the autocovariance function, in (4.6). The power spectrum of an MA process is:

$$p(f,g) = 2\sigma_a^2 |\Psi(\exp{-2\pi i f}, \exp{-2\pi i g})|^2 \tag{4.8}$$

$0 \leq |f| \leq 1/2$, $0 \leq |g_j| \leq 1/2$, $1 \leq j \leq m$.

Next AR processes are considered.

Theorem 4.5. The autocorrelation function of an AR process is found by multiplying (3.4) by $z_{x-\ell, t-k}$ and taking expectations. For $\ell = k = 0$

$$\sigma_z^2 = \sigma_a^2 \{ 1 - \sum_{n=-p}^{q} \sum_{k=1}^{r} \phi_{n,k} \rho_{n,k} \}^{-1} . \tag{4.9}$$

Note $\sigma_z^2 > 0$, and

$$\rho_{n,k} = \sum_{n=-p}^{q} \sum_{k=1}^{r} \phi_{n,k} \rho_{x+n, t-k} \tag{4.10}$$

Note $\rho_{n,k}$ satisfied the same form as (3.4) for all $\{n,k\}$, except $n = k = 0$. As $n,k \to \pm\infty$, $\rho_{n,k} \to 0$, provided the AR model is stationary.

Theorem 4.6. If conditions for stationarity (4.5) are satisfied, every finite AR process may be represented by an infinite MA process.

Proof: From (3.4)

$$z_{x,t} = \phi^{-1}(B_x, B_t) a_{x,t} = \sum_{d=0}^{\infty} (\sum\sum \phi_{n,k} F_x^n B_t^k)^d a_{x,t} \qquad (4.11)$$

Theorem 4.7. For invertibility of an MA model, set (4.3) equal to zero. We require that $|B_x| \leq 1$, $|B_t| \leq 1$, restricting $\theta_{n,k}(\psi_{n,k})$. For restrictions on $\phi_{n,k}$ the same method is used in the corresponding characteristic equation of AR model

$$\Phi(B_x, B_t) = (1 - \sum_{n=-q}^{p} \sum_{k=1}^{\infty} \phi_{n,k} B_x^n B_t^k) \qquad (4.12)$$

This theorem may also be stated for the MA(AR) models, or in fact for the ARMA model. The roots of the characteristic equation must lie outside the unit circle in each variable B_i when all other B's are set to one. These conditions yield stationarity in m dimensions, m+1 variables, and imply stationarity in every direction in the m+1 variables.

Theorem 4.8. The power spectrum for an AR process is:

$$p(f,g) = 2\sigma_a^2 |\Phi(\exp-2\pi i f, \exp-2\pi i g)|^{-2} \qquad (4.13)$$

$0 \leq |f| \leq 1/2$, $0 \leq |g_j| \leq 1/2$, $1 \leq j \leq m$.

Note the similarity to (4.8).

Theorem 4.9. Given an AR process to determine $\rho_{\ell,m}$, it is necessary to first find the corresponding MA expansion of the AR process. Then

$$\rho_{\ell,m} = \frac{C_{\ell,m}}{C_{00}} = \frac{\sum_{n=-\infty}^{\infty} \sum_{k=0}^{\infty} \psi_{n,k} \psi_{n+\ell, k+m}}{\sum_{n=-\infty}^{\infty} \sum_{k=0}^{\infty} \psi_{n,k}} \qquad (4.14)$$

where $C_{\ell,m}$ is the covariance $E(z_{x,t} z_{x-\ell, t-m})$. This is very important since the recurrence relationship of $\rho_{\ell,m}$ given by (4.10) does not provide a way of finding all the $\rho_{\ell,m}$.

5. THE PARTIAL AUTOCORRELATION FUNCTION

One of the most important questions that must be faced is the choice of m. Usually spatial considerations make clear the value of m. Otherwise choose that m which minimizes σ_z^2, but theoretical considerations should have the greater weight. If m is given, how is r the order determined? For the MA model use the cut-off property of the autocorrelation function. For the AR model the cut-off property is provided by the partial autocorrelation function. For m = 0, it is known that the last coefficient ϕ_r in an AR model with terms $\phi_1, \ldots, \phi_r$ is a partial coefficient of correlation and more importantly ϕ_{r+1}, i > r are all zero. For r = 1 in m dimensions m + 1 "last" coefficients ϕ will be partial coefficients of correlation, non-zero, but all other ϕ's would be exactly zero. In large samples these other ϕ's will tend to be small instead of being exactly zero as in the theoretical model.

The definition for the partial coefficient of correlation for AR models is $\phi_{ij} \neq 0$ for i = 1,2,...,m, j = 1,2,...,r, $-1 < \phi_{ir} < 1$ and $\phi_{ij} = 0$, i > m, j > r. For any AR models for ϕ_{ij}, i > m, j > r, one may prove $\phi_{ij} = 0$. See Taneja and Aroian (1980) for further details and proofs.

6. YULE-WALKER EQUATIONS, AR MODELS

The Yule-Walker equations for AR models in m dimensions will be given. Suppose we use (2.3), $z_{x,t} = \phi_1 z_{x,t-1} + \phi_2 z_{x-1,t-1} + a_{x,t}$. Multiply (2.3) by the coefficients of ϕ_1 and ϕ_2, take expected values and obtain the Yule-Walker equations

$$\rho_{01} = \phi_1 + \phi_2 \rho_{10}$$
$$\rho_{11} = \phi_1 \rho_{10} + \phi_2 \tag{6.1}$$

which, solved for ϕ_1 and ϕ_2, yield

$$\phi_1 = (\rho_{01} - \rho_{11}\rho_{10})/(1-\rho_{10}^2)$$
$$\phi_2 = (\rho_{11} - \rho_{01}\rho_{10})/(1-\rho_{10}^2) \tag{6.2}$$

These are the usual least squares equations. Hence the Yule-Walker equations in m dimensions for any order r may be found by the same method.

Thus the Yule-Walker equations will be of the form $\rho = P_\rho \phi$, ρ and ϕ are column vectors and P_ρ is the matrix of correlations. By the usual least squares theory

$$V(\hat{\phi}) \simeq n^{-1}(1-\rho'\hat{\phi})P_\rho^{-1}, \tag{6.3}$$

and if we substitute the sample values r for ρ then $V(\hat{\phi}) \simeq n^{-1}(1-r'\hat{\phi})R_\rho^{-1}$. For the example,

$$r' = (r_{01}, r_{11}), \quad R_2 = \begin{pmatrix} 1 & r_{10} \\ r_{10} & 1 \end{pmatrix}$$

$$\phi' = (\phi_1, \phi_2), \quad R_2^{-1} = (1-r_{10}^2)^{-1} \begin{pmatrix} 1 & -r_{10} \\ -r_{10} & 1 \end{pmatrix},$$ (6.4)

$$V(\hat{\phi}_1, \hat{\phi}_2) = n^{-1}(1 - r_{01}\hat{\phi}_1 - r_{11}\hat{\phi}_2) R_2^{-1}$$

$$\sigma^2_{\hat{\phi}_1} = \sigma^2_{\hat{\phi}_2} = n^{-1}(1-r_{10}^2)^{-1}(1-\hat{\phi}_1^2-\hat{\phi}_2^2-2\hat{\phi}_1\hat{\phi}_2 \, r_{10}),$$

and $\hat{\rho}_{\hat{\phi}_1, \hat{\phi}_2} = -r_{10}$. This method is general and if $a_{x,t}$ are distributed normally the estimates $\hat{\phi}$ are asymptotically unbiased, consistent, and approximately the maximum likelihood estimates; Perry and Aroian (1979), Aroian and Taneja (1980) have extended this method to the MA and ARMA models. The approximate maximum likelihood estimates for AR models are derived in a recent paper of Aroian (1981), and should be consulted. Detailed further properties of AR models and a typical example are given by Perry (1983).

7. ARMA MODELS

Denote the characteristic equation of the MA model by $\Theta(B_x, B_t)$ and that of the AR model by $\Phi(B_x, B_t)$, then

$$\Phi(B_x, B_t) z_{x,t} = \Theta(B_x, B_t) a_{x,t} \quad .$$ (7.1)

$B_x = (B_{x_1}, B_{x_2}, \ldots, B_{x_m})$, an m dimensional ARMA model of order $(r,s;p,q;u,v)$. Now (7.1) may be written as an infinite MA model:

$$z_{x,t} = \Theta(B_x, B_t) \Phi^{-1}(B_x, B_t) a_{x,t},$$ (7.2)

or as an infinite AR model

$$a_{x,t} = \Phi(B_x, B_t) \Theta^{-1}(B_x, B_t) z_{x,t} \quad .$$ (7.3)

Both results are important, particularly (7.2), since it is useful in finding $\rho_{m,n}$, and in forecasting $z_{x+\ell_1, t+\ell_2}$, $\ell_1, \ell_2 > 0$. The restrictions on ϕ_i and θ_i for invertibility and stationarity of (7.1) are exactly those of the MA and AR models jointly as given in theorem (4.7). An example is given for $m = 1$, $r_1 = r_2 = 2$ in Aroian and Taneja (1980). The autocovariance and partial autocorrelation functions are a combination of those of the MA and AR models.

The power spectrum is:

$$p(f,g) = 2\sigma_a^2 \frac{|\Theta(\exp-2\pi i f, \exp-2\pi i g)|^2}{|\Phi(\exp-2\pi i f, \exp-2\pi i g)|^2}$$

$0 \le |f| \le 1/2$, $0 \le |g_{x_j}| \le 1/2$, $1 \le x_j \le m$.

8. MULTIVARIATE AR MODELS

A concise description of multivariate AR time series models is given following Aroian (1979). Let

$$z^p_{x,t} = \sum_{p=1}^{s} \sum_{i=0}^{k} \sum_{j=1}^{\ell} \{\phi^{p1}_{ij} z^1_{x-i,t-j} + \phi^{p2}_{ij} z^2_{x-i,t-j} + \ldots + \phi^{pp}_{ij} z^p_{x-i,t-j}\} + a^p_{x,t}, \quad i \le j \quad (8.1)$$

where $x^p_{x,t}$ is the pth characteristic of event $z_{x,t}$; the mean vector is $\mu = (\mu_1, \mu_2, \ldots, \mu_s)$ which for convenience is assumed to be zero; the variance vector is $\sigma^2_z = (\sigma^2_1, \sigma^2_2, \ldots, \sigma^2_s)$; x, the space variable, is $(x_1, x_2, \ldots, x_m)$, t the time variable; stationarity is assumed over (x,t); the $a^p_{x,t}$'s are i.i.d. variables with mean zero, variances σ^2_{ap}, $a^p_{x,t}$ is independent of $z^p_{x,t-k}$ for $k \ne 0$. Further $w^p_{x,t} = z^p_{x,t} \sigma_p^{-1}$, and for convenience no change is made in the notation for $a^p_{x,t}$ under such a transformation. In actuality k may be infinite and m may in fact be infinite.

The AR process may be thought of as the output $z^p_{x,t}$ from a linear filter with transfer function $\phi_p^{-1}(B_x, B_t)$ when the input is white noise $a^p_{x,t}$. An example for $s = 2$, $\ell_1 = 1$, $k = 1$ is:

$$z^1_{x,t} = \phi^{11}_{01} z^1_{x,t-1} + \phi^{11}_{11} z^1_{x-1,t-1} + \phi^{12}_{01} z^2_{x,t-1} + \phi^{12}_{11} z^2_{x-1,t-1} + a^1_{x,t}$$
$$z^2_{x,t} = \phi^{21}_{01} z^1_{x,t-1} + \phi^{21}_{11} z^1_{x-1,t-1} + \phi^{22}_{01} z^2_{x,t-1} + \phi^{22}_{11} z^2_{x-1,t-1} + a^2_{x,t} \quad (8.2)$$

From (8.1) $z^1_{x,t}$ (and similarly for $z^p_{x,t}$):

$$z^1_{x,t} = (1 - \phi^{11}_{01} B_t - \phi^{11}_{11} B_x B_t - \phi^{21}_{01} B_t - \phi^{21}_{11} B_x B_t - \ldots - \phi^{p1}_{01} B_t - \phi^{p1}_{11} B_x B_t)^{-1} a^1_{x,t} \quad (8.3)$$

This may be expanded to obtain a p-variate moving average series in $a^1_{x,t}$.

The power spectra are:

$$p^p(f,g) = 2\sigma^2_a \left| \{1 - (\phi^{1p}_{01} + \phi^{1p}_{11} B_x) B_t - (\phi^{2p}_{01} + \phi^{2p}_{11} B_x) B_t - \ldots - (\phi^{pp}_{01} + \phi^{pp}_{11} B_x) B_t \} \right|^{-2} \quad (8.4)$$

The conditions on the ϕ's of $z^1_{x,t}$ are:

$$\left| \phi^{11}_{01} + \phi^{11}_{01} + \phi^{12}_{01} + \phi^{12}_{11} + \ldots + \phi^{1p}_{01} + \phi^{1p}_{11} \right| < 1$$

$$\left| \phi^{11}_{01} - \phi^{11}_{11} + \phi^{12}_{01} - \phi^{12}_{11} + \ldots + \phi^{1p}_{01} - \phi^{1p}_{11} \right| < 1 \quad (8.5)$$

Further results given in Aroian (1979), both for general s, and for $s = 2$, are the autocorrelation function, the cross correlation function between $z^1_{x,t}$ and $z^2_{x,t}$, solution of equations (8.2) for the ϕ's and σ^2_{ap}, determination of $\phi^p(B_x, B_t)$, the characteristic function, conditions for stationarity, the partial autocorrelation function, estimation, and extensions. Clearly similar results are readily obtained for MA and ARMA multivariate models after further very extensive researches.

9. ARMA MODEL EXAMPLES

The autoregressive moving average model (2.5) is defined for $m = 1$, $r_1 = r_2 = 2$ by:

$$z_{x,t} = \phi_1 z_{x,t-1} + \phi_2 z_{x-1,t-1} - \theta_1 a_{x,t-1} - \theta_2 a_{x-1,t-1} + a_{x,t} \qquad (9.1)$$

The characteristic $z_{x,t}$ in which one is interested is given as a linear combination of two past values of z at $(x,t-1)$ and $(x-1,t-1)$, $r_2 = 2$; and two past values of a variable $a_{x,t-1}$, $a_{x-1,t-1}$, random shocks, with an error $a_{x,t}$, $r_1 = 2$.

If $\phi_1 = \phi_2 = 0$, then

$$z_{x,t} = a_{x,t} - \theta_1 a_{x,t-1} - \theta_2 a_{x-1,t-1} \qquad (9.2)$$

is a moving average MA mdoel $m_1 = 1$, $r_1 = 2$; if $\theta_1 = \theta_2 = 0$

$$z_{x,t} = \phi_1 z_{x,t-1} + \phi_2 z_{x-1,t-1} + a_{x,t} \qquad (9.3)$$

is an autoregressive AR model $m = 1$, $r_2 = 2$.

If the t variable is omitted and if in (9.1) one replaces x and x-1 on the right hand side by x-1 and x-2 except in $a_{x,t}$ one gets a purely spatial model; while if the x variable is omitted, one gets the usual time series. A simpler ARMA model is obtained if one sets $\phi_1 = \theta_1 = 0$, or $\phi_2 = \theta_2 = 0$. The MA model, the AR model, and the ARMA model with their properties will be considered.

10. PROPERTIES OF THE MA MODEL

The autocorrelation function of the MA model (9.2), the values which θ_1 and θ_2 may have, and the corresponding AR model with an infinite set of coefficients will be given. From Voss et al (1980), the variance is

$$\sigma_z^2 = \sigma_a^2 (1 + \theta_1^2 + \theta_2^2) \qquad (10.1)$$

and the correlations are

$$\rho_{01}(1+\theta_1^2+\theta_2^2) = -\theta_1, \rho_{10}(1+\theta_1^2+\theta_2^2) = \theta_1 \theta_2, \rho_{11}(1+\theta_1^2+\theta_2^2) = -\theta_2 \qquad (10.2)$$

all other autocorrelations are zero. Note $\rho_{11} = \rho_{-1-1}$, $\rho_{01} = \rho_{0-1}$, $\rho_{10} = \rho_{-10}$. This cut-off property of the autocorrelation function aids in the identification of an MA model $m = 1$, $r_1 = 2$. A corresponding cut-off property is true for any MA model, (m,r). Rewrite (9.2) as

$$z_{x,t} = (1 - \theta_1 B_t - \theta_2 B_x B_t) a_{x,t} \qquad (10.3)$$

Rewrite (10.3) as

$$a_{x,t} = (1-\theta_1 B_t - \theta_2 B_x B_t)^{-1} z_{x,t} = \{\sum_{j=0}^{\infty} (\theta_1 + \theta_2 B_x)^j B_t^j\} z_{x,t} . \qquad (10.4)$$

This is an infinite AR model representation of (10.2) provided $|\theta_1|+|\theta_2|<1$, since $|B_x|<1$, $|B_t|<1$ for convergence of (10.4). Every MA process is stationary but for a representation as an infinite AR model, the condition $|\theta_1|+|\theta_2|<1$ is necessary. Note the coefficients of $z_{x-\ell,t-k}$ approach zero as $\ell, k \to \infty$. These results may also be obtained by successive substitutions in (10.2). We infer that

$$1+\theta_1^2+\theta_2^2 = -\theta_1/\rho_{01} = \theta_1\theta_2/\rho_{10} = -\theta_2/\rho_{11} , \qquad (10.5)$$

$$\theta_1 = -\rho_{10}/\rho_{11}, \ \theta_2 = -\rho_{10}/\rho_{01}, \ \theta_1 = (\theta_2\rho_{01})/\rho_{11} , \qquad (10.6)$$

$$\theta_2 = \frac{-\rho_{11} \pm \rho_{11}[1-4(\rho_{01}^2+\rho_{11}^2)]^{1/2}}{2(\rho_{01}^2+\rho_{11}^2)} . \qquad (10.7)$$

Hence $1-4(\rho_{01}^2+\rho_{11}^2) > 0$, or $\rho_{01}^2+\rho_{11}^2 < 1/4$, $\qquad (10.8)$

a circle of radius 1/2. From $\theta_2 = -\rho_{10}/\rho_{01}$, we conclude $\rho_{10} = -\rho_{01}\theta_2$.

From $|\theta_1|+|\theta_2|<1$, the added restrictions apply:

$$-1/\rho_{10}<1/\rho_{11}+1/\rho_{01}<1/\rho_{10} \text{ if } 0<\rho_{10}<1, \text{ or } 1/\rho_{10}<1/\rho_{11}+1/\rho_{01}<-1/\rho_{10} \qquad (10.9)$$

if $-1<\rho_{10} \leq 0$.

Thus the only values of $\{\rho_{01}, \rho_{10}, \rho_{11}\}$ possible are those in the intersection of (10.8) and (10.9).

In summary, given θ_1 and θ_2 then ρ_{01}, ρ_{10}, and ρ_{11} may be found from (10.2). If a representation of (9.2) as an infinite AR model (10.4) is desired, then $|\theta_1|+|\theta_2|<1$ and the coefficients of this AR model are given by (10.4). Additionally $\{\rho_{01}, \rho_{10}, \rho_{11}\}$ satisfy (10.6), (10.7), (10.8), and (10.9). Conversely, given $\rho_{01}, \rho_{10}, \rho_{11}$, which define an MA process (9.2), then if (10.4) is satisfied, all equations (10.5), (10.6), (10.7), and (10.8) are satisfied. Note ρ_{01} and ρ_{11} must satisfy (10.7) and consequently $\rho_{10} = -\theta_2\rho_{01}$. Hence given $\{\rho_{01}, \rho_{11}\}$, ρ_{10} is determined. If an arbitrary set $\{\rho_{01}, \rho_{10}, \rho_{11}\}$ are from some unknown time series in m dimensions the equations (10.5), (10.6), (10.7), (10.8) and $|\theta_1|+|\theta_2|<1$ will not be all satisfied. Thus, estimation proceeds iteratively from r_{01}, r_{10} and r_{11}. For instance, let $\theta_1 = .2$, $\theta_2 = -.5$, and $\sigma_z^2 = 1.29\sigma_a^2$, so that $\rho_{01} = -.1550$, $\rho_{10} = -.0775$, and $\rho_{11} = .3876$. All equations (10.5-10.9) are satisfied. Suppose the sample values are $r_{01} = -.14$, $r_{10} = -.10$, $r_{11} = .40$, searching iteratively to satisfy (10.5) as closely as possible we obtain $\hat{\theta}_1 = .22$, $\hat{\theta}_2 = -.55$, $1+\hat{\theta}_1^2+\hat{\theta}_2^2 = 1.3509$, and resulting $\hat{\rho}$'s from (10.5) are $r_{01} = -.163$, $r_{10} = -.090$, $r_{11} = .407$.

The MA model may be written in AR form as

$$a_{x,t} = \sum_0^\infty \sum_0^\infty \pi_{i,j} z_{x-i,t-j}, \quad \pi_{00} = 1, \quad \pi_{01} = \theta_1, \quad \pi_{11} = \theta_2, \tag{10.10}$$

$$\pi_{02} = \theta_1^2, \quad \pi_{12} = 2\theta_1\theta_2, \quad \pi_{22} = \theta_2^2, \quad \pi_{03} = \theta_1^3, \quad \pi_{13} = 3\theta_1^2\theta_2, \quad \pi_{23} = 3\theta_1\theta_2^2, \quad \pi_{33} = \theta_3^2,$$

$$\pi_{0j} = \theta_1 \pi_{0,j-1}, \quad \pi_{ij} = \theta_1 \pi_{i,j-1} + \theta_2 \pi_{i-1,j-1}, \quad \pi_{jj} = \theta_2 \pi_{j-1,j-1}.$$

11. SIMULATION OF THE MA MODEL

Given $\{\theta_1, \theta_2\}$, $|\theta_1| + |\theta_2| < 1$, we may simulate

$$z_{x,t} = a_{x,t} - \theta_1 a_{x,t-1} - \theta_2 a_{x-1,t-1} \tag{11.1}$$

by first generating a set of random numbers $(a_{x+i,t+j})$, $i = 0,1,2,\ldots,r+4$, $j = 0,1,2,\ldots,s+4$, from a distribution $\mu = 0$, $\sigma = 1$. This gives $(r+5)(s+5)$ values of $a_{x+i,t+j}$ random numbers. Then from these values of a's calculate $(r+5)(s+5)$ values of $(z_{x+i,t+j})$ with the chosen theoretical set $\{\theta_1,\theta_2\}$ using (11.1). All a's not generated as above are identically zero. Delete the first two rows, first two columns, the last three rows and the last three columns of the Z matrix. This produces a reduced Z matrix of rs values. Note the number five is chosen for convenience, so an inner Z matrix may represent the process properly. In the example $\theta_1 = .2$, $\theta_2 = -.5$, and the $a_{x,t}$'s are normally distributed with $\mu = 0$, $\sigma_a = 1$, $r = s = 10$, $z_{x,t}$ is given as the first number in each cell of Table 1. The $a_{x,t}$, the second number if each cell of Table 1 is not the 10x10 inner part of 15x15 original Z but is determined from the 10x10 Z matrix by using

$$a_{x,t} = z_{x,t} + \theta_1 a_{x,t-1} + \theta_2 a_{x-1,t-1} \tag{11.2}$$

with $\theta_1 = .2$, $\theta_2 = -.5$ where now $a_{x,t-1}$ and $a_{x-1,t-1}$ are taken as zero for initial starting values. Thus in the first row $a_{x,t} = z_{x,t}$. In the first column all unknown a's for $x < 1$ are zero. In this way a set of a's are built up slightly different from the inner 10x10 $a_{x,t}$'s obtainable from the 15x15 original Z matrix. This is done since in a sample of $z_{x,t}$ the $a_{x,t}$ must be determined in this way, although in practice the $\{\theta_1,\theta_2\}$ would be unknown and would be a set $\{\hat{\theta}_1,\hat{\theta}_2\}$ estimated from the sample $z_{x,t}$'s. In the second row all needed a's are known as soon as the first a is determined, and subsequently for each row, since the needed a's in any cell are found by using the $z_{x,t}$ in that cell and the $a_{x,t-1}$ in the cell directly above and the $a_{x-1,t-1}$ in the cell one to the left and above using (11.2). This set of a's is used in finding all further Z's for all subsequent sets $\{\theta_1,\theta_2\}$. From the 10x10 Z matrix determine $r_{m,n}$.

$\bar{z}$ (the sample mean), σ_z^2 (the sample variance of z), r_{01}, r_{10}, r_{11}, $\hat{\theta}_1$, $\hat{\theta}_2$, using (10.5-10.7) iteratively. New estimates $\hat{z}_{x,t}$ may be found using any desired $(\hat{\theta}_1, \hat{\theta}_2)$ particularly a set satisfying (10.5-10.7) and having a minimum prediction error variance $\sigma_\varepsilon^2 = \sum\sum (z_{x,t} - \hat{z}_{x,t})^2/(n-1)$. This is essentially a least squares procedure. The minimum variance procedure is to vary a set $(\hat{\theta}_1, \hat{\theta}_2)$ in the plane such that $\hat{\sigma}_\varepsilon^2$ is a minimum. The theoretical value of σ_ε^2 is

$$\sigma_\varepsilon^2 = \sigma_a^2 \{(\theta_1 - \hat{\theta}_1)^2 + (\theta_2 - \hat{\theta}_2)^2\}$$

where θ_1 and θ_2 are population values. This method is remarkably effective, working well for autoregressive models also (Perry and Aroian, 1979). For convenience in Table 1, $z_{x,t}$ is replaced by $z(x,t)$. Note $\hat{z}(x,t)_1$ are values from the minimum prediction error variance set $\hat{\theta}_1 = .208$, $\hat{\theta}_2 = -.493$ almost identical to the theoretical set $\{.2, -.5\}$; while $\hat{z}(x,t)_2$ are from $\{.333, -.418\}$ the least squares set most nearly satisfying (10.5-10.7), which leads to a theoretical $\{\rho_{01}, \rho_{10}, \rho_{11}\}$ set of $\{-.259, -.108, .325\}$ compared with an actual $\{r_{10}, r_{01}, r_{11}\}$ of $\{-.255, -.133, .299\}$ from $z_{x,t}$. The values of $r_{m,n}$ in Table 2 are those from $z_{x,t}$ (the notation of $z(x,t)$ in Table 1 is used for convenience). Further from (12.1) in the next section given $\{.2, -.5\}$, the 95% confidence interval for θ_1 is $(.036, .364)$ and for θ_2 is $(-.664, -.336)$ and $\{.333, -.418\}$ cuts these limits.

12. ESTIMATION

The variances and covariances of $\{\hat{\theta}_1, \hat{\theta}_2\}$ are approximated by use of the sample values r_{01}, r_{10}, r_{11} to estimate $\rho_{01}, \rho_{10}, \rho_{11}$ and the relationship of the variance-covariance matrix of MA models which are the same as those of AR models. From the results of Perry and Aroian (1979) given in Section 6:

$$\hat{\sigma}_{\hat{\theta}_1^2}^2 = \hat{\sigma}_{\hat{\theta}_2^2}^2 = n^{-1}(1-r_{10}^2)^{-1}(1-\theta_1^2-\theta_2^2-2\theta_1\theta_2 r_{10}), \text{ and } \hat{\rho}_{\hat{\theta}_1,\hat{\theta}_2} = -r_{10} . \quad (12.1)$$

These correspond to the least squares solution of the corresponding AR model and if the $a_{x,t}$ are assumed to be distributed normally, these are close approximations to the maximum likelihood estimates of $\{\sigma_{\theta_1}^2, \sigma_{\theta_2}^2\}$. (See Box and Jenkins (1976), p. 283, for similar case, m = 0.) The starting values of $a_{x,t}$ as noted in the simulation would need to be considered for the maximum likelihood solution. The method leading to (12.1) is general and applies to the m dimensional MA model. The minimum variance estimates $\{\hat{\theta}_1, \hat{\theta}_2\}$ and the resulting estimates of the correlation between θ_1 and θ_2 may be found as noted in the next section for any set of data.

TABLE 1
Values of $z(x,t)$, $a(x,t)$, $\hat{z}(x,t)_1$, $\hat{z}(x,t)_2$

t	x	1	2	3	4	5	6	7	8	9	10
1	$z(x,t)$	-1.478	0.815	-0.301	0.337	-1.277	0.527	0.210	-0.625	0.867	-1.668
	$a(x,t)$	-1.478	0.815	-0.301	0.337	-1.277	0.527	0.210	-0.625	0.867	-1.668
	$\hat{z}(x,t)_1$	-1.478	0.815	-0.301	0.337	-1.277	0.527	0.210	-0.625	0.867	-1.668
	$\hat{z}(x,t)_2$	-1.478	0.815	-0.301	0.337	-1.277	0.527	0.210	-0.625	0.867	-1.668
2	$z(x,t)$	-1.110	-1.410	1.541	0.126	-0.681	0.245	0.132	1.506	-1.602	-0.741
	$a(x,t)$	-1.406	-0.508	1.073	0.343	-1.105	0.989	-0.090	1.276	-1.116	-1.508
	$\hat{z}(x,t)_1$	-1.098	-1.406	1.538	0.125	-0.673	0.250	0.127	1.510	-1.605	-0.734
	$\hat{z}(x,t)_2$	-0.913	-1.397	1.514	0.106	-0.539	0.280	0.061	1.572	-1.666	-0.590
3	$z(x,t)$	-0.105	-1.199	-0.701	-0.649	1.620	0.425	0.148	-0.124	0.595	-0.060
	$a(x,t)$	-0.386	-0.598	-0.232	-1.116	1.228	1.175	-0.364	0.176	-0.266	0.196
	$\hat{z}(x,t)_1$	-0.094	-1.185	-0.706	-0.659	1.626	0.425	0.142	-0.134	0.595	-0.040
	$\hat{z}(x,t)_2$	0.082	-1.016	-0.802	-0.783	1.739	0.384	0.079	-0.286	0.639	0.232
4	$z(x,t)$	0.450	-0.484	-1.375	-0.278	-1.252	0.540	-0.343	0.767	-2.043	-0.903
	$a(x,t)$	0.372	-0.411	-1.123	-0.385	-0.448	0.161	-1.003	0.984	-2.184	-0.731
	$\hat{z}(x,t)_1$	0.453	-0.477	-1.369	-0.267	-1.254	0.522	-0.348	0.768	-2.042	-0.903
	$\hat{z}(x,t)_2$	0.501	-0.373	-1.295	-0.110	-1.324	0.283	-0.391	0.773	-2.022	-0.907
5	$z(x,t)$	-2.642	0.460	-0.264	-1.163	-0.555	2.482	1.121	-1.062	0.911	-1.187
	$a(x,t)$	-2.568	0.191	-0.284	-0.679	-0.452	2.739	0.840	-0.364	-0.018	-0.241
	$\hat{z}(x,t)_1$	-2.645	0.461	-0.252	-1.152	-0.549	2.484	1.128	-1.063	0.922	-1.166
	$\hat{z}(x,t)_2$	-2.692	0.484	-0.081	-1.020	-0.464	2.497	1.241	-1.111	1.121	-0.911
6	$z(x,t)$	1.355	0.040	-0.401	-0.857	2.023	-1.367	-0.648	0.602	-0.817	-1.329
	$a(x,t)$	0.842	1.362	-0.553	-0.851	2.272	-0.593	-1.850	0.109	-0.639	-1.368
	$\hat{z}(x,t)_1$	1.376	0.056	-0.400	-0.850	2.031	-1.386	-0.674	0.599	-0.814	-1.327
	$\hat{z}(x,t)_2$	1.696	0.225	-0.379	-0.743	2.139	-1.694	-0.984	0.581	-0.785	-1.295
7	$z(x,t)$	0.554	-0.191	1.936	0.196	-0.843	1.234	-0.742	0.557	0.775	0.542
	$a(x,t)$	0.722	-0.340	1.144	0.303	0.037	-0.021	-0.815	1.503	0.593	0.588
	$\hat{z}(x,t)_1$	0.547	-0.208	1.931	0.207	-0.855	1.223	-0.723	0.569	0.779	0.557
	$\hat{z}(x,t)_2$	0.442	-0.441	1.898	0.355	-1.075	1.127	-0.447	0.694	0.851	0.776
8	$z(x,t)$	0.012	2.509	-1.481	0.012	0.692	-0.702	0.124	-2.361	0.689	1.718
	$a(x,t)$	0.157	2.080	-1.082	-0.499	0.548	-0.725	-0.029	-1.653	0.056	1.539
	$\hat{z}(x,t)_1$	0.006	2.507	-1.488	0.002	0.690	-0.702	0.131	-2.367	0.674	1.709
	$\hat{z}(x,t)_2$	-0.084	2.495	-1.605	-0.122	0.662	-0.702	0.234	-2.494	0.487	1.591
9	$z(x,t)$	0.298	-1.145	2.832	0.688	-0.602	-1.427	-1.601	-0.279	-1.305	-0.124
	$a(x,t)$	0.329	-0.808	1.576	1.129	-0.243	-1.846	-1.244	-0.596	-0.467	0.156
	$\hat{z}(x,t)_1$	0.297	-1.163	2.826	0.700	-0.603	-1.425	-1.596	-0.266	-1.294	-0.137
	$\hat{z}(x,t)_2$	0.277	-1.434	2.805	0.843	-0.634	-1.376	-1.538	-0.057	-1.177	-0.333
10	$z(x,t)$	-1.572	1.603	-0.668	1.921	1.776	1.193	-1.140	-0.279	1.135	-1.252
	$a(x,t)$	-1.506	1.277	0.051	1.359	1.162	0.945	-0.466	0.224	1.339	-0.988
	$\hat{z}(x,t)_1$	-1.575	1.607	-0.675	1.901	1.770	1.209	-1.117	-0.266	1.143	-1.250
	$\hat{z}(x,t)_2$	-1.616	1.683	-0.811	1.642	1.716	1.458	-0.823	-0.098	1.246	-1.234

MEAN OF $z(x,t) = -0.0757732$ VAR. OF $z(x,t) = 1.29559$
MEAN OF $a(x,t) = -0.0620963$ VAR. OF $a(x,t) = 1.03742$
MEAN OF $\hat{z}(x,t)_1 = -0.0745114$ VAR. OF $\hat{z}(x,t)_1 = 1.29484$

The prediction error variance of $\hat{z}(x,t)_1 = .00009444634$
For $z(x,t)_1$, $\hat{\theta}_1 = .208$, $\hat{\theta}_2 = -.493$

MEAN OF $\hat{z}(x,t)_2 = -0.0575717$ VAR. OF $\hat{z}(x,t)_2 = 1.32942$

The prediction error variance of $\hat{z}(x,t)_2 = .0245026$
For $z(x,t)_2$, $\hat{\theta}_1 = .333$, $\hat{\theta}_2 = -.418$

Where $\hat{z}(x,t)_1$ is the prediction with minimum error without regard to (3-5) to (3-8)
$\hat{z}(x,t)_2$ is the prediction with minimum error with regard to (3-5) to (3-8).

TABLE 2
ESTIMATED CORRELATION COEFFICIENTS
$r_{m,n}$ from $z_{x,t}$

m \ n	0	1	2	3	4	5
-5	-.02440	-.08929	.01751	.09018	-.07126	.01399
-4	-.00683	-.01955	-.15060	.12190	-.06967	.06822
-3	.00744	-.21941	-.09758	-.16610	.14620	.06586
-2	-.04944	-.02406	.09758	.02137	.03338	.00839
-1	-.13263	.15886	-.05186	.07963	-.08132	.03068
0	1.0000	-.25518	.10146	.01626	-.05956	.02659
1	-.13263	.29921	.03112	-.13549	.03900	-.12985
2	-.04944	.05475	.02902	.05822	-.10431	.04176
3	.00744	-.00829	-.04685	.12587	.01059	.00946
4	-.00683	-.08132	.06358	-.00833	.07500	-.04081
5	-.02440	.00370	-.06670	-.05665	.05308	.01927

13. ARMA MODELS

The results of Oprian et al. (1980) are used freely in addition to some new results mainly in the determination of values of $\{\phi_1, \phi_2, \theta_1, \theta_2\}$ as related to the values of $\{\rho_{01}, \rho_{10}, \rho_{11}, \rho_{-1,1}\}$. The results are:

$$\sigma_z^2 = \sigma_a^2 \{1 - \theta_1(\phi_1 - \theta_1) - \theta_2(\phi_2 - \theta_2)\} (1 - \phi_1 \rho_{01} - \phi_2 \rho_{10})^{-1}$$

$$\rho_{01} = \phi_1 + \phi_2 \rho_{10} - \theta_1 \sigma_a^2 / \sigma_z^2; \quad \theta_1 \sigma_a^2 / \sigma_z^2 = \phi_1 + \phi_2 \rho_{10} - \rho_{01} \quad (13.1)$$

$$\rho_{11} = \phi_1 \rho_{10} + \phi_2 - \theta_2 \sigma_a^2 / \sigma_z^2; \quad \theta_2 \sigma_a^2 / \sigma_z^2 = \phi_1 \rho_{10} + \phi_2 - \rho_{11}$$

$$\rho_{10} = \phi_1 \rho_{1-1} + \phi_2 \rho_{01} - \theta_2 (\phi_1 - \theta_1) \sigma_a^2 / \sigma_z^2 \ .$$

The second and third equations may be solved for $\theta_i \sigma_a^2 / \sigma_z^2$ and these results substituted in the first and fourth equations to obtain $\{\phi_1, \phi_2, \theta_1, \theta_2\}$. If $\phi_1 = \phi_2 = 0$, or $\theta_1 = \theta_2 = 0$, the corresponding MA or AR models are obtained. For $m, n \geq 2$,

$$\rho_{m,n} = \phi_1 \rho_{m,n-1} + \phi_2 \rho_{m-1,n-1} \ . \quad (13.2)$$

The ARMA model may be represented as an infinite moving MA model

$$\sum_{i=0}^{\infty} \sum_{j=0}^{\infty} \psi_{ij} a_{x-i, t-j} \quad (13.3)$$

or an infinite AR model

$$\sum_{i=0}^{\infty} \sum_{j=0}^{\infty} \pi_{ij} z_{x-i, t-j} \ , \quad (13.4)$$

restrictions already noted $|\phi_1| + |\phi_2| < 1$ and $|\theta_1| + |\theta_2| < 1$, naturally restricting the values of $\{\rho_{01}, \rho_{10}, \rho_{11}\}$ also. The infinite MA model is given by

$$z_{x,t} = (1 - \theta_1 B_t - \theta_2 B_x B_t)(1 - \phi_1 B_t - \phi_2 B_x B_t)^{-1} a_{x,t}$$

$$= (1 - \theta_1 B_t - \theta_2 B_x B_t) \{ \sum_{i=0}^{\infty} (\phi_1 + \phi_2 B_x)^i B_t^i \} a_{x,t} \ , \quad (13.5)$$

with $\psi_{00} = 1$, $\psi_{01} = \phi_1 - \theta_1$, $\psi_{11} = \phi_2 - \theta_2$, $\psi_{02} = \phi_1$,

$$\psi_{12} = -\phi_1 \theta_2 + 2\phi_1 \phi_2 - \phi_2 \theta_1, \ldots, \psi_{0j} = \phi_1 \psi_{0,j-1} \ , \quad (13.6)$$

$$\psi_{ij} = \phi_1 \psi_{i,j-1} + \phi_2 \psi_{i-1,j-1}, \quad \psi_{ii} = \phi_2 \psi_{i-1,i-1} \ .$$

Similarly the ARMA model as an AR model:

$$\pi_{00} = 1, \quad \pi_{01} = \theta_1 - \phi_1, \quad \pi_{11} = \theta_2 - \phi_2, \quad \pi_{02} = \theta_1(\theta_1 - \phi_1)$$

$$\pi_{12} = -\theta_1 \phi_2 + 2\theta_1 \theta_2 - \theta_2 \phi_1, \quad \pi_{22} = \theta_2(\theta_2 - \phi_2) \ , \quad (13.7)$$

$$\pi_{0j} = \theta_1 \pi_{0,j-1}, \pi_{ij} = \theta_1 \pi_{i,j-1} + \theta_2 \pi_{i-1,j-1}, \quad \pi_{ii} = \theta_2 \pi_{i-1,i-1} \ .$$

For $\phi_1 = .2$, $\phi_2 = -.6$, $\theta_1 = .2$, $\theta_2 = -.5$, we have

$$\psi_{00} = 1, \quad \psi_{11} = -.1, \quad \psi_{12} = -.02, \quad \psi_{22} = -.6, \quad \psi_{13} = -.004, \quad (13.8)$$

$\psi_{23} = .024$, $\psi_{33} = -.036$, etc.

The autocorrelation function must be found by using (13.5)

$$\rho_{m,n} = (\sum\sum \psi_{ij} \psi_{i-m,j-n}) (\sum\sum \psi_{ij}^2)^{-1}, \quad m = n \neq 0 \qquad (13.9)$$

and

$$\sigma_z^2 = \sigma_a^2 \sum\sum \psi_{ij}^2. \qquad (13.10)$$

Thus $\sigma_z^2/\sigma_a^2 = 1.01741$, obtained from (13.9) and (13.10), must be used in (13.1) to solve for $\{\rho_{01}, \rho_{10}, \rho_{11}, \rho_{1-1}\} = \{.058, -.091, -.127, -.213\}$ as compared to the MA model $\{-.155, -.078, .389, 0\}$, and the AR model $\{.33, -.21, -.64, -.07\}$. In fact, had $\phi_1 = .2$, $\phi_2 = -.6$, $\theta_1 = .2$, $\theta_2 = -.6$, been chosen, then the ARMA model would have been reduced to $(1 - .2B_t + .6B_x B_t) z_{x,t} = (1 - .2B_t + .6B_x B_t) a_{x,t}$, or $z_{x,t} = a_{x,t}$!

Conversely, had a sample set of s_z^2/s_a^2, $\{r_{01}, r_{10}, r_{11}, r_{1-1}\}$ been given, estimates $(\hat{\phi}_1, \hat{\phi}_2, \hat{\theta}_1, \hat{\theta}_2)$ could be found from (13.1) as already given by substituting the sample estimates for the population values. These results, with the exception of (13.1) and the methods of solution given there, are due to Robert Perry in his Ph.D. thesis.

How may the estimates of $\{\phi_1, \phi_2, \theta_1, \theta_2\}$ in a sample of n be found? One method is to choose the set $\{\hat{\phi}_1, \hat{\phi}_2, \hat{\theta}_1, \hat{\theta}_2\}$ which minimizes s_ε^2, the sample estimates of σ_ε^2 as done in section 10 by varying the four constants numerically. The other method is to find the corresponding AR model though π_{22} and then apply the method given in section 12.

14. FORECASTING

Now that the models have been defined, their properties given, it is necessary to show how these results may be used for forecasting. It should be realized that in forecasting it is not necessary to assume that the future repeats the past. All that is needed is to assume that the errors of the past will be repeated in the future, although the events themselves will not necessarily be repeated. However, it is assumed that the model type holds true from the past to the future. Actually characteristics of many processes may change with time and those which change slowly or not at all are the easiest to model and the most useful to apply. If, however, a process changes from stationary to nonstationary, it is possible to use finite differences in two or more variables just as in zero dimensional processes, finite differences in one variable are used. It is assumed in the forecasting models that values are fixed and exact. Later these restrictions may be relaxed. The general case of forecasting in m dimensions is complicated, depending on whether the values of $z_{x,t}$ involving the space vector x are assumed known or unknown for x+i as well as x-i. However, it is assumed that $z_{x,t}$, $a_{x,t}$, $z_{x-\ell_1,t-\ell_2}$, $a_{x-\ell_2,t-\ell_2}$ are known for $\ell_1, \ell_2 \geq 0$, and in all cases if $\ell_2 > 0$, $z_{x-\ell_1,t+\ell_2}$ are unknown. Whether $a_{x+\ell_1,t-\ell_2}$ is known or unknown, $\ell_1 > 0$ depends on the particular process. Only processes one-sided in space or time are modeled.

The exact value of $z_{x+\ell_1,t+\ell_2}$ is given by

$$z_{x+\ell_1,t+\ell_2} = \sum_{i=0}^{\infty} \sum_{j=0}^{\infty} \psi_{i,j} a_{x-i+\ell_1, t-j+\ell_2}, \quad \ell_1, \ell_2 > 0. \tag{14.1}$$

The forecast of $z_{x,t}$ at $(x+\ell_1, t+\ell_2)$, given $z_{x,t}$, denoted by $\hat{z}_{x,t}(\ell_1,\ell_2)$, will involve terms in $a_{x-r,t-s}$, $r,s > 0$, and the error is denoted by

$$e_{x,t}(\ell_1,\ell_2) = z_{x+\ell_1,t+\ell_2} - \hat{z}_{x,t}(\ell_1,\ell_2). \tag{14.2}$$

Thus

$$\hat{z}_{x,t}(\ell_1,\ell_2) = \sum_{i=0}^{L_1} \sum_{j=0}^{L_2} \psi_{i,j} a_{x-i+\ell_1, t-j+\ell_2}, \quad L_1 \leq \ell_1, \, L_2 \leq \ell_2, \tag{14.3}$$

but L_1 and L_2 may not equal ℓ_1 and ℓ_2 simultaneously.

$$E\hat{z}_{x,t}(\ell_1,\ell_2) = 0$$

$$\sigma^2_{e_{x,t}}(\ell_1,\ell_2) = \sigma_a^2 \sum_0^{L_1} \sum_0^{L_2} \psi_{ij}^2, \quad \psi_{00} = 1, \, L_1 \leq \ell_1, \, L_2 \leq \ell_2. \tag{14.4}$$

If the more general ARMA model is first changed to an infinite MA model given by (13.5), then the previous formulas apply.

We consider the example

$$z_{x,t} = a_{x,t} + \psi_{01} a_{x,t-1} + \psi_{11} a_{x-1,t-1} + \psi_{02} a_{x,t-2} + \psi_{12} a_{x-1,t-2} + \psi_{22} a_{x-2,t-2}. \tag{14.5}$$

The true result at $(x+\ell_1, x+\ell_2)$ is

$$z_{x+\ell_1,t+\ell_2} = \sum_0^2 \sum_0^2 \psi_{ij} a_{x-i+\ell_1, t-j+\ell_2}, \quad \psi_{00} = 1, \, i \leq j.$$

Let $\ell_1 = 1$, $\ell_2 = 2$, $L_1 = 1$, $L_2 = 2$

$$\hat{z}_{x,t} = \sum_{i=0}^{2} \sum_{j=0}^{2} \psi_{ij} a_{x-i+\ell_1, t-j+\ell_2} - \sum_{i=0}^{1} \sum_{j=0}^{2} \psi_{ij} a_{x-i+\ell_1, t-j+\ell_2}$$

$$\hat{z}_{x,t} = \psi_{12} a_{x,t} + \psi_{22} a_{x-1,t} \tag{14.6}$$

since all other a's are unknown and assumed to be zero.

Thus

$$e_{x,t}(1,2) = a_{x+1,t+2} + \psi_{01} a_{x+1,t+1} + \psi_{11} a_{x,t+1} + \psi_{02} a_{x+1,t}, \quad \text{and} \tag{14.7}$$

$$Ee_{x,t}(1,2) = 0, \quad \sigma^2_{e_{x,t}}(1,2) = (1+\psi_{01}^2+\psi_{11}^2+\psi_{02}^2)\sigma_a^2 \tag{14.8}$$

In the MA case (9.2):

$$z_{x+\ell_1,t+\ell_2} = a_{x+\ell_1,t+\ell_2} - \theta_1 a_{x+\ell_1,t+\ell_2-1} - \theta_2 a_{x+\ell_1-1,t+\ell_2-1} , \tag{14.9}$$

$$e_{x+\ell_1,t+\ell_2} = a_{x+\ell_1,t+\ell_2} - \theta_1 a_{x+\ell_1,t+\ell_2-1} - \theta_2 a_{x+\ell_1,t+\ell_2-1} \tag{14.10}$$

$$\hat{z}_{x,t}(\ell_1,\ell_2) = 0 \tag{14.11}$$

$$\sigma^2_{e_{x,t}}(\ell_1,\ell_2) = \sigma_a^2(1+\theta_1^2+\theta_2^2) , \tag{14.12}$$

for $\ell_1 \geq 2$ or $\ell_2 \geq 2$. The other cases are

$$\hat{z}_{x,t}(0,1) = -\theta_1 a_{x,t} - \theta_2 a_{x-1,t}, \quad \hat{z}_{x,t}(1,0) = -\theta_2 a_{x,t-1}$$

$$e_{x,t}(0,1) = a_{x,t+1}, \quad e_{x,t}(1,0) = a_{x+1,t} - \theta_1 a_{x+1,t-1}$$

$$\hat{z}_{x,t}(1,1) = -\theta_2 a_{x,t}, \quad e_{x,t}(1,1) = a_{x+1,t+1} - \theta_1 a_{x+1,t}, \text{ and} \tag{14.13}$$

$$\sigma_e^2(0,1) = \sigma_a^2, \quad \sigma_e^2(1,0) = \sigma_a^2(1+\theta_1^2), \quad \sigma_e^2(1,1) = \sigma_a^2(1+\theta_1^2) .$$

Forecasts and forecast errors are correlated. In any MA model (and by extension any AR model or ARMA model) the forecast errors are:

$$e_{x,t}(0,1) = a_{x,t+1}, \quad e_{x,t}(1,0) = a_{x+1,t} - \theta_1 a_{x+1,t-1} , \text{ and}$$

$$e_{x,t}(1,1) = a_{x+1,t+1} - \theta_1 a_{x+1,t} . \tag{14.14}$$

Hence

$$\rho\{e_{x,t}(0,1), e_{x,t}(1,0)\} = \rho\{e_{x,t}(0,1), e_{x,t}(1,1)\} = 0 ,$$

$$\rho\{e_{x,t}(1,0), e_{x,t}(1,1)\} = -\theta_1(1+\theta_1^2)^{-1} . \tag{14.15}$$

Correlations among other forecast errors may be found. Forecasts are also correlated. From (14.13)

$$\rho\{\hat{z}_{x,t}(0,1), \hat{z}_{x,t}(1,0)\} = \rho\{\hat{z}_{x,t}(1,0), \hat{z}_{x,t}(1,1)\} = 0 , \quad \text{but}$$

$$\rho\{\hat{z}_{x,t}(0,1), \hat{z}_{x,t}(1,1)\} = \theta_1(\theta_1^2+\theta_2^2)^{-1/2} . \tag{14.16}$$

The autocorrelation function for forecasts and forecast errors may be found in a similar manner.

It has been assumed in the foregoing that all parameters θ_i are known where in fact estimates of θ_i are used, including the sample mean $\bar{z}_{x,t}$. A discussion of this matter is given in Box and Jenkins (1976), pages 267-268. For the mean, an additional source of variation, the σ_a^2/n is added to the formula for σ_e^2 given in (14.4) and is important if n is fairly small, $n \leq 100$, depending on the model. The power spectra of the discussed models may be found in Voss et al. (1980), Taneja and Aroian (1980), and Oprian et al. (1980).

15. APPLICATIONS

The first application of the theory by Campbell and Schaeffer (1979) used a nonstationary time series in one dimension to examine the changes occurring at sampling stations on the Chicago Sanitary and Ship Canal. They concluded that "Using data from upstream sampling sites, downstream levels of dissolved oxygen, total dissolved solids, nitrates, and ammonia are (more) accurately predicted (than if only the local data were used). The method is simple, insensitive to extreme values and responsive to changes in the system." Briefly, straightforward regression theory was used since the data were deemed not to be stationary. The model used was

$$z_{x,t} = \phi_{x-1,t-1} z_{x-1,t-1} + \phi_{x-1,t-2} z_{x-1,t-2} + \phi_{x-2,t-1} z_{x-2,t-1} + \phi_{x-2,t-2} z_{x-2,t-2} + a_{x,t}.$$

In another example the heights of the lower Mohawk River (above sea level) are estimated and predicted, Perry and Aroian (1979). Flooding occurs if the height of the ground surrounding the river is less than the height of the water due to the rise in the river. The model explained in (6.2) is used to predict the future heights from past data. Observations were taken at six locations on the river above and below each of three locks of the canal. Each observation was the extreme high in each six-month period, January-June, July-December, for the years 1967-76. The data at the top of each lock always had a much smaller standard deviation than the data below the lock, and the height of the river below the lock was always less than the height above the lock, since rivers always proceed downhill. Hence the transformed variable $w_{x,t} = (z_{x,t} - \bar{z})/s_x$ ($\bar{z}$ is the estimated mean, s_x the estimated σ_x) was used. A complete modeling using the partial autocorrelation function showed $w_{x,t} = -.12 w_{x,t-1} + .26 w_{x-1,t-1} + a_{x,t}$. Predicted and actual values for the first six months and the second six months were given for $z_{x,t}$ by reversing the transformations.

16. CONCLUSIONS

Many challenges remain: exact maximum likelihood results for estimation of ARMA models; χ^2 test and Kolmogarov-Smirnov tests of adequacy now in process of solution; intervention analysis; further results on confidence intervals for ϕ_{ij} and θ_{ij}; multivariate analysis; the frequency domain in m dimensions almost untouched, now under study; use of new techniques in regression analysis, particularly strong consistent estimates---more applications and examples; and nonlinear models. Perry's thesis (1981) includes nine computing programs useful for AR models, M = 1. The suggestions of the referees have been most useful. Special thanks are due Oliver D. Anderson and Keith Ord.

REFERENCES

ANSLEY, CRAIG F. (1979). An algorithm for the exact likelihood of a mixed autoregressive-moving average process. Biometrika 66, 1, 59-65.

AROIAN, L.A. (1979). Multivariate autoregressive time series in m dimensions. Proceedings of the Business and Economics Section, American Statistical Society Annual Meeting, 585-590.

AROIAN, L.A. (1980). Time series in m dimensions, definition, problems and prospects. Communications in Statistics, Simulation and Computation, B9, 5, 453-465.

AROIAN, L.A. (1981). Time series in m dimensions, maximum likelihood estimation for autoregressive models. Proceedings of the Business and Economics Section, American Statistical Society Annual Meeting, 334.

AROIAN, L.A. and GEBIZLIOGLU, OMER (1980). Time series in m dimensions, spatial models. Proceedings of the Statistical Computing Section, American Statistical Association Annual Meeting, 320-325.

AROIAN, L.A. and SCHMEE, JOSEF (1980). General results: time series in m dimensions. Proceedings of the Eleventh Annual Modeling and Simulation Conference, University of Pittsburgh, Pittsburgh, Pennsylvania, 1517-1522.

AROIAN, L.A. and TANEJA, VIDYA (1980). Some simple examples of time series in m dimensions. Modeling and Simulation, Instrument Society of America, Research Triangle Park, North Carolina, V. II, part 4, 1505-1516.

ATKINSON, A.C. and PEARCE, M.C. (1976). The computer generation of beta, gamma, and normal random variables. Journal of the Royal Statistical Society, A, 139, 4, 431.

BENNETT, R.J. (1979). Spatial Time Series. Pion Ltd., London.

BESAG, J.E. (1972). On the correlation structure of some two-dimensional stationary processes. Biometrika, 59, 1, 43-48.

BOX, G.E.P. and JENKINS, G.M. (1976). Time Series Analysis: Forecasting and Control. Holden-Day, Inc., San Francisco, California.

CAMPBELL, D. and SCHAEFFER, D.J. (1979). Pollution in the Chicago Sanitary and Ship Canal. Environmental Management 3, 283-288.

CHEN, G.J., LAI, T.L., and WEI, C.A. (1981). Convergence systems and strong consistency of least squares estimates in regression models. Journal of Multivariate Analysis, 11, 319-333.

CLIFF, A.D. and ORD, J.K. (1981). Spatial Processes: Models and Applications. Pion Ltd., London.

GEBIZLIOGLU, OMER (1981). Time Series in M Dimensions: Spatial Models. Ph.D. Thesis, Union College. Available from University Microfilms, Ann Arbor, MI, USA.

GEBIZLIOGLU, OMER (1983). On the modelling, estimation and hypothesis testing for spatial ARMA processes. Special Topics International Time Series Meeting, Toronto, Canada.

HAUGH, L.D. (1984). An introductory overview of some recent approaches to modelling spatial time series. In Time Series Analysis: Theory and Practice 5. Proceedings of the International Conference held at Nottingham University, England. Ed.: O.D. Anderson, North-Holland, Amsterdam and New York, 287-302.

LARIMORE, W.E. (1977). Statistical inference on stationary random fields. Proceedings of the IEEE 65 (6), 961-970.

LJUNG, GRETA (1982). Testing the adequacy of a fitted autoregressive-moving average model. American Statistical Association Meeting, Cincinnati, Ohio.

LJUNG, GRETA and BOX, G.E.P. (1979). The likelihood function of stationary autoregressive-moving average models. Biometrika 66, 2, 265-270.

NICHOLLS, D.F. and HALL, A.D. (1979). The exact likelihood function of multivariate autoregressive-moving average models. Biometrika 66, 2, 259-264.

OPRIAN, C., TANEJA, V., VOSS, D. and AROIAN, L.A. (1980). General considerations and interrelationships between MA and AR models, time series in m dimensions, the ARMA model. Communications in Statistics, Simulation and Computation, B9, 5, 515-532.

ORD, J.K. (1975). Estimation methods for models of spatial interaction. Journal of the American Statistical Association 70, 120-126.

PERRY, ROBERT (1981). Time Series in m dimensions, Autoregressive Models, M=1, Ph.D. Thesis, Union College. May be obtained from University Microfilms, Ann Arbor, MI, USA.

PERRY, R. and AROIAN, L.A. (1979). Of time and the river; time series in m dimensions, the one dimensional autoregressive model. Proceedings of Statistical Computing Section, American Statistical Association Meeting, 383-388.

PFEIFER, P.E. and DEUTSCH, S.J. (1980a). A three-stage iterative procedure for space-time modeling. Technometrics 22, 1, 35-47.

PFEIFER, P.E. and DEUTSCH, S.J. (1980b). Independence and sphericity tests for the residuals of space-time ARMA models. Communications in Statistics, Simulation and Computation, B9, 5, 533-549.

PFEIFER, P.E. and DEUTSCH, S.J. (1980c). Stationarity and invertibility regions for low order STARMA models. Communications in Statistics, Simulation and Computation, B9, 5, 551-562.

PHADKE, M.S. and KEDEM, G. (1978). Computation of the exact likelihood function of multivariate moving average models. Biometrika 65, 3, 511-519.

QUIGG, DAVID L. (1983). ARMA model identification--time series in m dimensions. Special Topics International Time Series Meeting, Toronto, Canada.

RIPLEY, BRIAN D. (1981). Spatial Statistics. New York: John Wiley and Sons.

SHARP, W.E. and AROIAN, L.A. (1983). Statistical space series on a square net. Special Topics International Time Series Meeting, Toronto, Canada.

TANEJA, V. and AROIAN, L.A. (1980). Time series in m dimensions, autoregressive models. Communications in Statistics, Simulation and Computation, B9, 5, 491-513.

VOSS, D., OPRIAN, C., and AROIAN, L.A. (1980). Moving average models, time series in m dimensions. <u>Communications in Statistics, Simulation and Computation</u>, B9, 467-489.

WANG, S.G. and WU, C.F. (1983). Further results on the consistent direction of least squares estimators. <u>Annals of Statistics</u>, 11-4, 1257-1262.

AUTOREGRESSIVE MODELS IN M DIMENSIONS, M=1, THEORY AND EXAMPLES

Robert James Perry
842 Hampton Avenue
Schenectady, NY 12309 USA

Leo A. Aroian
Institute of Administration and Management
Union College
Schenectady, NY 12308 USA

For autoregressive models, AR, M=1, formulas of estimation, prediction, forecasting, time spatial differences and transformations are provided, methods general for all M.

1. INTRODUCTION

The properties of autoregressive, AR, models in M dimensions $M \geq 1$ are stated, using the notation of Aroian (1983). The most general theoretical AR model is:

$$z_{x,t} = \sum_{n=-\infty}^{\infty} \sum_{k=1}^{\infty} \phi_{n,k} \, z_{x+n,t-k} + a_{x,t} \tag{1.1}$$

temporal order, r, if $1 \leq k \leq r$; spatial order, $p_i + q_i$, if $-q \leq n \leq p$, where x is a vector. The theorems governing AR processes given in section 4 of Aroian (1984) are used. The properties of the autocorrelation function, nonstationary models using differences, estimation and forecasting in general for space-time AR processes, are explained. One practical example is given, that of highway deterioration.

2. PROPERTIES OF AR, M=1

The AR model M=1 is:

$$z_{x,t} = \sum_{n=0}^{p} \sum_{k=1}^{r} \phi_{n,k} \, z_{x-n,t-k} + a_{x,t} \, , \tag{2.1}$$

a more convenient notation and special case of (1.1), x a vector, temporal order r, and spatial order p. It is assumed that $E(z_{x,t}) = 0$. Stationarity is assured by $\sum_{n=0}^{p} \sum_{k=1}^{r} | \phi_{n,k} | < 1$.

The variance is:

$$\sigma_z^2 = \sigma_a^2 \{1 - \sum_{n=0}^{p} \sum_{k=1}^{r} \phi_{n,k} \, \rho_{n,k} \}^{-1} , \tag{2.2}$$

the autocorrelation function, $\rho_{c,d}$ is:

$$\rho_{c,d} = \sum_{n=0}^{p} \sum_{k=1}^{r} \phi_{n,k} \, \rho_{c-n,d-k} \tag{2.3}$$

for all (c,d), except $c = d = 0$, from which the Yule-Walker equations are obtained.

For zero dimensions $\rho_s = \rho_{-s}$, (2.4)

for $m = 1$, $\rho_{r,s} = \rho_{-r,-s}$, $\rho_{-r,s}$

for $m = 2$, $\rho_{r_1,r_2,s} = \rho_{-r_1,-r_2,-s}$,

four cases where r_i may take any value $\pm r_i$, and for any m, the autocorrelation function has 2^m pieces. The corresponding infinite moving average process, MA, may be determined from the AR process in four ways: by successive substitutions eliminating the $a_{x+m,\,t-k}$; division of $a_{x,t}$ by function $\Phi(B_x,B_t)$ operating on $z_{x,t}$; by expansion of the inverse of the characteristic function $\Phi^{-1}(B_x,B_t) = \psi(B_x,B_t)$, and the binomial expansion; or by the recursive technique using $\Phi^{-1}(B_x,B_t)\psi(B_x,B_t) \equiv 1$, and equating like powers of B_x,B_t.

Thus,

$$z_{x,t} = \Phi^{-1}(B_x,B_t)a_{x,t} = \psi(B_x,B_t)a_{x,t},$$

$$z_{x,t} = \sum_{j=0}^{\infty}\sum_{i=0}^{\infty} \psi_{ij}\, a_{x-i,t-j},\ i \leq j,\quad (2.5)$$

the MA expansion of $z_{x,t}$ and $\psi(B_x,B_t) = \sum_{j=0}^{\infty}\sum_{i=0}^{\infty} \psi_{ij} B_x^i B_t^j$, $i \leq j$,

ψ_{ij} are found in terms of the ϕ_{ij}'s by any of the above four methods.

Theorem (4.9) of Aroian (1983) states: Given an AR process, $\rho_{\ell,m}$ is given by (2.6):

$$\rho_{\ell,m} = \frac{C_{\ell m}}{C_{00}} = \frac{\sum_{n=-\infty}^{\infty}\sum_{k=0}^{\infty} \psi_{n,k}\,\psi_{n+\ell,k+m}}{\sum_{n=-\infty}^{\infty}\sum_{k=0}^{\infty} \psi_{n,k}^2} \quad (2.6)$$

where $C_{\ell,m}$ is the covariance. These results are needed to find further $\rho_{c,d}$ using (2.3).

From (2.2) $\sigma_z^2/\sigma_a^2 > 0$, $\{1 - \sum_{n=0}^{p}\sum_{k=1}^{r} \phi_{n,k}\rho_{n,k}\} > 0$ implies that $\phi A \phi^T > 0$, a positive definite form where ϕ is a vector, and A is the correlation matrix of the $\rho_{n,k}$ given above.

A simple example illustrates these results:

$$z_{x,t} = \phi_1 z_{x,t-1} + \phi_2 z_{x-1,t-1} + a_{x,t},$$

$$\phi_1 = .2,\ \phi_2 = -.6\ |\phi_1| + |\phi_2| < 1, \quad (2.7)$$

then

$\sigma_z^2/\sigma_a^2 = 1.82$ from (2.5) $\rho_{01} = .33$, $\rho_{11} = -.64$, $\rho_{10} = -.21$ and from (2.3)

$\rho_{01} = \phi_1 + \phi_2\rho_{10}, \quad \rho_{11} = \phi_1\rho_{10} + \phi_2 ,$

$\phi_1 = (\rho_{01}-\rho_{10}\rho_{11})/(1-\rho_{10}^2), \quad \text{and} \quad \phi_2 = (\rho_{11}-\rho_{01}\rho_{10})/(1-\rho_{10}^2) .$ (2.8)

The complete autocorrelation function is best determined in general by using (2.6). For (2.7), $\quad |A| = \rho_{10}^2 + \rho_{01}^2 + \rho_{11}^2 - 2\rho_{01}\rho_{10}\rho_{11} - 1 < 0 ,$ (2.9)

from (2.7) and (2.8). Set (2.9) equal to zero, fix ρ_{10}; then we have an ellipse oblique to the axes in the (ρ_{01},ρ_{11}) plane. This defines the possible values for ρ_{01} and ρ_{11} given ρ_{10}, but not the permissible values determined by the two inequalities:

$-(1+\rho_{10}) < \rho_{01} + \rho_{11} < 1 + \rho_{10} \quad \text{and} \quad -1 + \rho_{10} < \rho_{01} - \rho_{11} < 1 - \rho_{10} .$

See Figure 1 with $\rho_{10} = -0.21$.

Note for (2.6) given C_{00}, σ_z^2/σ_a^2, (2.7), ϕ_1 and ϕ_2, then ρ_{01}, ρ_{10} and ρ_{11} are determined without (2.5). Confidence limits for ϕ_{ij} in any AR models may be determined approximately by formulas in Aroian (1983) and Perry's thesis (1981).

Figure 1

Permissible and Allowable Regions for ρ_{01}, ρ_{11}

$\rho_{10} = -0.21$

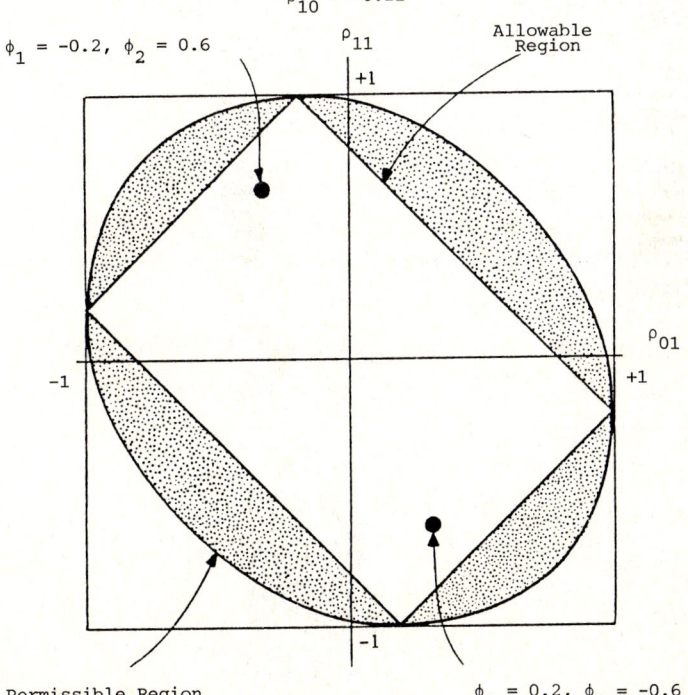

3. DIFFERENCES, NONSTATIONARY $z_{x,t}$

The general m-dimensional nonstationary autoregressive integrated moving average process ARIMA is defined with nonzero mean

$$\mu_{x,t}, \quad z_{x,t} - \mu_{x,t} = \tilde{z}_{x,t}, \quad \phi(B_x, B_t)\tilde{z}_{x,t} = \theta_0 + \theta(B_x, B_t)a_{x,t} \qquad (3.1)$$

m and x vectors,

$$\phi(B_x, B_t) = \Phi(B_x, B_t)\nabla^d_{m,n}, \quad \nabla^d_{m,n} = (1 - B_x^m B_t^n)^d, \qquad (3.2)$$

where components of m are either zero or one and n is either zero or one: $\phi(B_x, B_t)$ is the generalized autoregressive operator, $\Phi(B_x, B_t)$ is the autoregressive operator, $\theta(B_x, B_t)$ is the moving average operator and $\nabla^d_{m,n}$ is the difference operator. Note: $\nabla^{d_i}_{m,n} \nabla^{d_j}_{m,n} = \nabla^{d_i + d_j}_{m,n}$.

Also:

$$w_{x,t} = \nabla^d_{m,n}\tilde{z}_{x,t} = \nabla^d_{m,n}z_{x,t}, \quad E(w_{x,t}) = \mu_w = \theta_0\{1 - \sum_{n=-p}^{q}\sum_{k=1}^{r}\phi_{n,k}\}$$

if a deterministic trend exists. Hereafter μ_w is assumed zero unless otherwise indicated.

In Perry's thesis (1981) ten sets of data were generated with parameters $\phi_1 = 0.2$, $\phi_2 = -0.6$; each data set consisted of twenty linear and twenty time sets, n = 400, a comparison of methods of estimation: minimum mean square, ordinary least squares, and Yule-Walker estimation showed that the ordinary least squares method for estimation is preferred for AR models since it is the same as approximate maximum likelihood, Aroian (1981), while minimum mean square is most expensive because of more required computations. Recent results of Lai et al (1981), Chen et al (1981) and Wang and Wu (1983) support this view.

4. FORECASTING WITH AN AR(1;1,0) MODEL

A forecast made from $z_{x,t}$ for z at time t+m and place x+ℓ is denoted by $z_{x,t}(\ell,m)$, where $z_{x-i,t-j}$ are known for $(i,j) \geq 0$. Define

$$z_{x+\ell,t+m} - z_{x,t}(\ell,m) = e_{x,t}(\ell,m) \qquad (4.1)$$

where $z_{x+\ell,t+m}$ is the true value, $z_{x,t}(\ell,m)$ is the forecast value, and $e_{x,t}(\ell,m)$ is the forecast error. From (2.5)

$$z_{x+\ell,t+m} = \sum_{j=i}^{\infty}\sum_{i=0}^{\infty}\psi_{ij}a_{x+\ell-i,t+m-j}, \quad i \leq j, \qquad (4.2)$$

exactly the MA expression for an AR model.

Thus, for (2.7) $z_{x,t}(0,1) = \phi_1 z_{x,t} + \phi_2 z_{x-1,t}$, $e_{x,t}(0,1) = a_{x,t+1}$,

$z_{x,t}(1,0) = \phi_1 z_{x+1,t-1} + \phi_2 z_{x,t-1} - \phi_2 a_{x+1,t-1}$, $e_{x,t}(1,0) = \phi_1 a_{x+1,t-1} + a_{x+1,t}$,

and in this case if it is assumed that $a_{x+1,t-1}$ is known, then
$z_{x,t}(1,0) = \phi_1 z_{x+1,t-1} + \phi_2 z_{x,t-1}$, and $e_{x,t}(1,0) = a_{x+1,t}$.
Similarly $z_{x,t}(1,1) = \phi_1^2 z_{x+1,t-1} + \phi_1 \phi_2 z_{x,t-1} + \phi_2 z_{x,t}$ with
$e_{x,t}(1,1) = \phi_1^2 a_{x+1,t-1} + \phi_1 a_{x+1,t} + a_{x+1,t+1}$ or $e_{x,t}(1,1) = \phi_1 a_{x+1,t} + a_{x+1,t+1}$.
Note, $e_{x,t}(1,1)$ is correlated with $e_{x,t}(1,0)$, but not with $e_{x,t}(0,1)$,
and $e_{x,t}(1,0)$ and $e_{x,t}(0,1)$ are independent. Similarly,
$z_{x,t}(0,2) = \phi_1 z_{x,t}(0,1) + \phi_2 z_{x-1,t+1} + \phi_1^2 z_{x,t} + \phi_1 \phi_2 z_{x-1,t}$,
$e_{x,t}(0,2) = \phi_2 z_{x-1,t+1} + a_{x,t+1} + a_{x,t+2}$. For M=1, the four possible values
of a are $a_{x-1,t-j}$, $a_{x+i,t-j}$, $a_{x-i,t+j}$, and $a_{x+1,t+j}$. Now, $a_{x-i,t-j}$ is always
known, $i \geq 0$, $j \geq 0$, including $a_{x,t}$; $a_{x+i,t-j}$, $i,j > 0$ may be known or
unknown, known if at the time of observations the values of $z_{x+i,t-j}$ have been
obtained, otherwise unknown (actually $a_{x+i,t-j}$ may be found by predicting the
values of $a_{x+i,t-j}$ in x instead of t); clearly both $a_{x\pm i,t+j}$ for $i,j > 0$
are unknown.

Six cases, combinations of ℓ, m, and $a_{x\pm i,t\pm j}$ occur; $\ell=m$, $\ell>m$, $\ell<m$, and
whether $a_{x-i,t-j}$ only is known, or if $a_{x\pm i,t-j}$ are known. First, $\ell = m \neq 0$,
with $a_{x-i,t-j}$ known, then
$z_{x,t}(m,m) = \sum_{j=i}^{\infty} \sum_{i=m}^{\infty} (\quad)$, $e_{x,t}(m,m) = \sum_{j=i}^{\infty} \sum_{i=0}^{m-1} (\quad)$,
where $(\quad) = \psi_{ij} a_{x+\ell-i,t+m-j}$. (4.3)
Then $\sigma_z^2(m,m) = \sigma_a^2 \sum_{j=i}^{\infty} \sum_{i=m}^{\infty} \psi_{ij}^2$ and $\sigma_{e_{x,t}}^2 = \sigma_a^2 \sum_{j=i}^{\infty} \sum_{i=0}^{m-1} \psi_{ij}^2$.
If $a_{x\pm i,t+j}$ are known, $\ell=m$
$z_{x,t}(m,m) = \sum_{j=m+1}^{\infty} \sum_{i=0}^{\infty} (\quad) + \psi_{mm} a_{x,t}$
and for $i > m$, $j = i$ to ∞, $e_{x,t}(m,m) = \sum_{j=i}^{m} \sum_{i=0}^{m-1} (\quad)$
$\sigma_z^2(m,m) = \sigma_z^2(\psi_{m,m}^2 + \sum_{j=m+1}^{\infty} \sum_{i=0}^{\infty} \psi_{ij}^2)$, $\sigma_e^2(m,m) = \sigma_a^2 (\sum_{j=1}^{m} \sum_{i=0}^{m-1} \psi_{ij}^2)$. (4.4)
For $\ell > m$, and $a_{x-i,t-j}$ known, $i,j \geq 0$
$z_{x,t}(\ell,m) = \sum_{j=i}^{\infty} \sum_{i=\ell}^{\infty} (\quad)$
$e_{x,t}(\ell,m) = \sum_{j=i}^{\infty} \sum_{i=0}^{\ell-1} (\quad)$ (4.5)
$\sigma_z^2(\ell,m) = \sigma_a^2 \sum_{j=1}^{\infty} \sum_{i=\ell}^{\infty} \psi_{ij}^2$, $\sigma_e^2(\ell,m) = \sigma_a^2 \sum_{j=i}^{\infty} \sum_{i=0}^{\ell-1} \psi_{ij}^2$

If $a_{x\pm i, t-j}$ are known, $\ell > m$

$$z_{x,t}(\ell,m) = \sum_{j=m+1}^{\infty} \sum_{i=0}^{\infty} (\quad) \quad , \quad e_{x,t}(\ell,m) = \sum_{j=i}^{m} \sum_{i=0}^{m} (\quad)$$

$$\sigma_z^2(\ell,m) = \sigma_a^2 \sum_{j=m+1}^{\infty} \sum_{i=0}^{\infty} \psi_{ij}^2 \quad , \quad \sigma_e^2(\ell,m) = \sigma_a^2 \sum_{j=i}^{m} \sum_{i=0}^{m} \psi_{ij}^2 \quad , \qquad (4.6)$$

For $\ell < m$, $i,j \geq 0$, $a_{x-i, t-j}$ known,

$$z_{x,t}(\ell,m) = \sum_{j=m}^{\infty} \sum_{i=\ell}^{m-1} (\quad) + \sum_{j=i}^{\infty} \sum_{i=m}^{\infty} (\quad)$$

$$e_{x,t}(\ell,m) = \sum_{j=1}^{m-1} \sum_{i=0}^{m-1} (\quad) + \sum_{j=m}^{\infty} \sum_{i=0}^{\ell-1} (\quad)$$

$$\sigma_z^2(\ell,m) = \sum_{j=m}^{\infty} \sum_{i=\ell}^{m-1} \psi_{ij}^2 + \sum_{j=1}^{\infty} \sum_{i=m}^{\infty} \psi_{ij}^2$$

$$\sigma_e^2(\ell,m) = \sum_{j=i}^{m-1} \sum_{i=0}^{\ell-1} \psi_{ij}^2 + \sum_{j=m}^{\infty} \sum_{i=0}^{\ell-1} \psi_{ij}^2 \qquad (4.7)$$

For $\ell < m$, $a_{x\pm i, t-j}$ known, but $a_{x+i, t}$ unknown,

$$z_{x,t}(\ell,m) = \sum_{j=m+1}^{\infty} \sum_{i=0}^{\infty} (\quad) + \sum_{j=m-1}^{m-1} \sum_{i=m-\ell}^{m} (\quad)$$

$$\sigma_z^2(\ell,m) = \sigma_a^2 \{ \sum_{j=m+1}^{\infty} \sum_{i=0}^{\infty} \psi_{ij}^2 + \sum_{j=m-1}^{m-1} \sum_{i=m-\ell}^{m} \psi_{ij}^2 \}$$

$$e_{x,t}(\ell,m) = \sum_{j=i}^{m-1} \sum_{i=0}^{\ell-1} (\quad) + \sum_{j=1}^{m-1} \sum_{i=m-\ell}^{m-1} (\quad)$$

$$\sigma_e^2(\ell,m) = \sigma_a^2 \{ \sum_{j=i}^{m-1} \sum_{i=0}^{\ell-1} \psi_{ij}^2 + \sum_{j=1}^{m-1} \sum_{i=m-\ell}^{m-1} \psi_{ij}^2 \} \qquad (4.8)$$

If the $z_{x,t}$ are distributed normally, then the confidence limits for $z_{x,t} \pm t_\alpha \sigma_e$ (4.9) using the appropriate formulas (4.3) - (4.8) depending on (ℓ,m) and the particular assumptions used. Here t_α is given by $\int_{-t_\alpha}^{t_\alpha} y_t dt = \alpha$ where y_t is $N(0,1)$. The correlation between $z_{x,t}(\ell,m)$ and $z_{x,t}(\ell',m')$ indicated by $\rho_z(\ell'm')$, $\ell' = \ell + u$, $m' = m + v$, either or both $u > 0$, $v > 0$ and that of $e_{x,t}(\ell,m)$ and $e_{x,t}(\ell',m')$ indicated by $\rho_e(\ell',m')$, are found by using the appropriate formulas (4.3) - (4.8) and substituting ℓ' and m' for ℓ and m. Thus for $\ell' > m'$, $a_{x-i,t-j}$ known, use (4.5).

Hence:

$$\rho_z(\ell',m') = \{ \sum_{j=i}^{\infty} \sum_{i=\ell}^{\infty} \psi_{ij} \psi_{i+u, j+v} \} \{ (\sum_{j=i}^{\infty} \sum_{i=\ell}^{\infty} \psi_{ij}^2) (\sum_{j=i}^{\infty} \sum_{i=\ell}^{\infty} \psi_{i+u, j+v}^2) \}^{-\frac{1}{2}} \qquad (4.10)$$

Also for the same case:

$$\rho_e(\ell',m') = \{ \sum_{j=i}^{\infty} \sum_{i=0}^{\ell-1} \psi_{ij} \psi_{i+u, j+v} \} \{ (\sum_{j=i}^{\infty} \sum_{i=0}^{\ell-1} \psi_{ij}^2) (\sum_{j=i}^{\infty} \sum_{i=0}^{\ell-1} \psi_{i+u, j+v}^2) \}^{-\frac{1}{2}} . \qquad (4.11)$$

5. FORECASTING HIGHWAY DETERIORATION

The highway data conditions are given in Table 1 for sixteen locations from Route 9 in Clinton County, New York State, the higher the number the better the condition of the highway. Observations were taken every two-tenths of a mile; station 1 indicates the start of the highway and station 16 the end. One would expect the highway condition to become poorer since travel is from station 1 to

TABLE 1
Original Data and Predictions

LOC	1975	1976	1977	1978	1979	1980	1981
1	3.113	2.845	2.498	2.611	2.65 2.62	2.53 2.08	2.38 2.32
2	4.519	4.083	3.574	3.334	3.21 3.21	3.07 2.84	2.92 2.14
3	3.295	2.249	2.686	2.671	2.61 2.16	2.70 1.88	2.06 1.82
4	4.293	3.669	3.511	3.366	3.20 2.89	3.25 2.55	2.62 2.44
5	4.268	3.734	3.335	3.219	3.07 2.99	2.73 2.64	2.66 2.46
6	4.517	4.228	3.639	3.608	3.38 3.27	3.13 2.90	2.93 2.39
7	4.532	3.965	3.832	3.774	3.49 3.47	3.32 3.26	3.00 3.08
8	4.448	4.034	3.745	3.699	3.47 3.52	3.25 3.38	3.19 3.05
9	3.549	3.102	3.319	2.988	2.99 2.68	2.87 2.65	2.68 3.42
10	2.092	1.632	2.092	2.036	2.68 1.66	2.29 1.03	2.26 .54
11	4.440	3.914	3.452	3.582	3.42 3.49	3.09 3.20	2.90 3.18
12	4.392	3.839	3.450	3.415	3.31 3.11	3.21 2.83	2.82 2.46
13	3.569	3.153	2.874	3.116	3.02 2.72	2.67 2.88	2.61 3.30
14	3.756	3.205	3.123	3.163	3.01 2.65	2.94 2.16	2.73 1.34
15	4.787	4.268	3.952	3.952	3.54 3.55	3.24 3.34	2.93 3.17
16	4.742	4.172	3.881	3.669	3.42	3.24	3.11
17	4.700	4.000	3.700	3.500	3.35	3.20	3.00

Notes: LOC = location, the original data, 1975-78, the predictions 1979-81. The top figure in each cell 1979-81 is by regression (5.3), and the lower figure by use of the transformation $w_{x,t}$, using $w_{x,t} = \phi_1 w_{x,t-1} + \phi_2 w_{x-1,t-1} + a_{x,t}$ and returning to $z_{x,t}$ by reversing the difference formula, (5.1).

station 16, and to deteriorate also with time. Estimates were made graphically for station 17, needed to obtain theoretical estimates for station 16. Since the data is nonstationary, differencing

$$w_{x,t} = z_{x,t} - z_{x,t-1} - z_{x-1,t} + z_{x-1,t-1} \tag{5.1}$$

was used for Table 1, losing column 1 and row 16. The resulting $w_{x,t}$ matrix as an AR model is

$$w_{x,t} = -.661 \, w_{x,t-1} - .0326 \, w_{x-1,t-1} + a_{x,t} \tag{5.2}$$

and is used to forecast $w_{x,t}$'s for 1979, 1980, and 1981, the second entry in each cell in Table 1. From the predicted $w_{x,t}$, equation (5.1) is used to determine the predicted $z_{x,t}$, using graphical estimates where needed. The first entry in the cell of predictions for 1979 to 1981 was found by using the regression equation, which needs no differencing,

$$z_{x,t} = .846 + .757 \, z_{x,t-1} - .0592 \, z_{x-1,t-1} \tag{5.3}$$

using estimated starting values when needed, rows 16 and 17, and starting values in columns 5, 6, and 7. The regression technique for nonstationary time series is strongly supported by the recent results of Lai et al (1981), Chen et al (1981), and Wang and Jeff Wu (1983). These results correct those given in Perry (1981).

6. CONCLUSIONS

Perry's thesis contains additionally the partial autocorrelation, and further results on the bias and variances of r_{01}, r_{10}, and r_{11}, and the bias and variance of the ϕ's by use of a Taylor series expansion. He provides computer programs for all computations (but not for forecasting as given here) and for transformations of the data. The helpful comments of the referees and the help of Richard Gruberg for the results of Table 1 are appreciated.

REFERENCES

AROIAN, L.A. (1981). Time series in M dimensions, maximum likelihood estimation for autoregressive models. Proceedings of the Business and Economic Statistics Section, American Statistical Association, Washington, D.C. 20005, 334-336.

AROIAN, L.A. (1983). Time series in M dimensions, past, present, and future. Presented at a plenary session of the Time Series Analysis and Forecasting Society, Toronto, Canada, 1-22. To be published, Time Series Analysis: Theory and Practice 6 by North-Holland.

ANDERSON, T.W. (1971). The Statistical Analysis of Time Series. John Wiley & Sons, Inc.

BARTLETT, M.S. (1946). On the theoretical specification of sampling properties of autocorrelated time series. Journal of the Royal Statistical Society, B8:27-41.

BENNETT, R.J. (1975). The representation and identification of spatio-temporal systems, an example of population diffusion in north-west England. Transactions of the Institute of British Geographers, 66:73-94.

BOX, G.E.P. and JENKINS, G.M. (1976). Time Series Analysis, Forecasting and Control. Holden-Day, San Francisco, CA.

CHEN, G.C., LAI, T.L. and WEI, C.Z. (1981). Convergency system and strong consistency of least squares estimates in regression models. Journal of Multivariate Analysis, V.11:319-333.

HANNAN, E.J. (1970). Multiple Time Series. John Wiley & Sons, Inc.

LAI, T.L., ROBBINS, H. and WEI, C.Z. (1981). Strong consistency of least squares estimates in regression models. Journal of Multivariate Analysis, V.9:343-361.

LOMNICKI, Z.A. and ZAREMBA, S.K. (1957). On the estimation of autocorrelation in time series, Annals of Mathematical Statistics, 28:140-158.

PERRY, R.J. (1981). Time series in M dimensions, autoregressive models, M=1. Ph.D. thesis, April 1981, Union College and University. University Microfilms International Dissertation Copies, P.O. Box 1764, Ann Arbor, MI 48106.

PERRY, R.J. and AROIAN, L.A. (1979). Of time and the river; time series in M dimensions, the one dimensional autoregressive model. Proceedings of the Statistical Computing Section, American Statistical Association Annual Meeting, 383-388.

QUENOUILLE, M.H. (1949). Approximate tests of correlation in time series. Journal of the Royal Statistical Society, 11:68-84.

TANEJA, VIDYA S. and AROIAN, L.A. (1980). Time series in M dimensions: autoregressive models. Commun. Statist.-Simula. Computa., B9(5):491-513.

VOSS, D.A., OPRIAN, C.A. and Aroian, L.A. (1980). Moving average models - time series in M dimensions. Commun. Statist.-Simula. Computa., B9(5):453-465.

WANG, S.G. and WU, C.F.J. (1983). Further results on the consistent directions of least squares efficiency. Annals of Mathematical Statistics, V.11:1157-1262.

WEAVER, R.J. and CLARK, R.O. (1977). Psychophysical scaling of pavement serviceability. Soils Engineering Manual, Soil Mechanics Bureau, New York State Department of Transportation, SEM-9, 1220 Washington, Ave., Albany, NY 12232.

ESTIMATING MISSING VALUES IN SPACE-TIME DATA SERIES[1]

Daniel A. Griffith
Department of Geography, State University of New York at Buffalo, Amherst, New York, 14260, United States

Robert P. Haining
Department of Geography, University of Sheffield, Sheffield, S10 2TN, England

Robert J. Bennett
Department of Geography, University of Cambridge, Cambridge, CB2 3EN, England

> First, the missing data problem is briefly surveyed within a classical statistical context. Second, the missing data problem for a planar surface, given a first-order Markov process, is reviewed. Third, the missing data problem for a time series, given a first-order Markov process, is outlined. Next, a solution to the missing data problem for a spatial time series, given a first-order Markov process over space and through time, is presented. Finally, estimation considerations for this space-time situation are discussed. An empirical example for agricultural production in Puerto Rico is also reported. Throughout this exposition emphasis is placed on maximum likelihood estimation.

1. INTRODUCTION

The problem of missing data has received considerable attention in the statistics literature [see, for example, Robinson (1980), Barham and Dunstan (1982), and Tan (1982)]. Of the various ways developed to handle this problem, the Orchard-Woodbury (1972) solution will be explored here. This solution is for an n-dimensional vector $\mathbf{X}$ having a multivariate normal distribution with constant mean $\mu = \mu\mathbf{1}$ and covariance matrix $\underline{\Sigma}$, which may be partitioned such that

$$\mathbf{X} = \begin{bmatrix} \mathbf{X}_o \\ \cdots \\ \mathbf{X}_m \end{bmatrix} \quad , \quad \mu\mathbf{1} = \mu \begin{bmatrix} \mathbf{1}_o \\ \cdots \\ \mathbf{1}_m \end{bmatrix} \quad , \text{ and } \quad \underline{\Sigma} = \begin{bmatrix} \underline{\Sigma}_{oo} & \vdots & \underline{\Sigma}_{om} \\ \cdots & & \cdots \\ \underline{\Sigma}_{mo} & \vdots & \underline{\Sigma}_{mm} \end{bmatrix} \quad .$$

Here $\mathbf{X}_o$ is a subvector of o observed values and $\mathbf{X}_m$ is a subvector of m missing values, where $o + m = n$. Furthermore, $\underline{\Sigma}_{oo}$ and $\underline{\Sigma}_{mm}$ denote the covariance submatrices associated with the observed and missing values, respectively, and $\underline{\Sigma}_{mo} = \underline{\Sigma}_{om}^T$ denotes the submatrix of covariation between the observed and missing values, where superscript upper case 'tee' denotes the operation of matrix transpose.

The resulting estimate of $\mathbf{X}_m$ is given by

---

[1] This research was supported, in part, by Research Grant No. 059.81, from the NATO Scientific Affairs Division.

$$E(\mathbf{X}_m | \mathbf{X}_o) = \hat{\mathbf{X}}_m = \boldsymbol{\mu}_m + \Sigma_{mo} \Sigma_{oo}^{-1} (\mathbf{X}_o - \boldsymbol{\mu}_o) \ . \tag{1}$$

The primary objective of this paper is to rewrite equation (1) for space-time data series, and then to study properties of this new version of the equation. In so doing, several simplifying assumptions will be made. First, space-time interdependencies are assumed to be characterized by a first-order Markov structure in time as well as in space. Second, the process generating random variable ξ is assumed to be isotropic over space and through time, such that any realization $\zeta_{i,t}$ for areal unit i at time t is supposed to come from a normal distribution having mean μ and variance σ^2. Therefore, $\boldsymbol{\mu} = \mu \mathbf{1}$, where $\mathbf{1}$ is an n-by-1 vector of ones. Third, the planar surface under study is assumed to have the same partitioning for all points in space, whereas the points in time are uniformly spaced over some time horizon. While these assumptions may seem restrictive, they are consistent with those used for much of the work that has been completed on serial correlation and spatial autocorrelation.

2. BACKGROUND

The extension of equation (1) to spatial data series has been reported on by Haining, Griffith and Bennett (1984). First, these results will be summarized, then a parallel extension will be made for temporal data series.

All three of these extensions are based upon a simplification that can be derived from the equality

$$\Sigma^{-1} = \mathbf{A} = \begin{bmatrix} \mathbf{A}_{oo} & \vdots & \mathbf{A}_{om} \\ \cdots & \cdots & \cdots \\ \mathbf{A}_{mo} & \vdots & \mathbf{A}_{mm} \end{bmatrix} \ .$$

This equality is worth noting here because the $\mathbf{A}$ rather than the Σ matrices often are what are specified for models of spatial, temporal, and space-time data series. Further, usually these matrices are functions of only a few parameters. Moreover, writing the partitioned form of matrix Σ in terms of the corresponding partitions of matrix $\mathbf{A}^{-1}$, using standard results, yields for equation (2)

$$\Sigma_{mo} \Sigma_{oo}^{-1} = - \mathbf{A}_{mm}^{-1} \mathbf{A}_{mo} (\mathbf{A}_{oo} - \mathbf{A}_{om} \mathbf{A}_{mm}^{-1} \mathbf{A}_{mo})^{-1} [(\mathbf{A}_{oo} - \mathbf{A}_{om} \mathbf{A}_{mm}^{-1} \mathbf{A}_{mo})^{-1}]^{-1} = - \mathbf{A}_{mm}^{-1} \mathbf{A}_{mo} \ .$$

This simplification will prove to be extremely useful in the ensuing analyses, and will be employed in the derivation of all missing value estimation equations.

2.1. A SOLUTION FOR SPATIAL DATA SERIES

Let the multivariate normal realization be for a set of $n = o + m$ areal units or sites at a single point in time. Suppose the nearest neighbor juxtaposition of these areal units or sites is depicted by the n-by-n binary matrix $\mathbf{W}$, whose row and column labels correspond to the same sequence of these areal units, and whose $w_{ij} = 0$ entries denote that areal units i and j are not adjacent, whilst its $w_{ij} = 1$ entries denote that areal units i and j are adjacent with one another. Finally, let ρ denote the nature and degree of spatial autocorrelation latent in variable X.

For this spatial data series situation, let $X = D^{-1}\xi$, where $D^T D = (I - \rho W)$ [this is an example of a first-order spatial Markov random field (Besag, 1974)]. The covariance matrix Σ is given by $(I - \rho W)^{-1}$, or $\Sigma^{-1} = (I - \rho W)$. Accordingly, the partitioned form of matrix Σ^{-1} may be written as follows:

$$\begin{bmatrix} (I_o - \rho W_{oo}) & \vdots & -\rho W_{om} \\ \cdots\cdots\cdots\cdots\cdots\cdots\cdots\cdots\cdots\cdots\cdots \\ -\rho W_{mo} & \vdots & (I_m - \rho W_{mm}) \end{bmatrix} .$$

Consequently, the version of equation (1) that describes a purely spatial data series is given by

$$\hat{X}_m = \mu 1_m - (I_m - \rho W_{mm})^{-1}(-\rho W_{mo})(X_o - \mu 1_o) . \qquad (2)$$

Three distinct cases of equation (2) have been outlined by Haining, Griffith and Bennett (1984). First, if only a single missing value exists, for location i say, then $W_{mm} = 0$ and hence $(I_m - \rho W_{mm})^{-1}$ reduces to 1, W_{mo} becomes the column vector (w_{mj}, $j \neq i$, with w_{mi} set to 0) and equation (2) reduces to

$$\hat{x}_m = \mu + \rho \sum_{j=1}^{j=n} w_{mj}(x_j - \mu) . \qquad (3)$$

Clearly the estimate $\hat{x}_m$ is a linear combination of adjacent map values. Second, if m>1 missing values exist, but none of these missing values are adjacent, then once again $W_{mm} = 0$ and hence $(I_m - \rho W_{mm})^{-1}$ reduces to an identity matrix. Thus, equation (2) reduces to

$$\hat{X}_m = \mu 1_m + \rho W_{mo}(X_o - \mu 1_o)$$

which is a set of simultaneous equations, where each equation is of the form specified by equation (3). Third, if m>1 missing values exist, and some of these missing values are adjacent, then equation (2) cannot be reduced, and holds as stated.

2.2. A SOLUTION FOR TEMPORAL DATA SERIES

Let the multivariate normal realization be for a set of $T = o + m$ points in time for a single areal unit. Suppose the juxtaposition of these points in time is depicted by a T-by-T binary matrix C, whose row and column labels correspond to the same ascending time sequence. So, $c_{ij} = 1$ if $j = i + 1$, whilst $c_{ij} = 0$ otherwise. If the vector X also is arranged in ascending order of time, then C will be a non-symmetric matrix having lower off-diagonal entries of unity, and zero entries everywhere else. Except for the first row, then, matrix C would have row sums equal to unity. Finally, let τ denote the nature and degree of serial correlation latent in variable X.

For this time series let $X = (I - \tau C)\xi$. Then the covariance matrix Σ is given by $[(I - \tau C)^T (I - \tau C)]^{-1}$, or $\Sigma^{-1} = (I - \tau C)^T (I - \tau C)$. This is an example of a first-order temporal Markov process. Consequently, the version of equation (1) that describes a pure time series is

$$\hat{X}_m = \mu 1_m - [\tau^2 C_{om}^T C_{om} + (I_m - \tau C_{mm})^T (I_m - \tau C_{mm})]^{-1} \{-\tau [C_{om}^T (I_o - \tau C_{oo}) +$$
$$(I_m - \tau C_{mm})^T C_{mo}]\}(X_o - \mu 1_o) \quad . \tag{4}$$

Again, three distinct cases of equation (4) can be outlined. First, if only a single missing value exists (other than the initial or final value), then $C_{mm} = 0$, $I_m = 1$, and $C_{om}^T C_{om} = 1$. Thus, equation (4) reduces to

$$\hat{x}_m = \mu + [\tau/(1+\tau^2)][(x_{m-1} - \mu) + (x_{m+1} - \mu)] \quad . \tag{5}$$

Second, if m>1 missing values exist, but none of these missing values are for consecutive time periods, then once again $C_{mm} = 0$, $(I_m - \tau C_{mm})^{-1} = I_m$, $C_{om}^T (I_o - \tau C_{oo}) = C_{om}^T$, and $C_{om}^T C_{om} = I_m$. Hence, $[\tau^2 C_{om}^T C_{om} + (I_m - \tau C_{mm})^T (I_m - \tau C_{mm})]^{-1} = (I_m \tau^2 + I_m)^{-1} = [1/(1+\tau^2)] I_m$. Now matrix C_{mo} is not so sparse. Thus, equation (4) reduces to the set of simultaneous equations

$$\hat{X}_m = \mu 1_m + [\tau/(1+\tau^2)](C_{om}^T + C_{mo})(X_o - \mu 1_o) \quad , \text{ or}$$
$$\hat{x}_k = \mu + [\tau/(1+\tau^2)][(x_{k-1} - \mu) + (x_{k+1} - \mu)] \tag{6}$$

where k denotes the non-adjacent points in time for which values of x are missing.

Third, if m>1 missing values exist, and some of these missing values are adjacent, then equation (4) holds as stated. Moreover,

$$\hat{X}_m = \mu 1_m + \tau[\tau^2 C_{om}^T C_{mo} + (I_m - \tau C_{mm})^T (I_m - \tau C_{mm})]^{-1} (C_{om}^T + C_{mo})(X_o - \mu 1_o) \tag{7}$$

Again the matrix expression $(C_{om}^T + C_{mo})$ casts each missing value as a linear combination of past and future known values. Clearly, equations (5), (6) and (7) need to be modified slightly when x_1 is involved.

But if a single cluster of missing values exists, then $\tau^2 C_{om}^T C_{mo}$ ensures that every entry in the diagonal of the matrix to be inverted in equation (4) is $(1 + \tau^2)$. Therefore, this expression reduces to $[(1+\tau^2)I_m - \tau(C_{mm}^T + C_{mm})]^{-1}$, which may be approximated by the matrix

$$\begin{bmatrix} 1 & \tau & \tau^2 & \cdots & \tau^m \\ \tau & 1 & \tau & \cdots & \tau^{m-1} \\ \tau^2 & \tau & 1 & \cdots & \tau^{m-2} \\ \cdot & & & \cdots & \cdot \\ \cdot & & & \cdots & \cdot \\ \cdot & & & \cdots & \cdot \\ \tau^m & \tau^{m-1} & \tau^{m-2} & \cdots & 1 \end{bmatrix} \quad .$$

This approximation is useful when m is large, since numerical inversion of the matrix may well become dominated by rounding errors. Furthermore, equation (6) suggests that each distinct cluster could be treated separately in this manner.

3. A SOLUTION FOR SPACE-TIME DATA SERIES

Let the multivariate normal realization be for a set of

$$T \cdot n = \sum_{t=1}^{t=T} (o_t + m_t) \quad , \text{ where } o_t + m_t = n \text{ for all } t,$$

areal units or sites at points in time. The general problem of modelling these types of data has been examined by Bennett (1979) and Eby (1980), among others. Suppose the spatial adjacency of areal units still is represented by the matrix **W**. Assume that **W** is constant over time, and that the temporal adjacency still is represented by the matrix **C**. As before, let τ and ρ respectively represent the nature and degree of temporal and spatial autocorrelation. Finally, initially let matrix **X** be organized such that it is initially partitioned as follows:

$$\mathbf{X}_{Tn}^T = (\mathbf{X}_1^T \vdots \mathbf{X}_2^T \vdots \ldots \vdots \mathbf{X}_T^T) \quad ,$$

where $\mathbf{X}_i$ is an n-by-1 vector of areal unit values for time period i.

For this spatial time series situation, let $\mathbf{X} = \mathbf{D}^{-1}\xi$ where $\mathbf{D} = [\mathbf{I}_T \otimes \mathbf{D}_s - \tau \mathbf{C}_T \otimes \mathbf{I}_n]$ and $\mathbf{D}_s^T \mathbf{D}_s = (\mathbf{I} - \rho \mathbf{W})$. The covariance matrix is a partitioned matrix producing T block diagonal elements for $\mathbf{D}^T\mathbf{D}$ of $[(\mathbf{I}_n - \rho \mathbf{W}) + \tau^2 \mathbf{I}_n]$, except for the last spatial distribution, which has a diagonal block entry of $(\mathbf{I}_n - \rho \mathbf{W})$. Meanwhile the (T-1) lower off-diagonal block entries are $-\tau \mathbf{D}_s^T$, whereas the (T-1) upper off-diagonal block entries are the transposes of these entries (i. e., $-\tau \mathbf{D}_s$). Accordingly,

$$\Sigma = \{[\mathbf{I}_T \otimes (\mathbf{I}_n - \rho \mathbf{W}) + \tau^2 (\mathbf{C}_T \otimes \mathbf{I}_n)^T (\mathbf{C}_T \otimes \mathbf{I}_n)] - \tau[\mathbf{C}_T \otimes \mathbf{D}_s^T + \mathbf{C}_T^T \otimes \mathbf{D}_s]\}^{-1} \quad , \text{ or }$$

$$\Sigma^{-1} = \{[\mathbf{I}_T \otimes (\mathbf{I}_n - \rho \mathbf{W}) + \tau^2 (\mathbf{C}_T \otimes \mathbf{I}_n)^T (\mathbf{C}_T \otimes \mathbf{I}_n)] - \tau[\mathbf{C}_T \otimes \mathbf{D}_s^T + \mathbf{C}_T^T \otimes \mathbf{D}_s]\} \quad ,$$

where $\otimes$ denotes Kronecker matrix multiplication and the subscripts denote the order of the square matrices to which they are attached. The term $-\tau \mathbf{D}_s$ represents the presence of interdependencies in both space and time. Furthermore, the diagonal elements of matrix $\mathbf{D}_s$ will be non-zero, and are the coefficients that relate $x_{i,t}$ to $x_{i,t-1}$ and $x_{i,t+1}$. Here a simple representation of the partitioned form of matrix Σ^{-1} is difficult to construct, since the rows and columns containing missing values must be permuted to destroy their space-time sequence in order to isolate a missing value partition. The vector $\mathbf{X}_{Tn}$ must be permuted, too, in order to maintain the original relationship between Σ^{-1} and **X**.

Let $\mathbf{C}^* = \mathbf{C}_T \otimes \mathbf{I}_n$, $\mathbf{I}^* = \mathbf{I}_T \otimes \mathbf{I}_n$, $\mathbf{W}^* = \mathbf{I}_T \otimes \mathbf{W}$, $\mathbf{D}_s^* = \mathbf{I}_T \otimes \mathbf{D}_s$, $\mathbf{D}_s^{*T}\mathbf{D}_s^* = (\mathbf{I}^* - \rho \mathbf{W}^*)$ and $\mathbf{I}' = (\mathbf{C}_T^T \otimes \mathbf{I}_n)(\mathbf{C}_T \otimes \mathbf{I}_n)$. Then Σ^{-1} may be rewritten, using these resultant Kronecker products, as $(\mathbf{D}_s^* - \tau \mathbf{C}^*)^T (\mathbf{D}_s^* - \tau \mathbf{C}^*)$, or $\{[(\mathbf{I}^* + \tau^2 \mathbf{I}') - \rho \mathbf{W}^*] - \tau(\mathbf{C}^{*T}\mathbf{D}_s^* + \mathbf{C}^*\mathbf{D}_s^{*T})\}$. After permuting matrices $\mathbf{W}^*$, $\mathbf{D}_s^*$ and $\mathbf{C}^*$ so that the missing values components appear in the lower right hand partition, each of these matrices may be partitioned as follows:

$$\mathbf{W}^* = \begin{bmatrix} \mathbf{W}_{oo}^* & \vdots & \mathbf{W}_{om}^* \\ \cdots & & \cdots \\ \mathbf{W}_{mo}^* & \vdots & \mathbf{W}_{mm}^* \end{bmatrix} \quad , \quad \mathbf{C}^* = \begin{bmatrix} \mathbf{C}_{oo}^* & \vdots & \mathbf{C}_{om}^* \\ \cdots & & \cdots \\ \mathbf{C}_{mo}^* & \vdots & \mathbf{C}_{mm}^* \end{bmatrix} \quad , \text{ and } \quad \mathbf{D}_s^* = \begin{bmatrix} \mathbf{D}_{soo}^* & \vdots & \mathbf{D}_{som}^* \\ \cdots & & \cdots \\ \mathbf{D}_{smo}^* & \vdots & \mathbf{D}_{smm}^* \end{bmatrix}$$

where $\mathbf{W}_{pq}^*$, $\mathbf{D}_{spq}^*$ and $\mathbf{C}_{pq}^*$ denote the space-time counterparts to $\mathbf{W}_{pq}$, $\mathbf{D}_{spq}$ and $\mathbf{C}_{pq}$ (p,q = o, m). If only one time period exists, the $\mathbf{C}_{pq}^*$ matrices disappear, $\mathbf{I}' = 0$, $\mathbf{I}_T = 1$, and this covariance matrix reduces to that for the pure spatial data series. If only one areal unit exists the $\mathbf{W}_{pq}^*$ matrices disappear, $\mathbf{D}_s^* = \mathbf{I}_n = 1$, and this covariance matrix reduces to that for the pure temporal data series.

Accordingly, the partitoned form of Σ^{-1} is

$$\begin{bmatrix} [(I_o^* + \tau^2 I_o') - \rho W_{oo}^*] - \tau(C_{oo}^{*T}D_{soo}^* + C_{oo}^*D_{soo}^{*T}) & \vdots & -\rho W_{om}^* - \tau(C_{om}^{*T}D_{som}^* + C_{om}^*D_{som}^{*T}) \\ \cdots\cdots\cdots\cdots\cdots\cdots\cdots\cdots\cdots\cdots\cdots\cdots\cdots\cdots & & \cdots\cdots\cdots\cdots\cdots\cdots\cdots\cdots\cdots\cdots\cdots\cdots\cdots\cdots \\ -\rho W_{mo}^* - \tau(C_{mo}^{*T}D_{smo}^* + C_{mo}^*D_{smo}^{*T}) & \vdots & [(I_m^* + \tau^2 I_m') - \rho W_{mm}^*] - \tau(C_{mm}^{*T}D_{smm}^* + C_{mm}^*D_{smm}^{*T}) \end{bmatrix}.$$

Now the version of equation (1) that describes a space-time data series is given by

$$\hat{X}_m = \mu 1_m - \{[(I_m^* + \tau^2 I_m') - \rho W_{mm}^*] - \tau(C_{mm}^{*T}D_{smm}^* + C_{mm}^*D_{smm}^{*T})\}^{-1} \times$$
$$[-\rho W_{mo}^* - \tau(C_{mo}^{*T}D_{smo}^* + C_{mo}^*D_{smo}^{*T})](X_o - \mu 1_o) \ . \tag{8}$$

Now seven distinct cases of equation (8) can be outlined. First, if only a single missing value is present, then $m_t = 1$ for a single t, $I_m^* = 1$, $I_m' = 1$ (as long as the missing value is not at the beginning or end of an areal unit time series), $C_{mm}^* = 0$, $W_{mm}^* = 0$, $D_{smm}^* = 0$, the term in braces reduces to $1/(1 + \tau^2)$, and equation (8) reduces to, for the missing areal unit m value at time t,

$$\hat{x}_{m,t} = \mu + (1 + \tau^2)^{-1}[\rho \sum_{j=1}^{j=n} w_{mj}(x_{j,t} - \mu) + \tau \sum_{j=1}^{j=n} d_{mj}(x_{j,t-1} - \mu) + \tau \sum_{j=1}^{j=n} d_{mj}(x_{j,t+1} - \mu)] \tag{9}$$

which to some degree is a combination of equations (3) and (5). The terms $\rho w_{mj}/(1 + \tau^2)$ and $\tau d_{mj}/(1 + \tau^2)$ [recalling that $d_{mj} = f(\rho)$] represent prevailing space-time interdependencies. Next, a series of situations exist in which equation (8) reduces to a set of k (k=1,2,...,m) simultaneous equations, each being exactly like (9), when multiple missing values are present in either (a) non-adjacent areal units at the same point in time, or (b) a single areal unit for non-adjacent points in time, or (c) non-adjacent areal units as well as non-adjacent points in time. For each of these cases, too, $C_{mm}^* = 0$, $W_{mm}^* = 0$, and $D_{smm}^* = 0$, yielding

$$\hat{X}_m = \mu 1_m + (1 + \tau^2)^{-1}[\rho W_{mo}^* + \tau(C_{mo}^{*T}D_{smo}^* + C_{mo}^*D_{smo}^{*T})](X_o - \mu 1_o) \ . \tag{10}$$

These form the second, third and fourth cases.

The fifth case is where multiple missing values are present in adjacent areal units, at a single point in time. Here $C_{mm}^* = 0$, and therefore equation (8) reduces to

$$\hat{X}_m = \mu 1_m + [(1 + \tau^2)I_m^* - \rho W_{mm}^*]^{-1}[\rho W_{mo}^* + \tau(C_{mo}^{*T}D_{smo}^* + C_{mo}^*D_{smo}^{*T})](X_o - \mu 1_o) \ . \tag{11}$$

Equation (11) is, to some extent, a convolution of equations (2) and (6). The sixth case is where multiple missing values are present for a single areal unit at adjacent points in time. Here $W_{mm}^* = 0$, and equation (8) reduces to

$$\hat{X}_m = \mu 1_m + [(1+\tau^2)I_m^* - \tau(C_{mm}^{*T}D_{smm}^* + C_{mm}^*D_{smm}^{*T})]^{-1}[\rho W_{mo}^* + \tau(C_{mo}^{*T}D_{smo}^* + C_{mo}^*D_{smo}^{*T})](X_o - \mu 1_o) \ . \tag{12}$$

Equation (12) is somewhat of a convolution of equations (3) and (7). The seventh case arises when multiple missing values are present that are adjacent in both space and time, equation (8) cannot be reduced, and holds as stated.

Equations (8), (9), (10), (11) and (12) need to be modified when X_1 is involved.

4. SELECTED NUMERICAL EXAMPLES

Suppose that a region is partitioned into two juxtaposed areal units, and some variate is measured on each of these areal units for five time periods. The corresponding spatial covariance matrix would be

$$\begin{bmatrix} 1 & -\rho \\ -\rho & 1 \end{bmatrix},$$

which could be written in terms of its eigenvalues and eigenvectors as the matrix product $\mathbf{D}_s^T \mathbf{D}_s$, such that

$$\mathbf{D}_s = \begin{bmatrix} [(1+\rho^2)/2]^{1/2} & -[(1+\rho^2)/2]^{1/2} \\ [(1-\rho^2)/2]^{1/2} & [(1-\rho^2)/2]^{1/2} \end{bmatrix}.$$

Meanwhile, the temporal covariance matrix is

$$\begin{bmatrix} 1+\tau^2 & -\tau & 0 & 0 & 0 \\ -\tau & 1+\tau^2 & -\tau & 0 & 0 \\ 0 & -\tau & 1+\tau^2 & -\tau & 0 \\ 0 & 0 & -\tau & 1+\tau^2 & -\tau \\ 0 & 0 & 0 & -\tau & 1 \end{bmatrix}$$

which would decompose into

$$(\mathbf{I} - \tau\mathbf{C}) = \begin{bmatrix} 1 & 0 & 0 & 0 & 0 \\ -\tau & 1 & 0 & 0 & 0 \\ 0 & -\tau & 1 & 0 & 0 \\ 0 & 0 & -\tau & 1 & 0 \\ 0 & 0 & 0 & -\tau & 1 \end{bmatrix}.$$

Consequently, the space-time covariance matrix

$$\begin{bmatrix}
1+\tau^2 & -\rho(1+\tau^2) & -\tau & \tau\rho & 0 & 0 & 0 & 0 & 0 & 0 \\
-\rho(1+\tau^2) & 1+\tau^2 & \tau\rho & -\tau & 0 & 0 & 0 & 0 & 0 & 0 \\
-\tau & \tau\rho & 1+\tau^2 & -\rho(1+\tau^2) & -\tau & \tau\rho & 0 & 0 & 0 & 0 \\
\tau\rho & -\tau & -\rho(1+\tau^2) & 1+\tau^2 & \tau\rho & -\tau & 0 & 0 & 0 & 0 \\
0 & 0 & -\tau & \tau\rho & 1+\tau^2 & -\rho(1+\tau^2) & -\tau & \tau\rho & 0 & 0 \\
0 & 0 & \tau\rho & -\tau & -\rho(1+\tau^2) & 1+\tau^2 & \tau\rho & -\tau & 0 & 0 \\
0 & 0 & 0 & 0 & -\tau & \tau\rho & 1+\tau^2 & -\rho(1+\tau^2) & -\tau & \tau\rho \\
0 & 0 & 0 & 0 & \tau\rho & -\tau & -\rho(1+\tau^2) & 1+\tau^2 & \tau\rho & -\tau \\
0 & 0 & 0 & 0 & 0 & 0 & -\tau & \tau\rho & 1 & -\rho \\
0 & 0 & 0 & 0 & 0 & 0 & \tau\rho & -\tau & -\rho & 1
\end{bmatrix}$$

where the space-time sequence of observations is

$$\mathbf{x}^T = (x_{11}, x_{12}, x_{21}, x_{22}, x_{31}, x_{32}, x_{41}, x_{42}, x_{51}, x_{52}).$$

If the value x_{22} were missing (the case of a single missing value), then

$$\underline{\Sigma}_{mm} = 1 + \tau^2 \quad , \text{ and }$$

$$\underline{\Sigma}_{mo} = (\tau\rho \quad -\tau \quad -\rho(1+\tau^2) \quad \tau\rho \quad -\tau \quad 0 \quad 0 \quad 0 \quad 0).$$

Hence,

$$\hat{x}_{22} = \mu + (1+\tau^2)^{-1}[-\tau\rho(x_{11}-\mu) + \tau(x_{12}-\mu) + \rho(1+\tau^2)(x_{21}-\mu) - \tau\rho(x_{31}-\mu) + \tau(x_{32}-\mu)].$$

Similarly, if the values x_{22} and x_{42} were missing (the case of multiple non-adjacent missing values), then

$$\underline{\Sigma}_{mm} = \begin{bmatrix} 1+\tau^2 & 0 \\ 0 & 1+\tau^2 \end{bmatrix}, \text{ and}$$

$$\underline{\Sigma}_{mo} = \begin{bmatrix} \tau\rho & -\tau & -\rho(1+\tau^2) & \tau\rho & -\tau & 0 & 0 & 0 \\ 0 & 0 & 0 & \tau\rho & -\tau & -\rho(1+\tau^2) & \tau\rho & -\tau \end{bmatrix}.$$

Because $\underline{\Sigma}_{mm}^{-1} = (1+\tau^2)^{-1}I_2$, then

$$\hat{x}_{22} = \mu + (1+\tau^2)^{-1}[-\tau\rho(x_{11}-\mu) + \tau(x_{12}-\mu) + \rho(1+\tau^2)(x_{21}-\mu) - \tau\rho(x_{31}-\mu) + \tau(x_{32}-\mu)]$$

and

$$\hat{x}_{42} = \mu + (1+\tau^2)^{-1}[-\tau\rho(x_{31}-\mu) + \tau(x_{32}-\mu) + \rho(1+\tau^2)(x_{41}-\mu) - \tau\rho(x_{51}-\mu) + \tau(x_{52}-\mu)].$$

Parallel results are obtained if x_{31} also is missing.

Next, suppose that both x_{21} and x_{31} are missing (the case of multiple adjacent missing values). Now

$$\underline{\Sigma}_{mm} = \begin{bmatrix} 1+\tau^2 & -\tau \\ -\tau & 1+\tau^2 \end{bmatrix}, \text{ and}$$

$$\underline{\Sigma}_{mo} = \begin{bmatrix} -\tau & \tau\rho & -\rho(1+\tau^2) & \tau\rho & 0 & 0 & 0 & 0 \\ 0 & 0 & \tau\rho & -\rho(1+\tau^2) & -\tau & \tau\rho & 0 & 0 \end{bmatrix}.$$

Since

$$\underline{\Sigma}_{mm}^{-1} = (1+\tau^2+\tau^4)^{-1}\begin{bmatrix} 1+\tau^2 & \tau \\ \tau & 1+\tau^2 \end{bmatrix},$$

then

$$\hat{x}_{21} = \mu + (1+\tau^2+\tau^4)^{-1}\{\tau(1+\tau^2)(x_{11}-\mu) - \tau\rho(1+\tau^2)(x_{12}-\mu) + \rho[(1+\tau^2)^2-\tau^2](x_{22}-\mu) + \tau^2(x_{41}-\mu) - \tau^2\rho(x_{42}-\mu)\}, \text{ and}$$

$$\hat{x}_{31} = \mu + (1+\tau^2+\tau^4)^{-1}\{\tau^2(x_{11}-\mu) - \tau^2\rho(x_{12}-\mu) + \rho[(1+\tau^2)^2-\tau^2](x_{32}-\mu) + \tau(1+\tau^2)(x_{41}-\mu) - \tau\rho(1+\tau^2)(x_{42}-\mu)\}.$$

Parallel results are obtained if x_{21} and x_{22} are the clustered missing values, or if x_{12}, x_{22} and x_{31} are the missing values.

4.1. COMPARISON OF RESULTS

For comparative purposes, if the single missing value x_{22} is treated only within the context of a spatial data series, then

$$\hat{x}_{22} = \mu + \rho(x_{21} - \mu) \quad .$$

On the other hand, if x_{22} is treated solely within the context of a time series, then

$$\hat{x}_{22} = \mu + [\tau/(1 + \tau^2)][(x_{12} - \mu) + (x_{32} - \mu)] \quad .$$

The additional information derived from a space-time series context fails to be captured simply by combining these two results. Rather, it also involves the term $[-\tau\rho/(1 + \tau^2)][(x_{11} - \mu) + (x_{31} - \mu)]$, which is a convolution of space-time dependence and represents indirect effects of the separate spatial and temporal expressions. Once again, parallel results can be sketched for the remaining numerical examples outlined in this section.

5. CONCLUDING COMMENTS

The space-time model developed here was employed in an empirical study of sugar cane production in Puerto Rico. Five census periods (i. e., 1959, 1964, 1969, 1974 and 1978) were involved, and the island had been partitioned into 73 areal units (i. e., municipios) for the purpose of collecting agricultural census data. Six units had missing values in 1974, while 21 units had missing values in 1978, with some of these missing value units being the same for both time periods. The results of this study are reported in Griffith (1983). To briefly summarize, once the $\mathbf{X}_m$ vector had been estimated, the total data set was subjected to principal components, discriminant function and canonical correlation analyses to see whether or not one could distinguish between the observed and the estimated values. These analyses were unsuccessful in their attempts to statistically discriminate between $\mathbf{X}_o$ and $\mathbf{X}_m$. Furthermore, the EM algorithm converged in relatively few iterations. These results are encouraging, and suggest that those findings given in this paper should prove useful in the analysis of incomplete or censored spatial time series.

Future research needs to be conducted on variance estimates of $\hat{x}_m$, $\hat{\tau}$ and $\hat{\rho}$ when missing values are present, and situations in which ρ varies over time and τ varies over space.

REFERENCES

BARHAM, S. and DUNSTAN, F. (1982). Missing values in time series. In <u>Time Series Analysis: Theory and Practice 2</u> (Proceedings of the International Conference held in Dublin, Ireland, March 1982). Ed: O. Anderson, North-Holland, Amsterdam, 25-41.

BENNETT, R. (1979). *Spatial Time Series: Analysis-Forecasting-Control*. Pion, London.

BESAG, J. (1974). Spatial interaction and the statistical analysis of lattice systems. *Journal of the Royal Statistical Society* B **36**, 192-235.

DEMPSTER, A., LAIRD, N. and RUBIN, D. (1977). Maximum likelihood from incomplete data via the EM algorithm. *Journal of the Royal Statistical Society* **39B**, 1-38.

EBY, L. (1980). Prediction of autoregressive multivariate time series with an application to certain models of spatial time series. In *Analysing Time Series* (Proceedings of the International Conference held on Guernsey, Channel Islands, October 1979). Ed: O. Anderson, North-Holland, Amsterdam, 139-152.

GRIFFITH, D. (1983). Phasing-out of the sugar industry in Puerto Rico. In *Evolving Geographical Structures* (Proceedings of the NATO Advanced Studies Institute held at San Miniato, Italy, July 1982). Ed: D. Griffith and A. Lea, Martinus Nijhoff, The Hague, 196-228.

HAINING, R. (1978). Estimating spatial-interaction models. *Environment and Planning* A **10**, 305-320.

HAINING, R., GRIFFITH, D. and BENNETT, R. (1984). A statistical approach to the problem of missing spatial data using a first-order Markov model. *The Professional Geographer* **36**, in press.

MARTIN, R. (1983). Exact maximum likelihood for incomplete data from a correlated Gaussian process. Unpublished paper, Department of Probability and Statistics, University of Sheffield.

ORCHARD, R. and WOODBURY, M. (1972). A missing information principle: theory and applications. In *Proceedings* of the 6th Berkeley symposium on mathematical statistics and probability, vol. 1. Ed: L. le Cam, J. Neyman and E. Scott, University of California Press, Berkeley, 697-715.

ORD, K. (1975). Estimation methods for models of spatial interaction. *Journal of the American Statistical Assocation* **70**, 120-126.

ROBINSON, P. (1980). Estimation and forecasting for time series containing censored or missing observations. In *Time Series* (Proceedings of the International Conference held at Nottingham University, England, March 1979). Ed: O. Anderson, North-Holland, Amsterdam, 167-182.

TAN, S. (1982). Maximum likelihood estimation of stochastic linear difference equations with autoregressive moving average errors and with missing observations and observational errors. In *Applied Time Series Analysis* (Proceedings of the International Conference held at Houston, Texas, August 1981). Ed: O. Anderson and M. Perryman, North-Holland, Amsterdam, 397-408.

MAXIMUM LIKELIHOOD FITTING OF STARMAX MODELS TO INCOMPLETE SPACE-TIME SERIES DATA

D. S. Stoffer
Department of Mathematics and Statistics, University of Pittsburgh, Pittsburgh, Pennsylvania 15260, USA

In this paper we combine the spatial considerations of the space-time ARMA model and the parametrization of the ARMAX model to formulate a STARMAX model which can be used for modeling and forecasting the dynamics of multivariate populations which are functionally dependent upon spatial characteristics as well as time. Furthermore, due to the physical constraints imposed on a multivariate data collection system in both space and time, this model tolerates very general patterns of missing or incomplete data. As a consequence of Shumway and Stoffer (1982), the EM algorithm proposed by Dempster et al. (1977) is used in conjunction with modified Kalman smoothed estimators to derive a simple recursive procedure for estimating the parameters of the STARMAX model by maximum likelihood.

1. INTRODUCTION

Many problems which arise in the physical sciences require investigators to work with noisy data in both time and space. For example, in the general marine fisheries context, one observes environmental data at sensors which are distributed in space as well as time (cf. for example Mendelsohn (1982)). The fact that observations occurring contiguously in space can be expected to be as correlated as observations from adjacent time periods introduces several complications into traditional analyses. First of all, the volume of data increases as one may collect multivariate m×1 vector series observed at L locations for n time points, producing mLn intercorrelated data points. Secondly, physical constraints imposed on a data collection system of such magnitude almost guarantee that there will be stretches where observations are missed in time and/or space for certain components of the vector series. General techniques are needed both for smoothing and interpolating sections of the record with missing values and for constructing reasonable forecasts for future values.

Recently, several approaches to space-time modeling have been developed which depend on the completeness of the sample in space and time. If the vector process is spatially stationary and observed on a rectangular grid at each point in time, a rather detailed procedure for fitting a class of space-time models is described in Larimore (1977). If the spatial sampling is incomplete, structural models given in Cliff and Ord (1981) or in Pfeifer and Deutsch (1980) may be more appropriate.

In this paper we combine the spatial considerations of the space-time ARMA model of Pfeifer and Deutsch (1980) and the parametrization of the ARMAX model considered by Hannan (1976), to formulate a STARMAX model which can be used for modeling and forecasting the dynamics of multivariate populations which are functionally dependent upon spatial characteristics as well as time. Furthermore, due to the physical constraints imposed on a multivariate data collection system in both space and time, this model tolerates very general patterns of missing or incomplete data.

In view of the one-to-one correspondence between stationary ARMAX models and state-space models (cf. Hannan (1976)), the maximum likelihood fitting of STARMAX models to incomplete space-time data is a modification and generalization of the technique used in Shumway and Stoffer (1982). There, the EM algorithm (cf. Dempster el al. (1977)) was used to obtain a simple recursion for estimating the parameters of the state-space model. As a consequence of Shumway and Stoffer (1982), we are able to use the EM algorithm in conjunction with modified Kalman smoothed estimators to derive a recursive procedure for estimating the parameters of the STARMAX model by maximum likelihood. Moreover, analytical solutions are obtained under general spatial restrictions and general patterns of incomplete sampling.

In Section 2, we present the general form of the STARMAX model and discuss its relationship to the STARMA and ARMAX models. The spatial weighting of STARMAX models is discussed in Section 3. There, we give various mechanisms for representing the spatial orders among locations in regularly spaced systems as well as irregularly spaced systems. Since maximum likelihood fitting of STARMAX models involves an application of the EM algorithm, a brief discussion of the EM procedure and its properties is given in Section 4. The actual EM procedure as it pertains to STARMAX modeling is derived in Section 5 under the assumption that the sampling scheme is complete. Then, in Section 6, we given the missing data modifications to the results of Section 5. Finally, the recursive procedure for generating the modified Kalman smoothed estimators is given in the Appendix.

2. THE GENERAL STARMAX MODEL

We suppose that the $p \times 1$ population vector of some random field at time t denoted by x_t is of interest to an investigator. We may decompose x_t into components $x_j(t)$ denoting the true state $m \times 1$ vector at coordinate j and time t so that $x_t' = (x_1'(t), \ldots, x_L'(t))$ where L is the number of locations in the field and $p = mL$. We may further suppose that an $r \times 1$ environmental vector $z_t' = (z_1(t), z_2(t), \ldots, z_r(t))$ is measured concurrently. In the marine fisheries context for example, $x_j(t)$ could represent a certain fish population at coordinate j and time t, whereas z_t may measure sea surface temperature, sea level, pressure, and current fishing effort. In an air pollution context, $x_j(t)$ could represent the true average pollution index at certain location during day t, and z_t may parametrize wind speed and direction, temperature, and humidity. We note that $z_j(t)$ may be one environmental observation over the entire field at time t, or, $z_j(t)$ may be a vector of environmental observations at location j.

A model which describes the current state x_t in terms of the previous states $x_{t-1}, x_{t-2}, \ldots, x_{t-k}$ and the environment z_t (or z_{t-1}) may be expressed in the form

$$x_t = \sum_{j=1}^{k} \Lambda_j D_j x_{t-j} + \Psi z_t + w_t, \quad t \geq 1, \tag{2.1}$$

where Λ_j is the p×p diagonal space-time transition intensity matrix at lag j, D_j is a known p×p distance matrix which expresses the spatial relationship between the L locations in the random field at lag j. The p×r regression matrix Ψ expresses the relationship between the current population and the environmental effects vector, and w_t is p×1 white Gaussian noise: $w_t \sim N(0,Q)$. Equation (2.1) is a special case of the ARMAX model considered by Hannan (1976), and a modified multivariate version of the STARMA model of Pfeifer and Deutsch (1980).

For example, with k = 1 and $\Lambda = \text{diag}(\lambda_1, \ldots, \lambda_p)$, equation (2.1) can be rewritten for the ith sensor as

$$x_{ti} = \lambda_i \sum_{j=1}^{p} d_{ij} x_{t-1,j} + \sum_{j=1}^{r} \psi_{ij} z_{tj} + w_{ti}, \quad 1 \leq i \leq p,$$

which is essentially a constrained regression involving the present x_{ti} and the past values $x_{t-1,1}, \ldots, x_{t-1,p}$, as well as the present environmental effects $z_{t1}, \ldots, z_{tr}$. In this manner we see that the ith diagonal component of Λ, λ_i, describes the intensity of the system to or away from the ith sensor, and that d_{ij} is a measure of inverse distance from sensor i to sensor j.

Of course, it is not possible to observe the true state vector x_t and we must assume that it is measured implicitly by the surrogate q×1 vector y_t which may be incompletely observed at any given time. This may be modeled using the observation equation

$$y_t = M_t x_t + v_t, \quad t \geq 1, \tag{2.2}$$

where the q×p matrix M_t expresses the pattern which converts the unobserved stochastic vector x_t into the observed series y_t. Of course, M_t may represent patterns of missing observations. We assume that v_t is q×1 white Gaussian noise, $v_t \sim N(0,R)$.

It is convenient to rewrite the STARMAX model (2.1), (2.2) in the following manner to ease the notation and the calculations in the sequel. State equation (2.1) is now written as

$$X(t) = \Phi X(t-1) + \Psi z_t + W(t), \quad t \geq 1, \text{ or} \tag{2.3}$$

$$\begin{pmatrix} x_t \\ x_{t-1} \\ \vdots \\ x_{t-k+1} \end{pmatrix} = \begin{bmatrix} \Lambda_1 D_1 & \Lambda_2 D_2 & \cdots & \Lambda_k D_k \\ I & 0 \cdots 0 & & 0 \\ 0 & I & 0 & 0 \\ \vdots & \vdots & & \vdots \\ 0 & & I & 0 \end{bmatrix} \begin{pmatrix} x_{t-1} \\ x_{t-2} \\ \vdots \\ x_{t-k} \end{pmatrix} + \begin{pmatrix} \psi \\ 0 \\ \vdots \\ 0 \end{pmatrix} z_t + \begin{pmatrix} w_t \\ 0 \\ \vdots \\ 0 \end{pmatrix},$$

and the observation equation (2.2) is written as

$$y_t = M(t)X(t) + v_t, \quad t \geq 1, \text{ or} \tag{2.4}$$

$$y_t = [M_t, 0, \ldots, 0] \begin{pmatrix} x_t \\ x_{t-1} \\ \vdots \\ x_{t-k+1} \end{pmatrix} + v_t.$$

We shall assume that the initial state $X(0)$ is a $kp \times 1$ Gaussian vector, $X(0) \sim N(\mu, \Sigma)$, and that $X(0)$, $\{w_t\}$, and $\{v_t\}$ are mutually independent. Also, since the estimation procedure is carried out conditional on $\{z_t\}$, we shall assume that it is a deterministic sequence. This assumption is more of a convenience than a necessity. We note (cf. Hannan (1976)) that the minimal assumption is that $\lim_{N \to \infty} N^{-1} \sum_{t=1}^{N-n} z_t z'_{t+n}$, $n \geq 0$ exists and is finite with probability one.

The fitting of STARMAX models involves estimating by maximum likelihood, the covariance matrices Q and R, the diagonal space-time transition intensity matrices $\Lambda_1, \Lambda_2, \ldots, \Lambda_k$, and the environmental effects regression matrix ψ. One may compare the environmental effects and space-time components by the likelihood ratio tests of $\psi = 0$ and $\Lambda_j = 0$ for any or all $j = 1, 2, \ldots, k$, via the "innovations form" of the likelihood given in Section 5, equation (5.16). Forecasting of the space-time series x_t will involve calculating the conditional expectations $x_{t+n}^n = E\{x_{t+n} | y_1, \ldots, y_n, z_1, \ldots, z_n\}$ recursively for $t = 1, 2, \ldots$.

3. SPATIAL WEIGHTING OF STARMAX MODELS

The space-time autoregressive (STAR) model considered in Pfeifer and Deutsch (1980) is of the form

$$x_t = \sum_{j=1}^k \sum_{\ell=0}^{\nu_j} \phi_{j\ell} W^{(\ell)} x_{t-j} + w_t \tag{3.1}$$

where k is the order of the autoregression, ν_j is the spatial order of the jth autoregressive term, and $W^{(\ell)}$ is a $p \times p$ matrix of weights for spatial order ℓ. So for example, with $k = 1$ and $\nu_1 = 1$, equation (3.1) becomes

$$x_t = [\phi_{10} I + \phi_{11} W^{(1)}] x_{t-1} + w_t. \qquad (3.2)$$

Although the STAR models (3.1) and (3.2) are observation equations and not state equations as in the STARMAX model (2.1), we may still employ the basic mechanism for representing a hierarchical ordering of neighbors of each site, namely, the sequence of matrices $W^{(\ell)}$ described in (3.1).

To fix ideas, let us compare the first order STARMAX model to the first order STAR model in (3.2). Letting $D_1 = (I + W^{(1)})$ be a p×p matrix in (2.1), we obtain the state equation

$$x_t = \Lambda(I + W^{(1)}) x_{t-1} + \psi z_t + w_t. \qquad (3.3)$$

It is clear then that the spatial weighting of the two models might involve the same considerations. Hence, we could consider the spatial distance matrices in the STARMAX model (2.1), (2.2) to be of the form $D_j = I + W^{(1)} + W^{(2)} + \ldots + W^{(\nu_j)}$, $1 \leq j \leq k$ where, as in the STAR model, ν_j is the spatial order of the jth autoregressive term.

Indeed, specification of the distance matrices D_j, $1 \leq j \leq k$, must be left to the investigator of the space-time system in order that as much of the physical characteristics and constraints of the field as possible is employed in the model. For regularly spaced systems, a definition of spatial order (cf. Besag (1974); Pfeifer and Deutsch (1980)) is typically employed. Figure 3.1 shows the first four spatial order neighbors of a particular site in both a one-dimensional and two-dimensional Euclidean grid system. The weighting is a measure of inverse distance between neighbors in which "nearest neighbors" have the most effect on each other.

The weighting matrices adopted in the "nearest neighbor" concept are of the form:
$$W_{ij}^{(\ell)} = \begin{cases} 1/n_i^{(\ell)} & \text{if i and j are } \ell^{th} \text{ order neighbors} \\ 0 & \text{otherwise} \end{cases}$$
where $W_{ij}^{(\ell)}$ denotes the ijth element of $W^{(\ell)}$, and $n_i^{(\ell)}$ is the number of ℓth order neighbors possessed by site i. This weighting scheme is referred to as equal scaled weighting since all non-zero weights of a given site for a particular spatial order are equal and scaled so that $\sum_j W_{ij}^{(\ell)} = 1$.

For irregularly spaced systems (as well as regularly spaced systems) a reasonable method of spatial weighting might be based on the inverse of the Euclidean distance between each location. For example, if δ_{ij} is the distance from location i to location j, one could choose the distance matrices D_ℓ, $\ell = 1, 2, \ldots, k$ to be the p×p symmetric matrices with elements of the form

$$d_{\ell,ij} = [\delta_{ij} + g(\ell)]^{-h(\ell)}, \quad 1 \leq i, j \leq p, \qquad (3.4)$$

where g and h are positive non-decreasing functions of the autoregressive order

$\ell = 1,2,\ldots,k$. In this manner, the functions g and h may be fine tuned to the particular space-time characteristics of the sustem under investigation.

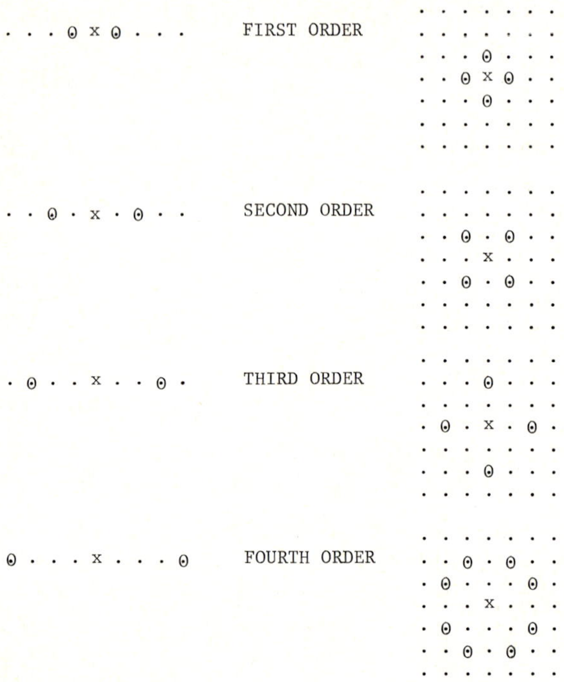

```
. . . ⊙ x ⊙ . . .      FIRST ORDER       . . . . . . .
                                         . . . . . . .
                                         . . . ⊙ . . .
                                         . . ⊙ x ⊙ . .
                                         . . . ⊙ . . .
                                         . . . . . . .
                                         . . . . . . .


. . ⊙ . x . ⊙ . .      SECOND ORDER      . . . . . . .
                                         . . ⊙ . ⊙ . .
                                         . . . x . . .
                                         . . ⊙ . ⊙ . .
                                         . . . . . . .
                                         . . . . . . .


. ⊙ . . x . . ⊙ .      THIRD ORDER       . . . ⊙ . . .
                                         . . . . . . .
                                         . ⊙ . x . ⊙ .
                                         . . . . . . .
                                         . . . ⊙ . . .
                                         . . . . . . .


⊙ . . . x . . . ⊙      FOURTH ORDER      . . . . . . .
                                         . . ⊙ . ⊙ . .
                                         . ⊙ . . . ⊙ .
                                         . . . x . . .
                                         . ⊙ . . . ⊙ .
                                         . . ⊙ . ⊙ . .
                                         . . . . . . .
```

Figure 3.1: One and two-dimensional spatial orders for regularly spaced systems

4. GENERAL THEORY OF THE EM ALGORITHM

Since the maximum likelihood fitting of STARMAX models involves an application of the EM algorithm, we shall give a brief discussion of the EM procedure and its properties here.

The EM algorithm (Dempster et al. (1977)) is a general approach to iterative computation of maximum likelihood estimates when observations can be viewed as incomplete data. The term incomplete data refers to the case where one may not be able to observe the stochastic process of interest, $\underline{x}$, but is only able to observe a function of the process $\underline{y} = \underline{y}(\underline{x})$. The EM algorithm is designed to find a value of $\underline{\theta}$ which maximizes the family of sampling densities $g(\underline{y}|\underline{\theta})$ which depend on parameter(s) $\underline{\theta}$ by making use of the associated family of sampling densities $f(\underline{x}|\underline{\theta})$.

Each iteration of the EM algorithm involves two steps called the expectation step (E-step) and the maximization step (M-step) which are described as follows:

Define

$$Q(\underline{\theta}|\underline{\theta}^*) = E\{\log f(\underline{x}|\underline{\theta})|\underline{y},\underline{\theta}^*\}. \quad (4.1)$$

The EM iteration $\underline{\theta}^{(r)} \to \underline{\theta}^{(r+1)}$ is defined as

E-step: Compute $Q(\underline{\theta}|\underline{\theta}^{(r)})$, $\quad (4.2)$

M-step: Choose $\underline{\theta}^{(r+1)} \, \epsilon\Omega$ to maximize $Q(\underline{\theta}|\underline{\theta}^{(r)})$, $\quad (4.3)$

where superscript r denotes the rth iteration, and Ω the parameter space. Heuristically, we would like to choose a $\underline{\theta}^* \epsilon\Omega$ to maximize $\log f(\underline{x}|\underline{\theta})$, but since $\log f(\underline{x}|\underline{\theta})$ is unknown, we maximize its current expectation given the data $\underline{y}$ and the current fit $\underline{\theta}^{(r)}$.

Some of the general properties of the EM algorithm (Dempster el al. (1977), Theorem 1 and its corollaries) are:

(1) $L(\underline{\theta}) = \log g(\underline{y}|\underline{\theta})$ is non-decreasing on each iteration of the EM algorithm, and is strictly increasing on any iteration for which $Q(\underline{\theta}^{(r+1)}|\underline{\theta}^{(r)}) > Q(\underline{\theta}^{(r)}|\underline{\theta}^{(r)})$.

(2) If the incomplete-data likelihood is bounded above, then a maximum likelihood estimate of $\underline{\theta}$ is a fixed point of the iterative process (4.2), (4.3).

The question of convergence of the EM algorithm to a maximum likelihood estimate of $\underline{\theta}$ is addressed in Wu (1983), Boyles (1980, 1983), and Dempster et al. (1977, Theorem 4), in the general case, and in Sundberg (1976) for the special case of a regular exponential family complete-data model. Dempster et al. and Sundberg consider the rate of convergence near a fixed point of the algorithm by obtaining a representation for the Jacobian matrix $DM(\underline{\theta}^*)$ at a fixed point $\underline{\theta}^*$ of M, where the mapping M: $\Omega \to \Omega$ satisfies (4.3), and D denotes the first partial derivative operator. It is noted that the largest eigenvalue of $DM(\underline{\theta}^*)$, if strictly less than one, gives the asymptotic $(r \to \infty)$ factor by which the error $||\underline{\theta}^{(r)} - \underline{\theta}^*||$ decreases at each iteration step, assuming that the sequence $\{\underline{\theta}^{(r)}\}$ converges to $\underline{\theta}^*$. Boyles gives conditions under which the sequence $\{\underline{\theta}^{(r)}\}$ will in fact converge for all starting points $\underline{\theta}^{(0)}$.

The most thorough discussion of the convergence of the EM algorithm given any starting value may be found in Wu (1983). There, it is shown that if the complete data likelihood is of the exponential family and certain mild conditions are met, the algorithm will locate a local maximum or global maximum if it is not trapped at a stationary point of the likelihood. It is his recommendation that since the convergence to stationary values, local maximum, or global maximum depends on the choice of starting points, several EM iterations should be tried with different starting points that are representative of the parameter space.

5. MAXIMUM LIKELIHOOD FITTING OF STARMAX MODELS

In order to develop the EM procedure for estimating the parameters of the STARMAX model (2.1), (2.2), by maximum likelihood, we note first that the joint log-likelihood of the complete data $X(0), x_1, \ldots, x_n, v_1, \ldots, v_n$ can be written in the form

$$\log L \stackrel{\circ}{=} -\frac{1}{2} \log |\Sigma| - \frac{1}{2}(X(0)-\mu)'\Sigma^{-1}(X(0)-\mu)$$

$$-\frac{n}{2} \log |Q| - \frac{1}{2} \sum_{t=1}^{n} (x_t - \phi^* X^*(t-1))' Q^{-1} (x_t - \phi^* X^*(t-1))$$

$$-\frac{n}{2} \log |R| - \frac{1}{2} \sum_{t=1}^{n} (y_t - M_t x_t)' R^{-1} (y_t - M_t x_t) \qquad (5.1)$$

where ϕ^* is the $p \times (kp+r)$ matrix $\phi^* = [\Lambda, \psi]$ and $\Lambda = [\Lambda_1, \Lambda_2, \ldots, \Lambda_k]$ is the $p \times kp$ matrix of transition intensity matrices. The $(kp+r) \times 1$ vector $X^*(t-1)$ is defined to be

$$X^*(t-1) = \begin{pmatrix} D\,X(t-1) \\ z_t \end{pmatrix}$$

where D is the $kp \times kp$ block diagonal matrix consisting of submatrices $D_1, D_2, \ldots D_k$, across the diagonal and $X(t-1)$ is defined in (2.3). It is clear then, that

$$\phi^* X^*(t-1) = \sum_{j=1}^{k} \Lambda_j D_j x_{t-j} + \psi z_t.$$

In order to calculate the E-step (4.2), it is convenient to define the $kp \times 1$ conditional mean

$$X^n(t) = E(X(t) | y_1, \ldots, y_n, z_1, \ldots, z_n) \qquad (5.2)$$

and the $kp \times kp$ covariance matrices

$$P^n(t) = \text{Cov}(X(t) | y_1, \ldots, y_n, z_1, \ldots, z_n) \qquad (5.3)$$

and

$$P^n(t,t-1) = \text{Cov}(X(t), X(t-1) | y_1, \ldots, y_n, z_1, \ldots, z_n). \qquad (5.4)$$

Then, by x_t^n we mean the $p \times 1$ subvector of $X^n(t)$ corresponding to $E(x_t | y_1, \ldots, y_n, z_1, \ldots, z_n)$. Similarly, by P_t^n and $P_{t,t-1}^n$ we mean the $p \times p$ submatrices of $P^n(t)$ and $P^n(t,t-1)$ respectively, which represent $P_t^n = \text{Cov}(x_t | y_1, \ldots, y_n, z_1, \ldots, z_n)$ and $P_{t,t-1}^n = \text{Cov}(x_t, x_{t-1} | y_1, \ldots, y_n, z_1, \ldots, z_n)$. A set of recursions for calculating (5.2), (5.3), and (5.4) is given in the Appendix where it is noted that they are modified Kalman smoothed estimates.

We shall also make use of the following matrices:

$$A = \sum_{t=1}^{n} [P^n(t-1) + X^n(t-1)X^{n'}(t-1)], \qquad (5.5)$$

$$B = \{\sum_{t=1}^{n} [P^n(t,t-1) + X^n(t)X^{n'}(t-1)]\}_{p \times kp}, \qquad (5.6)$$

where by $\{\cdot\}_{p \times kp}$ we mean the submatrix consisting of the first p rows of the $kp \times kp$ matrix within,

$$C = \sum_{t=1}^{n} [P_t^n + x_t^n x_t^{n'}], \tag{5.7}$$

$$F = \sum_{t=1}^{n} X^n(t-1) z_t', \tag{5.8}$$

$$G = \sum_{t=1}^{n} x_t^n z_t', \tag{5.9}$$

and

$$H = \sum_{t=1}^{n} z_t z_t'. \tag{5.10}$$

The dimensions of each matrix (5.5)-(5.10) are as follows: A $(kp \times kp)$, B $(p \times kp)$, C $(p \times p)$, F $(kp \times r)$, G $(p \times r)$, H $(r \times r)$, noting that p is the dimension of the state vector x_t, k is the order of the autoregression, and r is the dimension of the environmental vector z_t.

The E-step (4.2) in the case when vectors $y_1, \ldots, y_n$, $z_1, \ldots, z_n$ are fully observed is given by (cf. Shumway and Stoffer (1982))

$$Q(\underline{\theta}|\underline{\theta}^{(r)}) = -\frac{1}{2}\log|\Sigma| - \frac{1}{2}\text{tr}\{\Sigma^{-1}[P^n(0) + (X^n(0)-\mu)(X^n(0)-\mu)']\}$$

$$- \frac{n}{2}\log|Q| - \frac{1}{2}\text{tr}\{Q^{-1}(C - \Phi^* S_{10}^{*'} - S_{10}^* \Phi^{*'} + \Phi^* S_{00}^* \Phi^{*'})\}$$

$$- \frac{n}{2}\log|R| - \frac{1}{2}\text{tr}\{R^{-1}\sum_{t=1}^{n}[(y_t - M_t x_t^n)(y_t - M_t x_t^n)' + M_t P_t^n M_t']\}, \tag{5.11}$$

where S_{10}^* is a $p \times (kp+r)$ matrix and S_{00}^* is a $(kp+r) \times (kp+r)$ matrix composed of submatrices

$$S_{00}^* = \begin{bmatrix} DAD' & DF \\ F'D' & H \end{bmatrix} \quad \text{and} \quad S_{10}^* = [BD', G]. \tag{5.12}$$

The conditional means and covariances in (5.6)-(5.9) are evaluated at the current estimated values for the parameters $\underline{\theta}^{(r)} = \{\mu(r), \Sigma(r), \Lambda(r), \psi(r), Q(r), R(r)\}$.

The M-step (4.3) is now easily applied yielding

$$\Phi^*(r+1) = S_{10}^* S_{00}^{*-1}, \tag{5.13}$$

$$Q(r+1) = n^{-1}(C - S_{10}^* S_{00}^{*-1} S_{10}^{*'}), \tag{5.14}$$

and

$$R(r+1) = n^{-1}\sum_{t=1}^{n}[(y_t - M_t x_t^n)(y_t - M_t x_t^n)' + M_t P_t^n M_t']. \tag{5.15}$$

Note that since $\Phi^* = [\Lambda_1, \Lambda_2, \ldots, \Lambda_k, \psi]$, we may identify the updates $\Lambda_j(r+1)$, $1 \le j \le k$, and $\psi(r+1)$ by the corresponding submatrices in (5.13). In updating $\mu(r)$ and $\Sigma(r)$, one may either assume Σ to be known and take $\mu(r+1) = X^n(0)$, or assume μ to be known and take

$$\Sigma(r+1) = P^n(0) + (X^n(0)-\mu)(X^n(0)-\mu)'$$

or assume both μ and Σ are known. This consideration is due to the fact that only one observation can be used in estimating μ and Σ. It can be shown (Stoffer (1982)) however, that the recursive solutions to (5.5)-(5.9) are relatively

insensitive to variations in the initial conditions $\mu = X^0(0)$ and $\Sigma = P^0(0) > 0$.
To summarize the iterative procedure in the case when $y_1, \ldots, y_n$, $z_1, \ldots, z_n$ are fully observed:

(1) Start with initial estimates $\mu(0)$, $\Sigma(0)$, $\Lambda_1(0), \ldots, \Lambda_k(0)$, $\psi(0)$, $Q(0)$, and $R(0)$.

(2) On the rth iteration, run the modified Kalman smoother given in the Appendix on a STARMAX model specified by $\mu(r-1)$, $\Sigma(r-1)$, $\Lambda_1(r-1), \ldots, \Lambda_k(r-1)$, $\psi(r-1)$, $Q(r-1)$, and $R(r-1)$.

(3) Update the estimates by (5.13), (5.14), and (5.15), taking into consideration the previous discussion on updating $\mu(r)$ and $\Sigma(r)$.

(4) Stop when the incomplete-data log-likelihood and the parameter estimates stabilize. The incomplete-data log-likelihood can be written in the "innovations form" (cf. Gupta and Mehra (1974))

$$\log f(Y;\theta) \triangleq -\frac{1}{2} \sum_{t=1}^{n} \log |M_t P_t^{t-1} M_t' + R|$$
$$-\frac{1}{2} \sum_{t=1}^{n} (y_t - M_t x_t^{t-1})' [M_t P_t^{t-1} M_t' + R]^{-1} (y_t - M_t x_t^{t-1}) \qquad (5.16)$$

where x_t^{t-1} and P_t^{t-1} are the modified Kalman filter estimators given by the recursions in the Appendix.

6. MISSING DATA MODIFICATIONS

Suppose that at a given step, we define the partition of the $q \times 1$ observation vector $y_t' = (y_t^{(1)'}, y_t^{(2)'})$, where $y_t^{(1)}$ is the $q_1 \times 1$ observed portion and $y_t^{(2)}$ is the $q_2 \times 1$ unobserved portion. The overall complete data observation equation may now be expressed in the partitioned form

$$\begin{bmatrix} y_t^{(1)} \\ y_t^{(2)} \end{bmatrix} = \begin{bmatrix} M_t^{(1)} \\ M_t^{(2)} \end{bmatrix} x_t + \begin{bmatrix} v_t^{(1)} \\ v_t^{(2)} \end{bmatrix} \qquad (6.1)$$

where $M_t^{(1)}$ and $M_t^{(2)}$ are $q_1 \times p$ and $q_2 \times p$ matrices and

$$\text{cov} \begin{bmatrix} v_t^{(1)} \\ v_t^{(2)} \end{bmatrix} = \begin{bmatrix} R_{11} & R_{12} \\ R_{21} & R_{22} \end{bmatrix}. \qquad (6.2)$$

Using proofs similar to those in Stoffer (1982), it can be established that the equations in the Appendix hold for the missing data case given above if one makes the replacements $y_t' = (y_t^{(1)'}, \underline{0}')$, $M_t' = (M_t^{(1)'}, 0')$, and $R_{12} = R_{21} = 0$, at each step in the filter recursions, equations (5)-(9) of the Appendix. That is, if y_t is incomplete, the filtered and smoothed estimators can be calculated from the usual equations by entering zeroes in the observation vector y_t where data is missing, by zeroing out the corresponding rows of the design matrix M_t, and by zeroing out the off-diagonal elements of R corresponding to the covariance of $v_t^{(1)}$ and $v_t^{(2)}$ defined in (6.2). This leads to the smoothed estimators $x_t^{(n)}$ and the covariance functions $P_t^{(n)}$, $P_{t,t-1}^{(n)}$ in the missing data case. As previously mentioned we assume that the environmental vector z_t is a completely observable sequence. Otherwise, one may put some structure on z_t, say $z_t = g(z_{t-1},\ldots,z_1) + \varepsilon_t$ where $g(\cdot)$ is a function of the history of z_t and ε_t is a noise process, and predict z_t or any components of z_t which are missing using the usual techniques.

If the observation y_t is not completely observable, the maximum likelihood estimators as computed in the EM procedure now require that one take the conditional expectation of (5.1) under the assumption that y_t is incompletely observed. In this case, defining the incomplete data as $Y_n^{(1)} = (y_1^{(1)},\ldots,y_n^{(1)}, z_1,\ldots,z_n)$ we need only be concerned with the third term of (5.1), since the first two terms will have expectations which depend only on $x_t^{(n)}$, $P_t^{(n)}$, and $P_{t,t-1}^{(n)}$. The expectation of the third term can be computed by conditioning first on both $Y_n^{(1)}$ and x_t and then on $Y_n^{(1)}$ which leads to (cf. Stoffer (1982))

$$R(r+1) = n^{-1}\sum_{t=1}^{n} \Pi_t C_t \Pi_t' \tag{6.3}$$

where

$$C_t = \begin{bmatrix} S_t^{(1)} & S_t^{(1)} R_{2.1}' \\ R_{2.1} S_t^{(1)} & R_{2.1} S_t^{(1)} R_{2.1}' + R_{22.1} \end{bmatrix} \tag{6.4}$$

with

$$R_{2.1} = R_{21} R_{11}^{-1} \tag{6.5}$$

$$R_{22.1} = R_{22} - R_{21} R_{11}^{-1} R_{12}, \tag{6.6}$$

and

$$S_t^{(1)} = (y_t^{(1)} - M_t^{(1)} x_t^{(n)})(y_t^{(1)} - M_t^{(1)} x_t^{(n)})' + M_t^{(1)} P_t^{(n)} M_t^{(1)'}. \tag{6.7}$$

The matrix Π_t in (6.3) is a permutation matrix which reorders the variables in their original form. This is necessary because the application of (6.4) requires that the variables be ordered so that the observed values appear in $y_t^{(1)}$.

A simplification introduced in Shumway and Stoffer (1982) is for the case where the missing data part and the observed data part are uncorrelated so that R_{21} and hence $R_{2.1}$ are zero in (6.4) and the update for the missing part is R_{22}. If a vector observation y_t is missed completely, then the term C_t is simply R.

We may summarize by noting that in the missing data case, equations (5.5)-(5.14) are modified by replacing x_t^n, P_t^n, and $P_{t,t-1}^n$ by $x_t^{(n)}$, $P_t^{(n)}$, and $P_{t,t-1}^{(n)}$, and by replacing $X^n(t)$, $P^n(t)$, and $P^n(t,t-1)$ by $X^{(n)}(t)$, $P^{(n)}(t)$, and $P^{(n)}(t,t-1)$ where the superscript (n) denotes that a sub-component of y_t, the corresponding rows of M_t, and the appropriate off-diagonal elements of R have been zeroed out in the modified Kalman recursions given in the Appendix. Equation (5.15) is replaced by equations (6.3)-(6.7). The same convention is followed in computing the log-likelihood of the incomplete data given in the equation (5.16).

APPENDIX

We assume that the STARMAX model is written in the form of (2.3), (2.4), namely

$$X(t) = \Phi X(t-1) + \Psi z_t + W(t), \tag{1}$$

and

$$y_t = M(t) X(t) + v_t. \tag{2}$$

The modified Kalman smoother estimator

$$X^n(t) = E\{X(t) | y_1,\ldots,y_n, z_1,\ldots,z_n\} \tag{3}$$

for the model (1), (2) is obtained by minimizing the mean square error

$$P^n(t) = E\{(X(t) - X^n(t))(X(t) - X^n(t))' | y_1,\ldots,y_n, z_1,\ldots,z_n\} \tag{4}$$

and can be calculated recursively using the following equations. The derivation is based on techniques similar to those used in the derivation of the usual Kalman filters and smoothers (cf. Jazwinski (1970); pages 201, 217).

For $t = 1, 2, \ldots, n$

$$X^{t-1}(t) = \Phi X^{t-1}(t-1) + \Psi z_t, \tag{5}$$

$$P^{t-1}(t) = \Phi P^{t-1}(t-1)\Phi' + Q^*, \tag{6}$$

where $Q^* = \begin{bmatrix} Q & 0 \\ 0 & 0 \end{bmatrix}$, and

$$K_t = P^{t-1}(t)M'(t)[M(t)P^{t-1}(t)M'(t) + R]^{-1}, \tag{7}$$

$$X^t(t) = X^{t-1}(t) + K_t[y_t - M(t)X^{t-1}(t)], \tag{8}$$

$$P^t(t) = P^{t-1}(t) - K_t M(t) P^{t-1}(t). \tag{9}$$

The initial conditions for the recursion (5)-(9) are taken to be $X^0(0) = \mu$ and $P^0_0 = \Sigma > 0$. In order to calculate the smoothers (3) and (4), one performs the backward recursions $t = n, n-1, \ldots, 1$ on the equations

$$J_{t-1} = P^{t-1}(t-1)\Phi' [P^{t-1}(t)]^{-1}, \tag{10}$$

$$X^n(t-1) = X^{t-1}(t-1) + J_{t-1}[X^n(t) - X^{t-1}(t)], \tag{11}$$

and

$$P^n(t-1) = P^{t-1}(t-1) + J_{t-1}[P^n(t) - P^{t-1}(t)]J'_{t-1}. \tag{12}$$

We note that equation (5.6) in the text requires the covariance matrix $P^n(t,t-1)$ which can be calculated using the backward recursions

$$P^n(t-1,t-2) = P^{t-1}(t-1)J'_{t-2} + J_{t-1}[P^n(t,t-1) - \Phi P^{t-1}(t-1)]J'_{t-2}, \tag{13}$$

for $t = n, n-1, \ldots, 2$, where

$$P^n(n,n-1) = [I - K_n M(n)] \Phi P^{n-1}(n-1). \tag{14}$$

The derivation of (13) and (14) is lengthy and may be derived using techniques similar to those found in Shumway and Stoffer (1981).

ACKNOWLEDGMENTS

The author wishes to thank R.H. Shumway and the referees for their helpful comments. I am also greatly indebted to Mary Ann Kaditus for the superb typing of this manuscript.

This research is supported by a Faculty Research Grant from the Faculty of Arts and Sciences, University of Pittsburgh, and the Air Force Office of Scientific Research under contract F49620-82-K-001.

REFERENCES

BESAG, J.S. (1974). Spatial interaction and the statistical analysis of lattice systems. J. Roy. Statist. Soc., B, 36, 197-242.

BOYLES, R.A. (1980). Convergence results for the EM algorithm. Technical Report No. 13, Division of Statistics, University of California, Davis.

BOYLES, R.A. (1983). On the convergence of the EM algorithm. To appear, J. Roy. Statist. Soc., B.

CLIFF, A.D. and ORD, J.K. (1981). Spatial Processes, Models and Applications. Methuen, Inc., New York.

DEMPSTER, A.P., LAIRD, N.M. and RUBIN, D.B. (1977). Maximum likelihood from incomplete data via the EM algorithm. J. Roy. Statist. Soc., B, 39, 1-38.

HANNAN, E.J. (1970). Multiple Time Series, Wiley, New York.

HANNAN, E.J. (1976). The identification and parametrization of ARMAX and state-space forms. Econometrica, 44, 713-723.

JAZWINSKI, A.H. (1970). Stochastic Processes and Filtering Theory. Academic Press, New York.

JONES, R.H. (1980). Maximum likelihood fitting of ARMA models to time series with missing observations. Technometrics, 22, 389-395.

LARIMORE, W.E. (1977). Statistical inference on random fields. Proc. of the IEEE, Special Issue on Multidimensional Systems, 65, 961-970.

MENDELSSOHN, R. (1982). Environmental influences on fish population dynamics: A multivariate time series approach. Paper presented at Ann. Mtg., Amer. Statist. Assoc., Cincinnati, Ohio, August, 1982.

PFEIFER, P.E. and DEUTSCH, S.J. (1980). A three-stage iterative procedure for space-time modeling. Technometrics, 22, 35-47.

PFEIFER, P.E. and DEUTSCH, S.J. (1980). Identification and interpretation of first-order space-time ARMA models. Technometrics, 22, 397-408.

SHUMWAY, R.H. (1983). Some applications of the EM algorithm to analyzing incomplete time series data. Invited paper for ONR sponsored symposium on time series analysis of irregularly observed data, Institute of Statistics, Texas A&M University, College Station, TX 77843.

SHUMWAY, R.H. and STOFFER, D.S. (1981). Time series smoothing and forecasting using the EM algorithm. Technical Report No. 27, Division of Statistics, University of California, Davis.

SHUMWAY, R.H. and STOFFER, D.S. (1982). An approach to time series smoothing and forecasting using the EM algorithm. J. Time Series Anal., 3, 253-264.

SHUMWAY, R.H. and STOFFER, D.S. (1982b). An algorithm for parameter estimation and smoothing in space-time models with missing data. Prepared for NOAA Pacific Environmental Group, Final Report: P.O. 82-ABA-1198, September 1982.

STOFFER, D.S. (1982). Estimation of parameters in a linear dynamic system with missing observations. Ph.D. dissertation, University of California, Davis.

WU, C.F. (1983). On the convergence properties of the EM algorithm. Ann. Statist., 11, 95-103.

FORECASTING THE SPREAD OF AN EPIDEMIC

Andrew D. Cliff
Department of Geography, University of Cambridge, Cambridge CB2 3EN, England

J. Keith Ord
Division of Management Science, The Pennsylvania State University, University Park, PA 16802, USA

Disease histories in a given community often exhibit periods of fade-out followed by reintroductions from (neighboring) leading areas. Forecasting the onset of a new epidemic wave may be just as important as forecasting the magnitude of the whole epidemic. Logistic regression is used to develop forecasts of reintroduction, and the method is tested using data on measles for medical districts in Iceland. The results suggest that this approach is much better able to identify the possibility of a new wave than methods used previously.

1. INTRODUCTION

The mathematical modeling of epidemiological processes began some 75 years ago with the work of W. H. Hamer (1906) who developed the first version of what is now known as the Hamer-Soper model.

The total population at time t, $P(t)$, comprises three groups, susceptibles, infectives, and recovered. At time t, the numbers in each group are $S(t)$, $I(t)$, and $R(t)$, respectively, and

$$P(t) = S(t) + I(t) + R(t). \tag{1}$$

The original model was deterministic in form and treated the variables as continuous. This made it possible to formulate differential equations describing the rates of change of membership of the three groups. In the deterministic version, the solution yields damped oscillations despite several modifications by Soper (1929). Bartlett (1956) developed a stochastic version of the model which produces oscillations without damping, a condition he termed quasi-stationary. The stochastic formulation led to the Pandemic Threshold Theorem of Kendall (1957) which states that a small outbreak will result in an epidemic only when the number of susceptibles is "sufficiently large". For further details and a review of more recent work, see Bailey (1975) and Bartholomew (1982).

Most of the work to date has focused upon modeling rather than forecasting. However, the increased interest in cost-benefit evaluations of immunization programs (Ambrosch et al., 1978; Finkelstein et al., 1981) demonstrates the need for efficient forecasts of both the timing and the magnitude of disease outbreaks. In order to achieve any degree of predictability, we must know how the disease is transmitted and the likely paths it will follow once the first few cases have appeared. From the statistical viewpoint, it is necessary to study a disease where the medical reports of incidence are sufficiently accurate and timely to ensure that meaningful predictions can be made. The choice of location (Iceland) and disease (measles) are discussed in section 2.

Most forecasting effort is directed at the level of a process. The forecasts developed in Cliff et al. (1975), Cliff and Ord (1978), and Cliff et al. (1981) were of this type. However, a disease may spread very rapidly once it "catches hold", and a good forecasting method may be one which accurately predicts the onset of an epidemic rather than one which tracks the later stages of the spread. Thus, bearing in mind the Threshold Theorem, we now seek a procedure which will indicate when a major flare-up is likely to start rather than a prediction of the level. Professor Peter Haggett (pers. comm.) has suggested that such a forecast would indicate the potential for an epidemic, rather like the familiar indicator boards recording the level of risk for forest fires. Accordingly, in section 3, we develop the logistic regression model as a potential forecasting tool. In sections 4 and 5, the results of an application of this model to measles spread in Iceland is described. Directions for future research are briefly outlined in section 6.

2. MEASLES IN ICELAND

Measles has been widely studied since it is notifiable in many countries. Further, it is highly contagious and is usually transmitted in a single meeting between an infective and a susceptible. The virus does not noticeably mutate so that a single infection usually confers lifetime immunity upon the victim. Last, but by no means least, such a study is important since, as noted by Cliff et al. (1981, Chapter 8), measles remains one of the top killer diseases in the world.

The reasons for studying Iceland (an island community with excellent epidemiological records) are detailed in Cliff et al. (1981, Chapter 3). A data appendix to that text lists numbers of reported measles cases, by month and medical district, for 16 epidemics covering 1904-1974. These epidemics are summarized in Table 1. It can be seen that total fade-out occurred between epidemics. The characteristics of each wave are described in Cliff et al. (1981, Chapter 4).

Table 1: Timing and Magnitude of Measles Waves in Iceland (code numbers agree with those given by Cliff et al., 1981; the full data set appears on pages 201-229 of that volume)

Wave	Duration	Total Cases	Gap (in months)
I	April-November 1904	822	--
II	May 1907-August 1908	7397	39
III	April 1916-May 1917	4944	91
IV	June 1924-August 1926	6130	84
V	August 1928-December 1929	5317	23
VI	February 1936-March 1937	8408	73
VII	February 1943-May 1944	7155	70
VIII	November 1946-August 1948	4791	29
IX	January 1950-March 1952	6645	16
X	July 1952-April 1953*	1872	3 ⎫ 23*
XI	March 1954-November 1955	7787	10 ⎭
XII	March 1958-December 1959	7102	27
XIII	April 1962-May 1964	7405	27
XIV	November 1966-August 1968	6152	29
XV	October 1968-December 1970**	3625	1
XVI	October 1972-January 1974**	3953	21

*Wave X tended to hit only the areas missed by wave IX.
**The increase in immunizations from the mid-sixties has tended to change the pattern of the waves somewhat and to reduce the rate of incidence of the disease.

2.1. **Choice of Study Area.** The rapid growth in population since 1900 (see Table 2) and improved transport facilities have tended to reduce the time between epidemics, as shown in Table 1. However, the pattern of population growth has been very uneven, with several areas in the north declining in population. In this paper, we concentrate upon the medical district of Isafjarðar in the northwest of the country. This is a regional capital whose size remained fairly stable over the seventy-year period. Measles was consistently reintroduced from Reykjavikur, making the modeling task somewhat easier, although the lead time varied. Several of the empirical studies in Cliff et al. (1981) were carried out for the Isafjarðar district, allowing comparisons to be made between different forecasting approaches.

Table 2: Approximate Population of Iceland, 1750-1970 (in thousands)

	Year				
Region	1750	1850	1900	1950	1970
Reykjavikur (capital)	{50	{50	5	50	80
Rest of Iceland			60	80	130

2.2. **The Data Record.** In order to predict the onset of an epidemic, we need a clear statistical definition of what is being measured. Following Cliff et al. (1981, pp. 106-108), we define an <u>outbreak</u> as any period of time in which cases are reported, whereas an <u>epidemic</u> is an outbreak of sufficiently large volume. Several numerical procedures may be used to make this definition operational (see Cliff et al. (1981, pp. 108-110)). In the present study, we defined the district as "in-epidemic" at time t if $I(t) \geq 5$. Raising the cutoff point to 9 would reclassify two months in the period 1946-1948, whereas lowering it to 2 would reclassify seven months. Thus, the operational impact of the choice of cutoff is not pronounced.

The data record shows new cases on a monthly basis. Since measles has a latent period (time from initial contact to onset of disease) of about two weeks, all infectives will be removed or recovered within one month. For these reasons, it was decided to use $N(t)$, the number of new cases per month, as the key variable in trying to predict the onset of an epidemic (see Cliff and Ord (1978) for further discussion of this issue).

3. THE LOGISTIC MODEL

We assume that a system may be in one of two states, 0 or 1, at time t, described by the random variable Y_t. The general form of the model may be written as

$$P(Y_t = 1) = p_t = F(\underset{\sim}{x}_t' \underset{\sim}{\beta})$$
$$P(Y_t = 0) = 1 - p_t,$$
(2)

where $\underset{\sim}{x}_t$ is a vector of regressor variables, $\underset{\sim}{\beta}$ is a vector of unknown parameters, and F is a distribution function. Equation (2) is a form of the generalized linear model (GLIM) (see McCullagh and Nelder (1983)).

The earliest use of such models was in <u>probit</u> analysis (Finney, 1971), where F is taken to be the normal distribution function. However, the logistic distribution (<u>logit</u> analysis) is the canonical form for a binary response variable (McCullagh and Nelder, 1983, p. 24), and, in the absence of any clear arguments either way, the logistic was used in our study. Thus, if $z_t = x_t' \beta$,

$$F(z_t) = [1 + \exp(-z_t)]^{-1}. \qquad (3)$$

Estimation methods for logistic regression are described in Anderson (1972) and in McCullagh and Nelder (1983).

Previous studies using probits or logits have assumed cross-sectional data without any time dependence. Our present objective is to monitor a system through time and to develop a model to predict the state of the system. This means that we must take account of the dependence between the successive states of the system. That is, we need to examine the dependence of Y_t upon past values of Y_t and upon past and present values of z_t. Assume that z_t is a strictly stationary process with constant mean, μ, and variance, σ^2. Then we may specify

$$\text{corr}(z_t, z_{t-j}) = \rho_j, \quad j = 1, 2, \ldots \qquad (4)$$

Given that z_t is strictly stationary and that F is a well-behaved function, it follows that p_t is strictly stationary with, say,

$$E(p_t) = \pi = f(\mu) \qquad (5)$$

$$V(p_t) = \sigma_p^2 \qquad (6)$$

and

$$\text{cov}(p_t, p_{t-j}) = \theta_j, \quad j = 1, 2, \ldots \qquad (7)$$

Each θ_j will be some function of the corresponding ρ_j, the exact form depending upon the particular form adopted for F. For integer $k, m \geq 1$, we have

$$E(Y_t^k | z_t) = p_t \qquad (8)$$

$$E(Y_t^k Y_{t-j}^m | z_t, z_{t-j}) = p_t p_{t-j} \qquad (9)$$

and so on. It follows from (5) and (8) that

$$E(Y_t) = \pi \text{ and } V(Y_t) = \pi(1 - \pi).$$

Also, from (7) and (9), we have that

$$\text{cov}(Y_t, Y_{t-j}) = \theta_j.$$

Finally, from (6), we have that $\sigma_p^2 \leq \pi(1 - \pi)$, so that the autocorrelations of the Y_t are given by those for the p_t, scaled downwards by a constant factor. Exact results for the correlation between Y_t and z_t depend upon the form of F and appear to be intractable in general. However, it does follow that

$$\rho_j = 0 \rightarrow \theta_j = 0 \text{ and } \rho_j = 0 \rightarrow \text{corr}(Y_t, z_{t-j}) = 0.$$

Clearly, further work is required in order to develop satisfactory tools for model identification in this case, but the auto and cross-correlations do give some guidance. In the empirical work in section 5, the latent variable z_t is replaced by observables $\underset{\sim}{x}_t$, and so we rely upon a stepwise approach to model identification.

4. IMPLEMENTING THE LOGISTIC APPROACH

It was stated in section 2 that the number of new cases in month t, N(t) would be used as the primary explanatory variable. However, two other factors must be taken into account:

(i) the disease may be reintroduced from outside after a period of fade-out;

(ii) by the Threshold Theorem, the probability of an epidemic depends upon the number of susceptibles, S(t). Unfortunately, it is not feasible to measure S(t) directly, and some surrogate must be used.

To account for (i), we include $N_L(t)$ and a constant in the vector of regressor variables, where $N_L(t)$ refers to the number of cases in neighboring leading areas through which reintroduction is typically effected. The constant allows for other sources of reinfection, usually overseas.

One approach to (ii) is to develop a cohort model of the numbers of susceptibles through time; this was done in Cliff et al. (1981, appendix). In this study, we decided not to use the cohorts for two reasons. First, the number of reported cases tends to underestimate the actual number, although, around a 40% reporting rate, Iceland's records are better than most. Secondly, we sought a simple method which could be implemented by public health authorities with minimal extra data requirements. A reasonable first approximation is to assume that recruitment of new susceptibles (births) is a linear function of time with periodic depletions during epidemics. Thus, the estimated number of susceptibles at time t was defined as

$$S^*(t) = \delta_2 \{\delta_0 + \delta_1 t - C(t)\}, \tag{10}$$

where C(t) denotes the <u>total</u> number of cases reported since the beginning of the study period. We allow $\delta_2 \geq 1$ to cover under-reporting of cases. The constant δ_1 was chosen so that $S^*(0) \doteq S^*(T)$, where 0 and T represent the start and end points of the data record, but selected in the middle of interepidemic phases. Recall that the population of the study area was relatively stable during 1946-1968. A rapidly changing population would require some assumption about the proportion of the population ultimately infected.

In general, in a stochastic epidemic model,

$$E[n\{t + 1\}] \propto N(t)S(t) \tag{11}$$

(see Cliff et al. (1981, pp. 174-176)). That is, we assume homogeneous mixing within a district and that the number of new cases is proportional to the number of infective-susceptible contacts. Modifying N(t) and S(t) in light of the arguments presented above suggests the following form for

$$z_{t+1} = \underset{\sim}{x}_{t+1}' \underset{\sim}{\beta} = N^*(t)S^*(t) = \delta_2[\gamma_0 + \gamma_1 N(t) + \gamma_2 N_L(t)][\delta_0 + \delta_1 t - C(t)]. \tag{12}$$

We may reparametrize (12) to arrive at

$$z_{t+1} = \beta_0 + \beta_1 N(t) + \beta_2 N_L(t) + \beta_3 S^*(t) + \beta_4 N(t)S^*(t) + \beta_5 N_L(t)S^*(t), \tag{13}$$

where now $S^*(t) = \delta_1 t - C(t)$ as we allow δ_0 to be determined by the model. It would have been possible to enter t and C(t) as separate variables, but these tend to be highly correlated which may pose numerical difficulties. Thus, it was decided to retain $S^*(t)$ as stated since the epidemiological "trigger" depends upon (11). The value $\delta_1 = 4$ was used in the study since this keeps initial and final susceptibles approximately equal. The results are not sensitive to moderate changes in δ_1.

In (13), we would expect all the β_j to be positive, although β_3 and β_5 may be very small. Since the time taken to transmit the disease from one district to another varies, different lags for $N_L(t)$ were also used.

5. EMPIRICAL RESULTS

The logistic model was fitted using data for the period January 1946–December 1956 beginning and ending in a non-epidemic state. The fitted model was then used to make one-month-ahead predictions over the period January 1956–August 1968 (see Table 1 for choice of dates).

The autocorrelation function for Y_t, shown in Table 3, shows an exponential decay over lags 1-8 and then a steady negative value (around -0.1) thereafter. This is typical of the quasi-stationary nature of the in and out of epidemic phases of such series, intensified by use of the binary variable. The cross-correlations between Y_t and $N(t)$, $N_L(t)$ are shown in Table 4. The relationships suggested by (13) are consistent with these values. The higher cross-correlation for Y_t with $N_L(t)$ compared to $N(t)$ with $N_L(t)$ is indicative of the non-linear nature of the process.

Table 3: Autocorrelation Function for Y_t
(monthly data for Isafjarðar, 1/46-12/56)

Lag	1	2	3	4	5
Autocorrelation	0.724	0.518	0.379	0.241	0.171
Lag	6	7	8	9	10
Autocorrelation	0.101	0.099	0.030	-0.040	-0.110
Lag	11	12	13	14	15
Autocorrelation	-0.180	-0.182	-0.183	-0.175	-0.166
Lag	16	17	18	19	20
Autocorrelation	-0.148	-0.130	-0.112	-0.104	-0.105

Table 4: Cross-Correlation Functions for Y_t with N(t), Numbers of New Cases in Some District (Isafjarðar) and with $N_L(t)$, Numbers of New Cases in Lead District (Reykjavíkur) and N(t) with $N_L(t)$

Lag (k)	-6	-5	-4	-3	-2	-1	0
corr{Y_t, N(t + k)}	0.094	0.179	0.337	0.372	0.477	0.578	0.675
corr{Y_t, N_L(t + k)}	0.016	0.013	0.099	0.221	0.366	0.521	0.473
corr{N(t), N_L(t + k)}	-0.019	-0.008	0.044	0.110	0.171	0.229	0.221

Lag (k)	1	2	3	4	5	6
corr{Y_t, N(t + k)}	0.405	0.194	0.120	0.043	0.019	-0.004
corr{Y_t, N_L(t + k)}	0.342	0.220	0.041	-0.093	-0.136	-0.143
corr{N(t), N_L(t + k)}	0.170	0.107	0.017	-0.060	-0.097	-0.108

Two versions of the logistic model were fitted, one based upon equation (13), the other by stepwise search using all the variables in (13) and their lagged counterparts up to four months back. The results are summarized in Table 5. The postulated model shows β_3 and β_5 to be not significantly different from zero, as expected; these variables were not included by the final stepwise version. The stepwise version also showed changes in new cases in the lead area to be important, that is, $N_L(t) - N_L(t - 1)$. For these reasons the stepwise model was used in later work.

The final chi-square criterion for the fitted model had a value of 23.74 on 11 degrees of freedom, with p = 0.0139. None of the other variables tried was significant at even the 10% level. Although this is not wholly satisfactory, the model with only a constant term had a chi-square value of 88.09 on 15 degrees of freedom, so a worthwhile reduction was achieved. Since Y_t = 0 or 1, residuals are readily computed from Tables 6 and 8.

The behavior of the observed and fitted values is summarized in Table 6, where the one-step-ahead forecasts are given during the epidemic periods. The start of the first epidemic is accurately gauged although its revival is not. However, the clear-out of susceptibles allows the model to predict correctly that the revival is short lived. Since Y_t is either zero or one, we term forecasts as "accurate" if the P value jumps dramatically at the onset of the epidemic.

The second cycle recorded is a minor affair with only 34 cases in all. The model predicts an epidemic two months early but misses the correct start. The third wave is predicted correctly, both for onset and decline. Not shown in the table is a minor epidemic in 1952 which led to only 12 cases in Reykjavíkur and five cases in Isafjarðar. Following four cases in December 1952, the forecast value rose to 473/1000 for the following month, then promptly dropped back to its non-epidemic level in the next month. The complete data record for the third wave is shown in Table 7, so the patterns may be better understood.

In order to judge the performance of the fitted model, one-step-ahead forecasts were then generated for the period 1957-1968. These results are presented in Table 8. The start of the 1958/1959 wave is flagged accurately, although the end is missed due to there being a considerable number of susceptibles still available. In the second epidemic, an outbreak is predicted for the beginning of 1963 which fails to materialize until 7/63. It is unlikely that any forecasting method would have picked up the long delay. When the epidemic finally gets underway, it is missed for the first month (six cases) but quickly recognized thereafter. The pool of susceptibles is rapidly depleted and the fade-out accurately predicted. The wave in early 1967 is picked up fairly well, although the low number of susceptibles leads to an early prediction for fade-out. The minor peak in Reykjavikur has no impact since the susceptible population is still very low.

Table 5: Comparison of Fitted Logistic Models (the coefficients for each regressor variable are given, with their estimated standard errors in parenthesis) (variables are defined in the text)

Regressor Variable	Postulated Model	Stepwise Model
$N(t)$	0.2389 (0.0773)	0.4746 (0.1280)
$N_L(t)$	0.0079 (0.0037)	0.0158 (0.0048)
$N_L(t-1)$	--	-0.0150 (0.0047)
$S^*(t)$	-0.0065 (0.0068)	--
$N(t)S^*(t)$	0.0028 (0.0010)	0.0052 (0.0015)
$N_L(t)S^*(t)$	0.0001 (0.0001)	--
Intercept	-3.144 (0.586)	-3.987 (0.686)

Notes: (1) All calculations were performed using the SAS procedure LOGIST.
(2) Variables were included in the final stepwise model only if the reduction in the chi-square goodness-of-fit criterion was significant at the 5% level.
(3) Standard errors are based upon asymptotic theory (see McCullagh and Nelder (1983, p. 85)).

We feel that this study has illustrated the importance of including some measure of the number of susceptibles in any epidemiological forecasting procedure. Better measurement of the susceptible population may further improve the forecasting performance.

6. CONCLUSIONS

The empirical limitations of this approach are evident in that it gives only an indication of the onset of an epidemic and no idea of its magnitude. Nevertheless, it could prove an important signalling device for those involved in immunization programs. In this respect, the model performs better than the approaches described in Cliff et al. (1981) which tend to predict numbers quite well but to miss the onset of a new wave. Thus, the forecasts obtained by this approach might be used as inputs to a decision model weighing the costs and benefits of immunization policies.

Table 6: Comparison of Fitted and Actual Values for Ísafjarðar, 1946-1956 (the fitted P out of epidemics was 18/1000)

Dates					
11/46-3/48		12/49-10/51		3/54-6/55	
New Cases	Fitted P (x 1000)	New Cases	Fitted P (x 1000)	New Cases	Fitted P (x 1000)
0	18	0	18	0	18
0	20	0	18	0	18
1 (Start)	17	0 (Minor Peak)	18	0	18
15*	570	0	38	1 (Start)	19
13*	999	1	9	3	91
10* (Peak)	996	0	30	15*	514
5*	966	0	21	12*	1000
1	15	0	18	18* (Peak)	1000
6*	14	0	18	11*	1000
7*	467	0	17	10*	1000
13*	557	0	18	4	162
5*	973	0 (Start)	18	2	27
50*	269	0	23	0	107
70*	1000	0	140	0	18
4	15	0	938	0	20
0	18	0 (Peak)	153	0	17
0	18	10*	86		
		16*	961		
		3	511		
		4	7		
		0	104		

Notes: (1) All months with ≥ 5 new cases are termed in-epidemic.
(2) The fitted P values are estimated for month $t + 1$ given the data collected for month t. The P values are multiplied by 1000 for ease of display.
(3) The months designated minor peak, start, and peak refer to epidemic behavior in the lead area, Reykjavíkur.
(4) $Y_t = 1$ in months marked *, $Y_t = 0$ otherwise.

As noted in Cliff et al. (1981), population growth and improved communications have led to a speeding up of epidemic spread. This would suggest the possibility of coefficients varying slowly over time, but such models remain unexplored at present. Data on a finer time grid (e.g., weekly) would be valuable in identifying lag structures more accurately and improving the quality of estimates.

Table 7: Complete Data Record for 3/54-6/55

Isafjarðar			Reykjavikur
New Cases N(t)	Susceptible Trend $S^*(t)$	Indicator Y(t)	New Cases $N_L(t)$
0	153	0	0
0	157	0	0
0	161	0	3
1	164	0	26
3	165	0	28
15	154	1	91
12	146	1	380
18	132	1	917
11	125	1	878
10	119	1	290
4	119	0	24
2	121	0	1
0	125	0	0
0	129	0	6
0	133	0	2
0	137	0	1

Table 8: Comparison of Forecast and Actual Values for Isafjarðar, 1957-1968

Dates					
6/58-9/59		4/62-6/64		10/66-5/68	
New Cases	Forecast P (x 1000)	New Cases	Forecast P (x 1000)	New Cases	Forecast P (x 1000)
0	18	0	18	0	18
0	18	0 (Start)	19	0	18
0	19	0	22	0 (Start)	20
0 (Start)	25	0	16	0	107
0	64	0	21	3	339
0	999	0	20	24* (Peak)	487
5* (Peak)	1000	0	18	35*	1000
6*	358	0	227	16*	992
27*	1000	0 (Peak)	875	23*	99
85*	1000	0	984	12*	334
166*	1000	0	1	7*	73
51*	1000	0	0	7*	60
4	1000	0	1	0	61
0	183	0	11	0	17
0	18	0	9	0	20
0	18	6*	17	0 (Minor Peak)	19
		10*	997	0	31
		31*	1000	0	12
		24*	1000	0	15
		24*	1000	0	18